METHODS IN MOLECULAR BIOLOGY

Series Editor
John M. Walker
School of Life Sciences
University of Hertfordshire
Hatfield, Hertfordshire, AL10 9AB, UK

For further volumes:
http://www.springer.com/series/7651

Mitosis

Methods and Protocols

Edited by

David J. Sharp

Department of Physiology and Biophysics, Albert Einstein College of Medicine, Bronx, NY, USA

 Humana Press

Editor
David J. Sharp
Department of Physiology and Biophysics
Albert Einstein College of Medicine
Bronx, NY, USA

ISSN 1064-3745   ISSN 1940-6029 (electronic)
ISBN 978-1-4939-0328-3   ISBN 978-1-4939-0329-0 (eBook)
DOI 10.1007/978-1-4939-0329-0
Springer New York Heidelberg Dordrecht London

Library of Congress Control Number: 2014931984

Printed on acid-free paper

Humana Press is a brand of Springer
Springer is part of Springer Science+Business Media (www.springer.com)

Preface

Mitosis is an extraordinarily complex and dynamic process that must be tightly regulated but sufficiently flexible to ensure the faithful segregation of genetic material from mother to daughter cells. How this is achieved has fascinated biologists for more than a century, both for its role in the generation of life as we know it and its contributions to diseases such as cancer that arise from mitotic defects. Over the past several decades our understanding of mitosis has grown by leaps and bounds driven by the application of an increasingly diverse array of methodological techniques in a variety of experimental systems. Our goal in this volume is to provide a state-of-the-art overview of some of the most important approaches currently used in mitosis research spanning from the analysis of single molecules in isolation to their utilization within the complex environment of the cell.

To aid the reader, this volume has been divided into three parts, each focused on methods pertaining to distinct aspects of mitosis research. The chapters in Part I (Chapters 1–5) present approaches for visualizing and analyzing the dynamic behaviors of the spindle apparatus, the microtubule-based machine that drives chromosome segregation. A particular goal of this section is to arm the researcher with tools to exploit diverse cell types—each with their own particular strengths—to understand how the fluidity of the spindle and the proteins from which it is composed are harnessed to accurately segregate chromosomes. Part II (Chapters 6–8) focuses more generally on methods for studying and manipulating the microtubule cytoskeleton in cells and complex cell-free extracts. Although not necessarily specific to mitosis, these approaches are highly relevant to mitosis researchers since microtubules and microtubule-associated proteins are the primary structural and mechanical elements within the spindle. In this same vein, Part III (Chapters 9–12) provides state-of-the art biophysical and high resolution microscopy approaches for assessing complex interactions between microtubules and microtubule-associated proteins in isolation (Chapters 9–11) as well as microtubule structure in cells (Chapter 12). Finally, Part IV provides two "extras," the first of which provides methods for studying the effects of cell shape on cell division (Chapter 13) while the second describes methods for quantifying aneuploidy (aberrant chromosome number) which frequently results from mitotic defects and has been linked to human maladies ranging from birth defects to cancer (Chapter 14).

In sum, this volume is meant to serve two purposes. First and most obviously, it is for researchers who already have an experiment in mind, or at least a specific question that they want to answer, and want a tested and successful protocol for carrying it out. But secondly and somewhat less obviously, we hope that this volume and the diverse methods presented herein inspires some readers to expand their methodological repertoire by employing new techniques to address new (or old) questions related to the mechanisms of mitosis.

Bronx, NY   *David J. Sharp*

Contents

Part I

Imaging and Measurement of Mitotic Spindle Dynamics In Vivo

Chapter 1

Analysis of Mitotic Protein Dynamics and Function in *Drosophila* Embryos by Live Cell Imaging and Quantitative Modeling

Ingrid Brust-Mascher, Gul Civelekoglu-Scholey, and Jonathan M. Scholey

Abstract

Mitosis depends upon the mitotic spindle, a dynamic protein machine that uses ensembles of dynamic microtubules (MTs) and MT-based motor proteins to assemble itself, control its own length (pole–pole spacing), and segregate chromosomes during anaphase A (chromosome-to-pole motility) and anaphase B (spindle elongation). In this review, we describe how the molecular and biophysical mechanisms of these processes can be analyzed in the syncytial *Drosophila* embryo by combining (1) time-lapse imaging and other fluorescence light microscopy techniques to study the dynamics of mitotic proteins such as tubulins, mitotic motors, and chromosome or centrosome proteins; (2) the perturbation of specific mitotic protein function using microinjected inhibitors (e.g., antibodies) or mutants to infer protein function; and (3) mathematical modeling of the qualitative models derived from these experiments, which can then be used to make predictions which are in turn tested experimentally. We provide details of the methods we use for embryo preparation, fluorescence imaging, and mathematical modeling.

Key words *Drosophila melanogaster*, Mitosis, Time-lapse microscopy, Fluorescence speckle microscopy, Fluorescence recovery after photobleaching, Quantitative modeling, Spindle dynamics, Microtubules, Mitotic motors

1 Introduction

Mitosis, the process by which the replicated genetic material is separated and distributed to the daughter cell products of every cell division, depends upon the action of the mitotic spindle. Mitosis has been extensively studied for over a century in many different systems, of which the fruit fly, *Drosophila melanogaster*, is an especially useful model system [1] and is a focus of the current chapter.

The syncytial blastoderm embryo carries out 13 rounds of mitosis very rapidly and synchronously (cell cycle duration, ~10 min) without any intervening cytokinesis. Of these, mitoses

David J. Sharp (ed.), *Mitosis: Methods and Protocols*, Methods in Molecular Biology, vol. 1136,
DOI 10.1007/978-1-4939-0329-0_1, © Springer Science+Business Media New York 2014

10–13 occur in a single monolayer of mitotic spindles lying just underneath the cortex at the surface of the embryo that are easily visualized by light microscopy. Transgenic flies expressing mutated, or fluorescently (e.g., green fluorescent protein (GFP)) tagged, mitotic proteins have been and are being generated. Targeted gene expression is possible using the GAL4/UAS system [2]. In addition, the microinjection of fluorescently labeled proteins, function-blocking antibodies, or inhibitors to specific mitotic proteins is relatively straightforward. Injected molecules diffuse throughout the embryo, producing a concentration gradient, which, for an inhibitor, results in a gradient of defects, equivalent to an allelic series of mutants, which allows the range of functions of the protein of interest to be dissected.

The synchronous mitoses occurring during cycles 10–13 can be easily imaged using time-lapse confocal fluorescence microscopy. The resulting movies allow measurement of pole–pole distance as a function of time which is characteristic of each cycle and is very reproducible from embryo to embryo, so even small defects are quantitatively measurable. With the appropriate proteins tagged, kinetochore-to-pole or chromosome-to-pole movements are also reproducibly measurable. These measurements provide precise descriptions of spindle dynamics that can be used as a basis for analyzing the effect of protein inhibition on specific mitotic events.

Spindle dynamics can be further studied by microscopy techniques such as fluorescence recovery after photobleaching (FRAP) and fluorescence speckle microscopy (FSM). In FRAP experiments, which report on the bulk dynamics of proteins, the protein of interest is fluorescently labeled, the fluorescence in a certain region of the spindle is photobleached, and the movement of fluorescent and non-fluorescent proteins into and out of the bleached (i.e., dark) region will give rise to fluorescence recovery, yielding information about the diffusion and transport properties of the tagged protein. For example, if the proteins move in a coordinated manner, the bleached region itself will move as a unit, e.g., toward the pole as in poleward MT flux [3]. If proteins diffuse freely, fluorescence recovers without directional movement of the bleached region. Generally, a combination of both transport and diffusion occurs, and the fluorescence may recover before significant movement occurs. In the "inverse" technique of photoactivation, the protein is tagged with a photoactivatable dye, so that following stimulation, the region is now fluorescent which leads to clearer visualization of the region's movement. Poleward flux of spindle tubulin was first observed using this technique [4].

FSM has most commonly been used to study the behavior of polymerized proteins [5, 6]. Low concentrations of fluorescent monomers incorporate into polymers in sub-stoichiometric amounts relative to the non-fluorescent subunits which give rise to the appearance of fluorescent "speckle"-labeled polymers. The resulting fluorescent speckles will move, or appear and disappear as dynamic filaments assemble, disassemble, and undergo steady-state turnover. In the mitotic spindle, for example, MT speckles tend to move persistently from the equator to the pole due to poleward flux. More recently, similar FSM techniques have been applied to protein ensembles to study their movement [7], though the interpretation of these speckles may not be straightforward.

Mathematical or computational modeling can be used to augment the aforementioned experimental techniques in order to test the resulting qualitative/cartoon models quantitatively. This usually leads to predictions which can help design new experiments to validate, refine, or revise the qualitative model, establishing a feedback between quantitative model prediction and experimental testing. Mathematical/computational models are also used to reconstitute the dynamics of one or several of the identified components of the mitotic spindle in silico, to identify the kinetics underlying experimental observations such as those obtained using FRAP, FSM, or PAF of dynamic spindle components. These models have shown, for example, that highly dynamic MTs can drive robust, rapid, and steady spindle or chromosome movements [8–10] and have also identified areas of uncertainty that need further experimentation [11].

Below, we present details of the methods that we currently use to collect and prepare *Drosophila* syncytial embryos for microinjection and in vivo imaging [12], as well as techniques for time-lapse imaging, FRAP, FSM, and the quantitative modeling of mitotic spindle dynamics (*see* **Note 1**).

2   Materials

1. *Grape juice plates*: Mix 5.5 g bactoagar, 14.5 g dextrose or glucose, 7.15 g sucrose, 204.5 ml H_2O, 45 ml grape juice concentrate (100 % juice), and 625 µl 10 N NaOH in a 2 L glass beaker. Cover with parafilm and microwave until boiling. Stir occasionally to mix. Add 2.8 ml of acid mix (acid mix: 20.9 ml propionic acid, 2.1 ml phosphoric acid, and 27 ml H_2O). Mix and pour onto 35 mm Petri dishes. Let solidify at room temperature. Dry for 1 or 2 days, and then seal with parafilm and keep at 4 °C. Let come to room temperature before using.

2. *Yeast paste*: Dissolve about one teaspoon of yeast (Sigma YSC2, yeast from *Saccharomyces cerevisiae* type II) in water to form a thick paste. Place a small amount of this on each grape juice plate just prior to use.

3. *G-PEM buffer*: 80 mM Na-PIPES pH 6.9, 1 mM $MgCl_2$, 1 mM EGTA, 1 mM GTP.

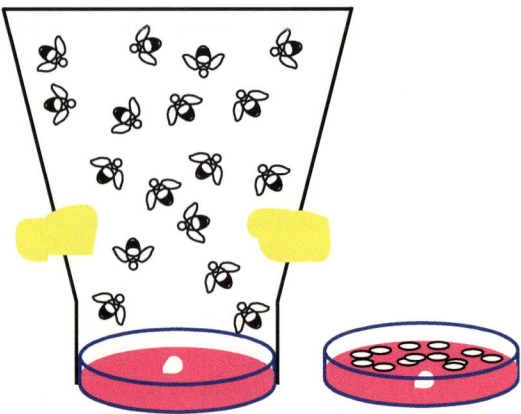

Fig. 1 Lay cage for embryo collection. Young flies are placed in a plastic bottle with cotton plugged holes and covered with a grape juice plate with fresh yeast paste, on which flies will lay their embryos (see text for details)

4. *Injection buffer*: 150 mM K-Aspartate, 10 mM K-phosphate, 20 mM imidazole, pH 7.2.

5. *PBS buffer*: 137 mM NaCl, 10 mM Phosphate, 2.7 mM KCl, pH 7.4.

6. *Heptane glue*: Unroll double sticky tape and place it in a 100 ml bottle. Add about 50 ml heptane, seal the bottle, and rock for several days. To use pour ~1 ml into a 2 ml tube, add heptane if the glue is too thick.

7. *Dehydration chamber*: Place bottom of a 35 mm dish inside a 100 mm Petri dish with the open side up. Add Drierite (anhydrous calcium sulfate) around it. It is recommended to use at least some indicating Drierite and change it when the color has changed.

8. *Lay cage*: For best laying, make the lay cage 1 or 2 days before use with young flies. Flies will lay embryos for about 10 days. Take a 6 oz., round bottom plastic bottle (Applied Scientific *Drosophila*), cut two small holes (about 1 cm²) on opposite sides, and fill with a cotton piece (this will allow air to flow). Tap the flies into the bottle and cover with a grape juice plate with a small amount of yeast paste (Fig. 1). Set on a flat surface with plate down at 25 °C or room temperature.

9. *Other materials needed*:

   Forceps (Dumont #5), separated into two arms; use one of them.

   Fine tipped paintbrush.

   Microinjection needles.

   Halocarbon oil 700.

3   Methods

3.1   Preparation of Needles for Microinjection

Note: All work with tubulin should be done at 4 °C to prevent polymerization. Needles should be at 4 °C before loading. We use lyophilized fluorescent tubulin from Cytoskeleton (Denver, CO).

3.1.1   Preparation of Lyophilized Tubulin

1. Take tubulin out of freezer and let stand on ice for a few minutes.

2. Centrifuge at $13800 \times g$ for 1–2 min so the powder accumulates at the bottom of the tube.

3. Resuspend in 2 μl cold water to make a 10 mg/ml solution.

4. Dilute in cold G-PEM buffer 5–8-fold. The dilution will depend on the amount to be injected and the microscopy technique to be used.

5. Centrifuge at $13800 \times g$ for 8–10 min.

6. Load cold needles and keep needles on ice when not injecting.

3.1.2   Preparation of Antibody or Chemical Inhibitor

1. The buffer in which an inhibitor or antibody is dissolved in should be injected as a control to ensure it has no effect on embryo mitosis. Tested buffers are the injection buffer, G-PEM buffer, and PBS (recipes in Subheading 2, items 3–5).

2. Centrifuge inhibitor/antibody at $13800 \times g$ for 5–7 min to remove aggregates before loading the needles.

3. The appropriate concentration varies depending on the specific inhibitor or antibody. Antibody concentrations are usually in the 5–20 mg/ml range.

4. To find the correct concentration, start with a concentration in this range. If no effect is observed, increase the concentration; if strong effects are observed, decrease the concentration. The ideal concentration will be such that the most severe phenotype is observed close to the injection site and the severity decreases with increasing distance away from the injection site.

5. If both a fluorescent protein and an inhibitor need to be injected, they can be mixed if both are at high enough concentrations and in compatible buffers. Mix and centrifuge before loading needles.

6. If they cannot be mixed, then inject fluorescent protein and wait at least 5 min before injecting the inhibitor. For double injections, start with more embryos and dehydrate 1 or 2 min longer.

3.2   Coverslip Preparation

1. Put a 50×22 mm coverslip on one side of a microscope slide. Tape the four corners to the slide so it doesn't move.

2. Using a cotton tipped applicator, streak a layer of heptane glue on the coverslip.

3. Put a piece of double sticky tape on the other side of the slide (Fig. 2a top).

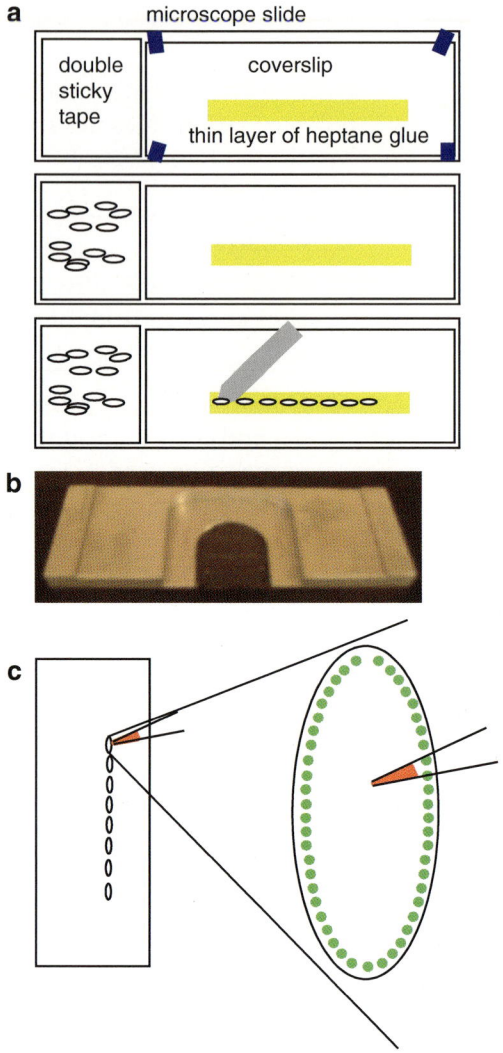

Fig. 2 Coverslip and embryo preparation. (**a**) Schematic of coverslip preparation and embryo dechorionation. (**b**) Microscopy chamber, coverslip is on bottom side with oil covering the embryos. (**c**) Schematic of embryo injection

3.3 Embryo Collection and Preparation for Microscopy

3.3.1 Embryo Collection

1. Put a new grape juice plate with yeast on the lay cage.

2. Change this plate after 30–60 min and discard.

3. Change the plate every 30–60 min and allow to mature so they are imaged at 2 h from start of collection (*see* **Notes 2** and **3** and next section).

3.3.2 Embryo Preparation

1. With a moistened brush, carefully pick up the embryos from the grape juice plate and place them on the double sticky tape on the slide (Fig. 2a middle).

2. Using one arm of the forceps, roll the embryos over the double sticky tape until the chorion (the outer membrane) breaks open.

3. Pick up the embryo by gently rolling it over the chorion so it sticks to the forceps and place it on the heptane glue on the coverslip with the long side of the embryo parallel to the long side of the coverslip. Set up 10–20 embryos in one row (Fig. 2a bottom) (*see* **Notes 4–6**).

4. Remove the coverslip and place it in the dehydration chamber for 3–8 min (this time depends on the humidity of the room and the amount to be injected). If embryos will not be injected, skip this dehydration step.

5. Place on a metal chamber with vacuum grease. Cover the embryos with halocarbon oil 700 to avoid further dehydration. Embryos are now ready for injection (Fig. 2b) (*see* **Note 7**).

3.3.3 Embryo Microinjection

1. Find the embryos under the microscope (*see* **Note 8**).

2. Move the embryos away, and find the needle without moving the focal plane.

3. Center the needle and move it up without moving it in *x* or *y* direction.

4. If the needle is not open, put the edge of the coverslip in the edge of the field of view, and lower the needle to the same focal plane. Very carefully move the coverslip until it hits the needle and gently breaks it open. Move the needle up.

5. Put the embryos in view (out of the needle's path), lower the needle into the oil, and make sure you obtain small liquid drops from the needle.

6. Put the embryos and needle in the same focal plane and carefully but steadily move the embryo into the needle, inject a drop into the embryo, and move the embryo away (or follow the instructions of your injection apparatus) (*see* **Notes 9** and **10**).

3.4 Time-Lapse Imaging

Imaging of embryos should be done on a confocal or deconvolution microscope, since the embryos are thick and there is a lot of background. Multiple wavelengths can be used to image different proteins (Fig. 3).

3.4.1 Acquisition

1. We image embryos on an Olympus (Melville, NY) microscope equipped with an Ultra-View spinning disk confocal head (PerkinElmer-Cetus, Boston, MA) and an Orca ER CCD camera (Hamamatsu).

2. We use either a 100× 1.35 NA or a 60× 1.4 NA objective. A 60× objective is useful for initial inhibition experiments when a larger part of the embryo is imaged to see more of the gradient. The 60× is necessary to inject on the confocal microscope for specific timing or fast-acting inhibitors.

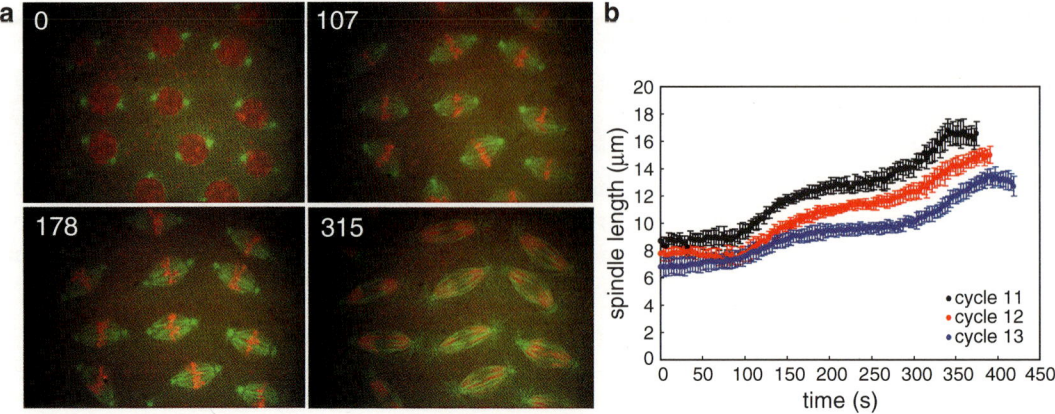

Fig. 3 Time-lapse microscopy and control measurements. (**a**) Images from a time-lapse series of an embryo expressing GFP-tagged tubulin and RFP-tagged histone. (**b**) Pole–pole distance measured for three cycles in wild-type embryos. These plots are very reproducible (adapted from [12])

3. GFP-tagged proteins are imaged with a 488 nm laser, RFP-tagged proteins can be imaged with a 546 nm or a 568 nm line, and far red fluorescence can be imaged with a 647 nm laser line.

4. The quality of the image is determined by the detector sensitivity, laser power, and exposure time. These will need to be set empirically to obtain the best images without photodamage to the embryos (*see* **Note 11**).

5. Increasing the gain on the detector will increase the signal, but also the noise. It is important to set the gain so that the response is linear. This can be checked with prepared slides that have different concentrations of fluorescent molecules. A plot of fluorescence intensity as a function of concentration should be linear. Decrease the gain if it is not.

6. Increasing the laser power will increase the fluorescence intensity. However, this may lead to photobleaching and photodamage. Use the lowest laser power that still allows a good image. Also, make sure that the image is not saturated, that is, the highest intensity should be less than the maximum measurable intensity. On a 12 bit camera, the highest intensity should be less than 4,095.

7. Increasing exposure time also increases the fluorescence intensity, but this may also lead to photobleaching and photodamage. Generally, samples will tolerate a longer exposure time at a lower laser power better than shorter exposure times and higher laser powers. It is important to keep the exposure time smaller than the time in which changes are expected. For example, if the spindle is elongating and the exposure time is long, the spindle will appear blurred.

8. Since mitosis in the embryo is fast, the time interval on the time-lapse series should be at most a few seconds, depending on the exact process to be studied. For measurements of pole–pole distance, a frame rate of 3–5 s is adequate; however, for FRAP or FSM discussed below, the frame rate needs to be much faster.

9. If the whole spindle needs to be imaged, for example, for localization studies, a 3D series is needed. The thickness of the imaging plane depends on the particular setup, for example, with a 100× objective and an optimized pinhole, it is 0.5 μm. Recording a 3D series will reduce the frame rate. The tradeoff between temporal and spatial resolution will depend on the process being studied—is it more important to study the whole spindle or to get fast images?

3.4.2 Analysis

Time-lapse movies can be used to measure spindle length, chromosome-to-pole movement, protein redistributions, or changes in relative concentration (*see* **Note 12**).

Length/Distance Measurements

1. Log the positions of each pole and chromosome or kinetochore if applicable in MetaMorph, MATLAB, or other image processing software.

2. Calculate the distance between poles, kinetochores, or kinetochore and pole.

3. Plot these distances as a function of time. Pole-to-pole distance as a function of time is very consistent in embryos and should be reproducible in wild-type embryos (Fig. 3b, **Note 4**).

Relative Protein Concentrations

The amount of protein in the spindle can be measured relative to the amount of tubulin within one embryo [13].

1. Use embryos expressing the protein of interest tagged with GFP.

2. Inject with rhodamine tubulin immediately after collection and wait before imaging to allow diffusion of rhodamine tubulin to an even distribution.

3. Image both channels making sure not to change any settings while imaging and that neither channel is saturated.

4. Measure the intensity of GFP and that of rhodamine on the spindle and in the background.

5. Subtract the background intensity and ratio the GFP with the rhodamine value. This allows comparison of the amount of protein on different spindles within single embryos, for example, to look at the effect of an inhibitor gradient [13].

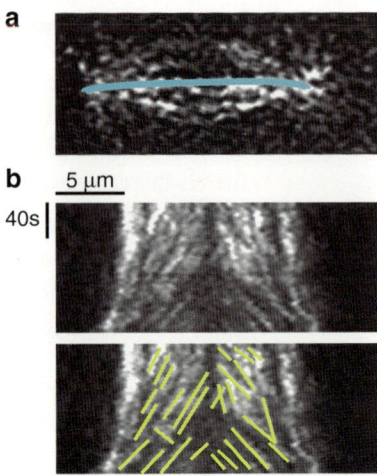

Fig. 4 Fluorescence speckle microscopy (FSM). (**a**) Still of a time-lapse movie after processing (sharpen high- and low-pass commands) showing the line used to create the kymograph. (**b**) Kymograph generated over a spindle MT bundle. *Bottom* shows the same kymograph with lines overlaid for clarity

3.5 Fluorescence Speckle Microscopy

FSM [5, 6] is used to look at tubulin dynamics within the spindle. Embryos are injected with small concentrations of fluorescent tubulin to allow the formation of fluorescent speckles as opposed to uniform labeling of the microtubules.

1. Obtain time-lapse images as described above with a low concentration of fluorescent tubulin at the fastest rate possible. Speckles should be clear, if the spindle is too bright and uniformly labeled, dilute the labeled tubulin further.

2. Images are processed as follows: no neighbors deconvolution or sharpen high filter followed by a low-pass filter in MetaMorph imaging software (Universal Imaging, West Chester, PA).

3. Generate a kymograph by tracing a line on the MT bundle of interest (Fig. 4a) and following this line as a function of time (MetaMorph has an integrated kymograph command). Note: If the spindle moves too much, align it before making the kymograph as explained in Subheading 3.6.2, FRAP data analysis.

4. Moving particles appear as oblique lines whose slopes correspond to their rate of movement (Fig. 4b). Measure this by logging the position (x_i, t_i) of the starting and ending points, note that the y dimension represents time. The distance traveled is calculated as (pixel size) $\times (x_{i+1} - x_i)$, while the elapsed time is given by (time/frame) $\times (t_{i+1} - t_i)$. The velocity is (distance traveled)/(elapsed time).

5. A histogram of observed velocities should have a Gaussian shape. A one-tailed histogram toward faster velocities may

indicate that faster particles are lost in the measurements. Repeat the experiment with a faster acquisition rate.

6. The rate of flux is obtained by subtracting the movement of the pole from that of the speckle.

Note: Motors may also form speckles [7], and their movement can be measured in the same way; however, it is important to remember that speckles are not individual motors and that motors may bind and unbind to the spindle. Careful interpretation of speckle movement is needed and can benefit from the application of modeling.

3.6 Fluorescence Recovery After Photobleaching

FRAP is used to measure bulk dynamics of proteins [14]. In this technique, the fluorescent tag on a protein of interest is photobleached, thus creating a dark region. As fluorescent proteins move into the region and bleached ones move out of the region, the fluorescence recovers, reflecting the bulk mobility of the protein. Any fluorescently labeled protein can be photobleached and the recovery used to analyze the protein dynamics. There are different FRAP setups that can be used; embryos should be imaged on a confocal microscope. We use a laser scanning Olympus confocal microscope with a 60× 1.4 NA objective.

3.6.1 Acquisition

1. Prepare embryos as above for time-lapse imaging.

2. Find a spindle in the stage of mitosis of interest. In embryos, this has to be done quickly as they progress through mitosis rapidly. Use a fast, low-resolution scan to find a spindle at the right stage.

3. Set the zoom and region so that only one spindle is imaged with a pixel size of about 0.1 μm. Again use a fast, low-resolution scan to set this up.

4. Draw the region you wish to bleach, it could be a circular or rectangular region (Fig. 5), and choose the area of the spindle. There could be differences depending on the region and mitotic phase [15].

5. Set up the bleach time and bleach laser power so that the bleach is deep (70–90 % bleach) but minimizes damage.

6. Acquire 3–5 prebleach images with the same imaging conditions as postbleach images.

7. Bleach an area of the spindle. If the bleach is larger than the selected region, lower the bleach time and/or laser power.

8. Time-lapse series acquisition should start immediately after the bleach. Because dynamics in the embryo are very fast, the acquisition time should be very rapid. For tubulin, we acquire at rates of 2–4 frames/s.

9. GFP-tagged proteins are visualized with a 488 nm laser. In our system, photobleaching is done with a separate 405 nm laser. This allows imaging to begin immediately after photobleaching.

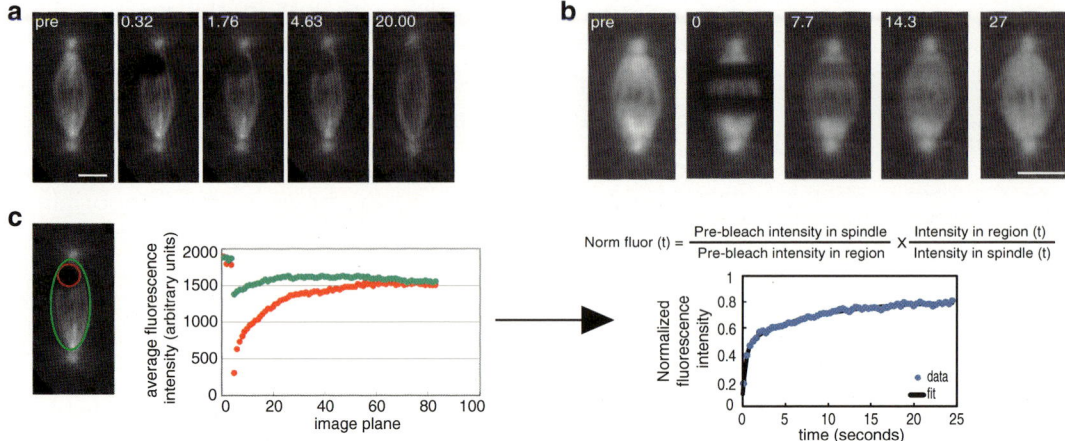

Fig. 5 Fluorescence recovery after photobleaching (FRAP). (**a**) Images from a time-lapse series of GFP-tubulin with a small circular region photobleached at time 0. (**b**) Images from a time-lapse series of GFP-tubulin with rectangular regions simultaneously bleached at equator and near poles to compare the dynamics in different regions of the spindle (adapted from [15]). (**c**) Measurement of fluorescence recovery. After spindle alignment, the fluorescence in the bleached region (*red circle*) and the spindle (*green outline*) are measured, and the recovery is calculated as shown to yield the corrected fluorescence intensity as a function of time (*right graph*). The experimental data (*blue*) is fitted to an exponential recovery. In this example, the recovery was fitted to a double exponential

Because the recovery in the spindle is very fast, it is important to minimize the time between bleach and acquisition of the first image.

10. This system also allows for photoactivation of PAGFP in the same manner. In this case, the decay of fluorescence is analyzed. This technique is especially useful for following movement, since it is a bright signal on a dark background [4].

3.6.2 Analysis

1. Correct spindles for movement:

   Log the pole positions in each time frame.

   Center the spindles at the midpoint between the two poles in each time frame.

   Rotate the image so that the angle of the pole–pole axis is constant.

2. Measure the fluorescence intensities of the bleached region ($F(t)$) and the spindle ($T(t)$) in each time frame (Fig. 5c).

3. Normalize the data by $I(t) = \left[F(t) T_{pre} \right] / \left[T(t) F_{pre} \right]$, where F_{pre} and T_{pre} are the mean fluorescence in the FRAP region and in the entire spindle just before the bleach (average 3–5 pre-bleach images) [16, 17].

4. Fit the normalized data to an exponential recovery curve Fit $(t) = F_0 + (F_{inf} - F_0) \times (1 - \exp(-\ln(2) \, t/t_{1/2}))$ [18] (Fig. 5c)

5. This fit yields the half time $t_{1/2}$; the percentage recovery is calculated as $(F_{inf} - F_0)/(F_{pre} - F_0)$.

6. If this exponential fit is not a good fit, a double exponential may be needed. In our hands, tubulin FRAP in small circular regions requires a double exponential fit.

3.7 Modeling

Quantitative modeling of mitosis is used to test the plausibility of qualitative models and to make predictions that can then be tested experimentally to validate, refine, or revise the model. Conversely, new experimental data can be used to restrict the parameters and to test and improve the model. This loop between experimental data and modeling data can be repeated to advance our understanding of the process (*see* **Note 12**).

The exact formulation of the model will vary depending on the particular aspect of mitosis being studied. Different methods can be applied to test the model (1) in a force-balance approach or (2) in a biochemical/kinetic approach, or (3) in a combination of both approaches, as described below (details for the implementation are in the following subsections).

1. In the force-balance (FB) approach, the model is developed by setting a coupled set of force-balance equations describing the motion of spindle parts (centrosomes, chromosomes, MTs) in low Reynolds number conditions. The kinetic and biophysical properties and concentrations of the proteins involved are used to describe the magnitude of forces exerted on the parts by force-generating enzymes/MTs. The sum of the forces exerted on each part, in turn, determines its rate of motion. The resulting coupled set of ODEs (rate equations, describing the velocity of the spindle part(s), $v = dx/dt$, Fig. 6a) are then analyzed to identify how variations in concentration, activity, and kinetic and biophysical properties of the proteins, MTs, or spindle parts affect the dynamic process studied. This analysis not only makes predictions on the key players that drive the process but also predicts how individual or synchronous changes in the properties of each molecule or its concentration would affect the outcome. Such predictions are then used to design new experiments and to validate, or refine and revise the proposed model.

2. In the biochemical/kinetic approach, models for binding/detachment dynamics of motor proteins in the spindle and/or the polymerization/depolymerization of MT ends can be quantitatively accounted for without considering the magnitude of the generated forces, simply by assuming the observed average rate of movements of the spindle parts to identify the kinetic properties that could match the observed experimental FRAP or FSM data. This method requires the development of in silico FRAP and in silico kymographs. The analysis of the results helps us narrow down the range of unknown parameters and helps us identify what type of changes in the kinetic

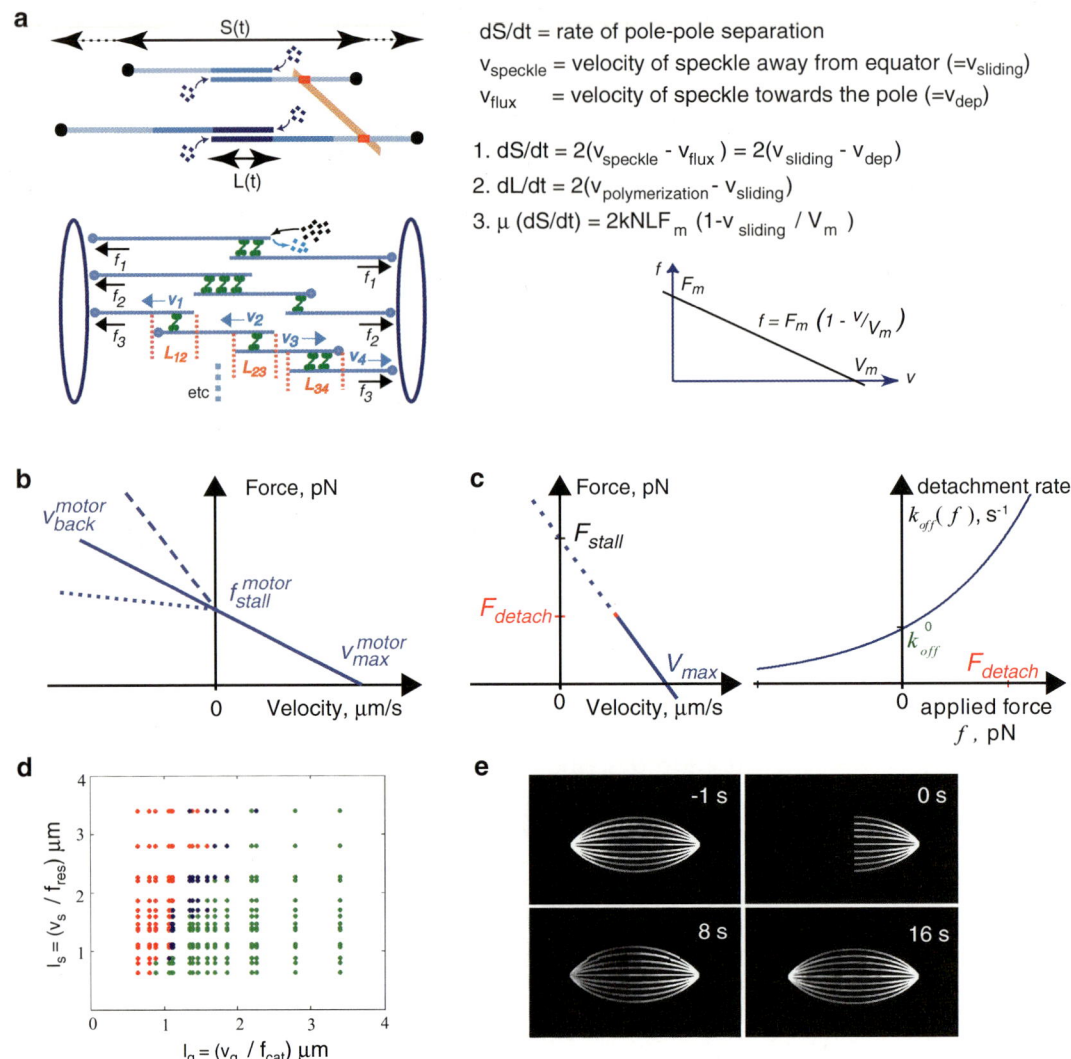

Fig. 6 Modeling. (**a**) (*Top*) Dynamics of spindle poles (*black dots*), ipMTs (*overlapping blue lines*), and tubulin speckles (*orange*). In preanaphase B, pole–pole spacing remains constant, and opposite end assembly–disassembly is associated with poleward flux (*orange*). In anaphase B, depolymerization at the poles ceases, and ipMT sliding drives pole–pole separation; thus, speckles move away from the equator at the same rate as the poles. (*Bottom*) A realistic spindle with ipMT arrays composed of two, three, and four overlapping MTs. In the lower ipMT array, the overlap length between the parallel (L_{12} and L_{34}) and antiparallel (L_{23}) MTs and the sliding velocity of each MT (V_1, V_2, V_3, and V_4) is indicated. The dynamic instability of the plus ends, resulting in an average net polymerization, is shown only for the *left* MT in the *top* ipMT array. Forces generated by bipolar motors in different ipMT arrays (e.g., f_1, f_2, and f_3) are different. (*Right*) Force-balance equations. (**b**) Force-velocity relationship. Different motor behaviors in backward velocity regime: (1) *solid line*, linear FV relationship, extends into backward regime; (2) *dashed line*, motor resists strongly to backward movement, exerts superstall forces; (3) *dotted line*, motor continues to exert stall force in response to backward movement. (**c**) Force-sensitive detachment of motor proteins from MTs. *Left panel* shows a linear FV curve, F_{detach} (a motor specific, measurable force) is the critical force at which the detachment rate of the motor increases e-fold. *Right panel* shows a typical force-sensitive MT-detachment rate of a (motor) protein. (**d**). High MT dynamic parameters explain rapid MT turnover. Color-coded summary of in silico FRAP results in terms of the mean growth (l_{grow}) and shrinkage lengths (l_{short}). *Red dots* show combinations of MT dynamics where the antiparallel

parameters can account for the experimentally observed changes in spindle organization. These predictions can then be tested experimentally.

Finally, the FB approach and the biochemical approach can be combined to further test the qualitative model by using all the constraints and knowledge available based on experimental observations.

3.7.1 Formulation of the Force-Balance Model

Assumptions

In formulating the model, a number of simplifying assumptions are made based on the architecture/organization of the mitotic spindle and the identity and properties of the force generators.

1. The spindle has symmetry around the pole–pole axis; therefore, the simplest model can be in one dimension. Under this assumption, elastic bending forces on MTs are neglected; furthermore, all movements and forces or their projections are assumed to be along this unique dimension.

2. MTs are polar filaments with their minus ends toward the pole and their plus ends away from the pole. All minus ends can be clustered at the pole or minus ends can be distributed along the spindle with MT bundles formed by cross-linked shorter MTs.

3. Forces generated by active enzymes on the same spindle part (e.g., centrosome, kinetochore, or MT) are assumed to be additive, i.e., the active force generators share the load equally and are synchronous.

4. Force-generating enzymes are assumed to obey a linear or a piecewise linear force-velocity relationship. The binding/dissociation kinetics of the motors can be explicitly included in the model, requiring individual representation of each enzyme in the model, or assumed to be equilibrated in a time scale faster than the motility time scale, and an average number of enzyme per length is represented in the equations. If the binding/dissociation kinetics of motor enzymes is explicitly considered, force sensitivity of the dissociation can also be included.

5. We assume that MT plus ends exhibit dynamic instability, fully characterized by growth velocity v_g, shrinkage velocity v_s, rescue frequency f_r, and catastrophe frequency f_c [19].

Fig. 6 (continued) overlap could not be maintained and vanished completely during preanaphase B; *blue dots* indicate combinations that gave rise to FRAP half times <10 s during preanaphase B (in agreement with our experimental observations); and *green dots* indicate combinations of parameters that led to FRAP half times >10 s. The values of MT dynamic parameters used were between 0.1 and 0.3 µm/s for v_g and v_s and 0.05 and 0.25 s^{-1} for f_{cat} and f_{res}, in steps of 0.06 µm/s for the velocities and 0.05 s^{-1} for the frequencies. (**d**) Snapshots from a typical in silico FRAP of a half-spindle in preanaphase B (**a** adapted from [30], **d** and **e** adapted from [15])

1. The model typically consists of a few core equations that describe the forces exerted on one or several spindle parts, including (1) the centrosome, (2) the MTs, and (3) the chromosomes, as necessary.

2. Each equation simply states that the sum of forces exerted on each component is balanced by the viscous drag forces (proportional to its time-dependent velocity and its drag coefficient) in the viscous medium the motility occurs. For example, if we assume that only outward sliding motors on antiparallel MTs exert forces on the centrosome, the FB equation on the centrosome is $F_{drag} = \sum_{\#MT} F_{sliding}$, where $F_{drag} = \mu_{pole} v_{pole}$, $F_{sliding} = LNf_{motor}$, and μ_{pole}, L, and N are the viscous drag coefficient of the centrosome, the total antiparallel overlap length, and the number of sliding motors per unit overlap, respectively.

3. In general, forces exerted on a spindle part may include (a) the forces generated by plus- and minus-end-directed motor enzymes along the length of, on the parallel or antiparallel overlaps of, or anchored to the plus/minus ends of MTs and to the cortex and/or the kinetochore or the chromosome arms; (b) the forces generated by polymerization/depolymerization of MT plus/minus ends; and (c) elastic forces due to stretching/compression of cohesion bonds between the sister chromatids. In models in one dimension, elastic forces resulting from the bending of the MTs are neglected.

4. If spindle parts change position/length in response to relative movement between these spindle parts and not by force generation or stochastic MT dynamics (i.e., the dynamic instability of MT plus ends), and if this length is an explicit part of one or more of the FB equation(s), kinematic equations describing the change in length (or position) should be coupled to the FB equations. For example, if an MT minus end interacting with the centrosome depolymerizes by removal of subunits at its minus end at the rate the MT is pushed into the pole by sliding motors, then the depolymerization rate of the MT minus end (which determines the length of the MT along with the dynamics of its plus end) is described by $v_{depoly}^{minus} = v_{depoly}^{minus} - v_{depoly}$, or equivalently $v_{pole} = v_{sliding}^{MT} - v_{depoly}^{minus}$.

5. Such kinematic equations, however, impose further assumptions on the FB equations, for example, in the case described above, the coupling of the kinematic equation to FB equations in which only the outward pushing of the spindle poles by the action of the motors on antiparallel MT overlaps is described implicitly implies that $v_{pole} \geq 0$ (equivalently, $v_{sliding}^{MT} \geq v_{depoly}^{minus}$), as there is no force or force generator to account for an inward movement of the spindle pole ($v_{pole} < 0$).

6. For example, to test the qualitative model suggesting that the outward sliding of antiparallel overlapping MTs, generated by kinesin-5 motors, drives the observed rapid and linear anaphase B spindle elongation when MT minus-end depolymerization at the poles is turned off, we set up a model based on three core equations (Fig. 6a). The first two are kinematic equations that describe the relationship between the spindle and overlap lengths as a function of time in terms of ipMT sliding and MT polymerization and depolymerization, while the third equation sets up a force balance on the spindle pole.

7. Assuming that all MTs behave uniformly and synchronously, the core system of FB-kinematic equations includes only FB equations for one MT. However, in the stochastic/full description of the model, an FB equation for each MT should be included, yielding a large set of coupled systems of equations [8].

3.7.2 Model Parameters

The parameters in the FB equations should be determined or at least estimated from experimental observations.

1. The number of MT bundles and the general geometry can be determined from 3D deconvolved images [10].

2. The number of MTs and the length of the overlap region can be determined from electron microscopy reconstructions [20].

3. The number of kinetochore MTs could also be obtained from electron microscopy reconstructions; although this has not been done for embryos, we can estimate this number based on reconstructions from insect spermatocytes and S2 cells [21, 22].

4. For motors acting on overlapping antiparallel MTs, the number of motors is assumed to be proportional to the overlap length, based on electron microscopy reconstructions [20].

5. The force-velocity relationships, motor gliding velocities, and other motor characteristics have been measured in vitro for some mitotic motors [23, 24].

6. A range for the polymerization rate of MT plus ends in the spindle can be estimated from experimental observation of EB1-GFP dynamics in the spindle.

7. Viscous drag coefficient of the centrosome, the MT, or the chromosome is either estimated experimentally [25] or can be estimated based on the geometry of the part and the estimate of the cytoplasmic viscosity [25].

8. For parameters which have not been measured inside the spindle, for example, MT catastrophe frequency, a range can be estimated from measurements of astral MTs, interphase MTs, observations from other organisms, or in vitro experiments (Table 1).

Table 1
Range of parameter values for typical model variables

Symbol	Meaning	Parameter values	References
v_g	MT plus end growth rate	0.1–0.4 µm/s	[31, 32]
v_s	MT plus end shrinkage rate	0.1–0.4 µm/s	[31, 32]
f_{res}	MT plus end rescue frequency	0.01–0.5 s^{-1}	[31, 32]
f_{cat}	MT plus end catastrophe frequency	0.01–0.5 s^{-1}	[31, 32]
N	Number of overlapping pairs of ipMTs	200–600	[8, 20]
k, n	Number of bound motors per unit AP MT length	up to 800 µm^{-1}	[20]
μ	Effective spindle pole viscous drag coefficient	1,000 pNs/µm	[25]
Motor parameters (Ncd or KLP61F)			
F_{stall}	Stall force	1–10 pN	[24, 33]
F_{det}	Detachment force	1–10 pN	[24]
V_{max}	Maximal velocity	0.01–0.5 µm/s	[34]
V_{back}	Backward velocity	0.001–5 µm/s	[35]
k_{on}	MT binding rate	0.01–10 s^{-1}	[36]
k_{off}^0	MT load-free detachment rate	0.01–10 s^{-1}	[36]

9. The depolymerization rates and the rescue frequency of MT plus ends are estimated and determined by fitting the experimental FRAP data to in silico FRAP (see below).

3.7.3 Solving the Force-Balance Model

1. The core set of coupled FB and kinematic equations, where the velocities of the spindle parts are their time derivatives, form a set of coupled ODEs and can be solved using built-in ODE solvers available in MATLAB, Mathematica, or similar packages, with user-defined initial conditions and a set of values for the parameters in the system of equations [26].

2. The solutions describe the behavior of components in an idealized spindle where all MTs and all force generators work synchronously and uniformly and can be plotted over time to compare with experimental data.

3. This deterministic approach is valuable to screen for the plausibility of the proposed mechanism in a coarse-grain approach, for example, to decide if the proposed system could produce the movements in a homogeneous, uniform spindle with fully synchronized motors and microtubules. In addition to this model testing, the deterministic solutions can also be used to screen for model conditions and parameters which have the potential to account for the experimental observation [26].

4. However, such a screen may overlook or falsely identify conditions and parameter sets in a realistic spindle in which MTs and motors behave stochastically.

5. The stochasticity of MT dynamics and the stochastic and force-sensitive kinetics of key spindle proteins affect the outcome profoundly; therefore, numerical solutions of the system of FB equations fully describing the spindle MTs including stochasticity provide the most realistic quantitative test of the qualitative model.

6. The large system of coupled FB equations (typically hundreds of equations), in which a separate FB equation describes each MT, can be solved numerically using a custom-made, explicit forward Euler algorithm. To this end, starting with user-defined initial conditions at $t = t_0$, at each time step $t_n = t_{n-1} + \Delta t$, the following steps are executed:

   6.1 The algebraic set of equations corresponding to the system of FB equations is solved based on the position and the current state of the spindle components (for MTs the state is either growth or shrinkage, and for motor proteins, the state is either bound or free).

   6.2 The solutions yield the velocities of the spindle parts (the poles, the MTs, and the kinetochores if applicable) as well as all the forces generated by and/or exerted on each spindle component.

   6.3 These velocities are used to compute the new positions of the parts for $t_{n+1} = t_n + \Delta t$. All positions are stored.

   6.4 Then, within the same time step, the stochastic MT plus-end dynamics and the binding/dissociation of the motors are invoked using the built-in pseudorandom number generator function of MATLAB: a random number is generated for each MT plus end and motor protein in the system.

   6.5 MT plus ends can switch from growth to shrinkage or shrinkage to growth, and motor proteins can switch from bound-to-free or free-to-bound states. To determine the new state of each MT plus end and protein motor in the next time step, the probability of each switch event is computed based on the defined rates using the current state of the MTs and motors at $t = t_n$ and the newly computed positions of all spindle parts in **step 6.3**. The rates are f_{cat}, f_{res}, k_{on} or k_{off} and $P_{switch} = 1 - \exp(-k_{switch})$, where k_{switch} is equal to f_{cat}, f_{res}, k_{on}, or k_{off}.

   6.6 Some examples: (1) the switch rate of an MT plus end may depend on its position within the spindle (i.e., length-dependent catastrophe) or on the force exerted on it; (2) the dissociation of a motor protein can depend on the applied force (i.e., force-sensitive dissociation rate) and can be different in the forward and backward velocity regime.

6.7 Next, for each MT plus end and motor protein, the probability of the switch event is compared with the random number, r, generated in **step 6.4**: if $r < P_{switch}$, the switch event is executed, whereas if $r \geq P_{switch}$, the current state is unchanged. In all cases, the new state (changed or unchanged) of each MT plus end and motor protein is stored for the next time step, $t = t_{n+1}$.

6.8 The polymerization or depolymerization rate is used to update the position of each MT plus end according to the MT's new state computed in the previous step.

6.9 Finally, the time step is increased by Δt, and the sequence of events is repeated until $n = t_{final}$, typically hundreds to thousands of steps.

7. The numerical solutions of the model yield the positions and states of all spindle parts as well as the time-dependent forces acting on these spindle parts. They can be plotted at each time step and be presented in graphs or stored as a sequence of frames which can be displayed as a movie to be compared with experimental data.

8. Due to the stochastic aspects of the numerical algorithm, the solutions should be computed tens of times with the same set of parameters, to evaluate the variance in the dynamics.

3.7.4 Analysis of the Force-Balance Model

The analysis of the model comprises the solution of the equations with different parameters within the biologically reasonable estimated range or known values and comparison with the available experimental data. The aims of this modeling are:

1. To quantitatively reproduce the experimental observation based on the properties of the molecules proposed to drive the motility event

2. To identify the parameters which alter the dynamics of the spindle (sensitive parameters)

3. To make predictions on the outcome in response to changes in these parameters

4. To identify parameters which do not affect the outcome within a reasonable biological range

5. To design new experiments to verify, revise, or falsify the model

For example, we modeled the formation of the spindle at nuclear envelope breakdown in an FB approach based on the action of dynamic MTs and two antagonistic motors, kinesin-5 generating outward force and kinesin-14 generating inward force. In this case, nonlinear (piecewise linear) force-velocity relationships were used for the motors in both the forward and backward direction based on in vitro data (Fig. 6b). The resulting model agreed with the experimental data but also suggested the existence of an additional elastic force for a robust steady-state length, and experiments showed that this was the nuclear lamina [27].

3.7.5 Formulation and Analysis of the Biochemical/Kinetic Model

To test if the observed dynamic properties of the MTs, inferred from FRAP and FSM data, and the distribution of the MT plus ends, inferred from EB1-GFP imaging, are consistent with the generally accepted model of spindle MT organization with minus ends near the spindle poles and plus ends extending toward the spindle equator and to identify the MT dynamic instability parameters that could account for the observed FRAP recovery rates, we developed a stochastic modeling approach.

1. In this model, the stochastic dynamic instability properties of MT plus ends and MT poleward flux are accounted for, without explicitly incorporating force generation, by assuming that as long as there is sufficient substrate (i.e., antiparallel overlap) for motor binding, the average spindle MT poleward sliding movement is maintained [15].

2. A custom-made MATLAB script describing the stochastic dynamics of MTs plus and minus ends in the spindle from metaphase through anaphase B was developed (*see* Subheading on "The Stochastic MT Dynamics Model").

3. Each run of the model yields a solution, comprised of the time-dependent positions of the MT plus and minus ends, and the (growth/shrinkage) state of the MT plus ends, associated with the specific set of parameters and initial conditions.

4. All solutions are stored in a large data file for further evaluation.

5. Each solution is used to perform multiple in silico FRAP [8] in different regions and/or mitotic phases, and in silico FSM or EB1 kymographs [15] (*see* Subheadings 3.7.6 and 3.7.7 below).

6. The in silico results are compared with the corresponding experimental data to determine the adequacy of the MT organization and parameters of the specific run. We have used this setup to study what causes the change in MT dynamics observed at anaphase B onset [15].

The Stochastic MT Dynamics Model

1. A set of model parameters (MT DI parameters, number of MTs) and initial conditions (state of MT plus ends and positions of the plus and minus ends) are defined.

2. At each time step, the stochastic switch probability of each MT plus end is computed, and a switch event is realized if appropriate (using the methodology explained above in Subheading 3.7.3).

3. The growth or shortening length of the MT end is computed, as applicable. For example, if the MT is in the growth state, the length of the MT is extended by $l = v_g \Delta t$.

4. Each MT's plus and minus ends are moved poleward by simultaneous depolymerization of the minus end at the experimentally observed average poleward flux rate, augmented by a stochastic value randomly selected to match the experimental variance in the poleward flux rate.

5. The new position and state of each MT's plus and minus ends are stored for the next time step, and the sequence of events is repeated until a user-defined final time step.

6. The model run is repeated (typically ten times) for each parameter set and initial condition, and the parameters are varied systematically in small increments to cover the biologically relevant range in the four-dimensional space of MT dynamic instability parameters (v_g, v_s, f_{cat}, and f_{res}).

7. During a run, if the total antiparallel overlap length of the MTs in the spindle drops below a threshold value (close to zero), the solutions are discarded and the parameter set and initial conditions are retained and labeled as "overlap deficient." All other solutions, together with the initial conditions and parameters, are stored for further evaluation (Fig. 6d).

Motor Protein Dynamics in the Spindle

Similarly to the stochastic model describing the dynamics of the MTs, the binding and unbinding of motors to spindle MTs and their free diffusion around the spindle can be modeled in a stochastic approach to study motor dynamics [10].

1. In this case, only the binding and dissociation of the motors on the spindle MTs (whose positions are already defined in the solution data file by the algorithm described above) have to be executed in a stochastic approach, similarly to the switch events of the MT plus ends between growth and shrinkage states.

2. The bound motors are moved along the MTs at their stepping rate, and the free motors are allowed to diffuse with a diffusion coefficient defined in the model parameters.

3. In silico FRAP of GFP-tagged motors and in silico kymographs of the motors can be generated (see next two sections) to compare with experimental data.

3.7.6 In Silico FRAP

1. A "bleach time" and a rectangular area (corresponding to an interval $[x_L, x_R]$ in the one-dimensional model, where x_L and x_R are the bleach boundaries proximal to the opposing poles) are defined.

2. The total length of MT segments within the bleach zone $[x_L, x_R]$ is computed and stored in a variable "fluo(t_n)" at each time step prior to the desired bleach time.

3. At the bleach time, the boundaries (left and right, proximal to the left and right pole, respectively) of each MT segment within the bleach interval are determined and stored as the left and right bleach boundary, for each MT, and "fluo($t_{bleach\text{-}time}$)" is defined as zero.

4. In the subsequent time steps, for each MT, the left and right bleach boundaries are moved at the known poleward flux rate, and if the MT depolymerizes past a bleach boundary, the boundary is updated as the new position of the MT plus end.

5. New fluorescent segments appear in the bleach zone as MTs slide poleward and depolymerize/polymerize (Fig. 6e).

6. For each MT, the newly appearing fluorescent segment length within the bleach zone is calculated, and the sum of all fluorescent MT segments within the bleach interval [x_L, x_R] is determined and recorded in the array "fluo(t_j)."

7. When the final time is reached, the array "fluo(t_j)" is normalized by the average of 10 consecutive prebleach time steps.

8. The normalized array is plotted as a function of time—the in silico FRAP.

9. As with experimental data, the in silico FRAP curve can be fit to a first-order exponential curve, which gives the half time and the percent of recovery.

3.7.7 In Silico FSM and In Silico Kymograph

The script developed for the stochastic MT dynamics model algorithm is augmented with the following steps:

In Silico FSM of Tubulin

1. An average "number of tubulin speckles," N_{tub}, per micron MT length is defined in the set of model parameters.

2. At the initial time step, the length of each MT (L_{MT}) is computed from its plus and minus end positions. The number of speckles on this MT is generated as f_tub$_{MT}$ = round($L_{MT} \times N_{tub}$) using the round function in MATLAB.

3. The positions of fluorescent tubulin subunits incorporated in the lattice are selected randomly and stored in array f_tub$_{MT}$ positions with the MT length.

4. In each subsequent time step for each MT, the MT poleward flux rate and the polymerization/depolymerization rates are determined and executed, and then the existing positions of f_tub$_{MT}$ fluorescent tubulin subunits are moved at the same rate as the MT in that time step and stored as the new speckle positions for the following time step.

5. If the position of a tubulin mark is further than the MT plus end (i.e., the MT depolymerized past the tubulin mark), this mark is discarded in the next time step, and the number of marks f_tub$_{MT}$ for that MT is decreased. Depending on the depolymerization rate, more than one mark could disappear in one time step.

6. The distance between the MT tip and the mark closest to it, l_{tip}, is computed. If round($l_{tip} \times N_{tub}$) is equal or larger than one (i.e., the MT has polymerized enough), a new tubulin mark is added along the new length and f_tub$_{MT}$ for that MT is increased by one. The new number of tubulin marks and its positions are stored for the next time step.

7. The positions of all tubulin marks on all spindle MTs are saved in an array.

8. The maximal distance between the spindle poles throughout the solution (i.e., in any time step) is determined and divided into 200 nm bins (the outer bins are allowed to extend slightly beyond the spindle poles' positions to keep the bin size fixed).

9. At each time step, the number of marks in each bin is computed (bin boundaries are assumed to be closed on the left and open on the right) and recorded in a time-dependent array.

10. The time-dependent array of the bins is plotted using the *imagesc* function, with *colormap(gray)* property in MATLAB, giving rise to an in silico FSM kymograph.

In Silico FSM of EB1

1. A model solution data file resulting either from the FB or the stochastic MT dynamics model is loaded.

2. The maximum distance between the spindle poles over time is computed and divided into 200 nm bins as described above.

3. At each time step, the number of MT tips in the growth state in each bin is computed and recorded in a time-dependent array.

4. This array is then plotted using the *imagesc* function, with *colormap(jet)* property in MATLAB to yield the corresponding EB1-GFP kymograph.

4   Notes

1. This protocol is relatively straightforward; however, each step requires practice to make sure the embryos are not damaged.

2. Embryos should be visualized about 2 h after laying, and they are usually collected for 1 h and allowed to mature for another hour, so that embryos are between 1 and 2 h "old." Shorter collection times (with an appropriate longer maturation time) will give embryos in a smaller age gap and thus, a narrower range of cycles.

3. Embryos should be imaged about 2 h after the start of the collection, so allow yourself sufficient time to complete all steps.

4. First image uninjected embryos to make sure the physical manipulation of picking up and dechorionating are not

damaging the embryo. Pole–pole plots are very reproducible, they are specific to each cycle, but the same cycle in different embryos should be essentially the same (Fig. 3b).

5. If the physical manipulation is not causing problems, dehydrate the embryos and inject them with buffer. The pole–pole plots should be the same as those obtained without injection. Spindles should be evenly distributed on the cortex. Multipolar spindles and connections between spindles usually result from too much dehydration. If you observe this, reduce the time in the dehydration chamber.

6. Free centrosomes indicate that nuclei have been lost to nuclear fallout; this indicates a problem in the manipulation or injection.

7. Once control embryos are reproducible and reliable, inject a fluorescent protein or inhibitor. Embryos injected with a fluorescent wild-type protein should behave like wild type—otherwise the protein is having an effect.

8. Embryos are injected on an inverted microscope usually with a 16× objective (Fig. 2c). However, if the injected inhibitor acts very fast or the time of injection needs to be determined precisely with respect to the time in mitosis, embryos can be injected under a 60× objective on the final imaging microscope (inverted).

9. When injecting inhibitors and antibodies, several controls are needed. First the buffer alone should not cause any defects. Second, the optimum concentration should be determined as explained in Subheading 3.1.2.

10. Ideally, two forms of inhibition should be tried and shown to give the same results, for example, an antibody and a dominant negative construct [28].

11. For all microscopy techniques, bleaching should be minimized by minimizing the laser power and exposure.

12. We have used these methods to study different stages of mitosis. For example, we have used all of these techniques to study the function of the kinesin-5 KLP61F (Fig. 7). We generated fly stocks expressing a functional KLP61F-GFP transgene [7]. Microinjection of a function-blocking antibody led to spindle collapse at the injection site [13, 29] and less severe defects further from the injection site [13, 29] (Fig. 7a). The gradient within one embryo can be quantified by measuring the relative concentration of KLP61F-GFP in each spindle and correlated to the severity of the defects in pole–pole distance plots (Fig. 7b). FSM of tubulin in partially inhibited spindles showed that KLP61F function is required for tubulin poleward flux [13] (Fig. 7c). KLP61F-GFP also forms speckles in the spindle (Fig. 7d), and FRAP of KLP61F-GFP (Fig. 7e) demonstrates

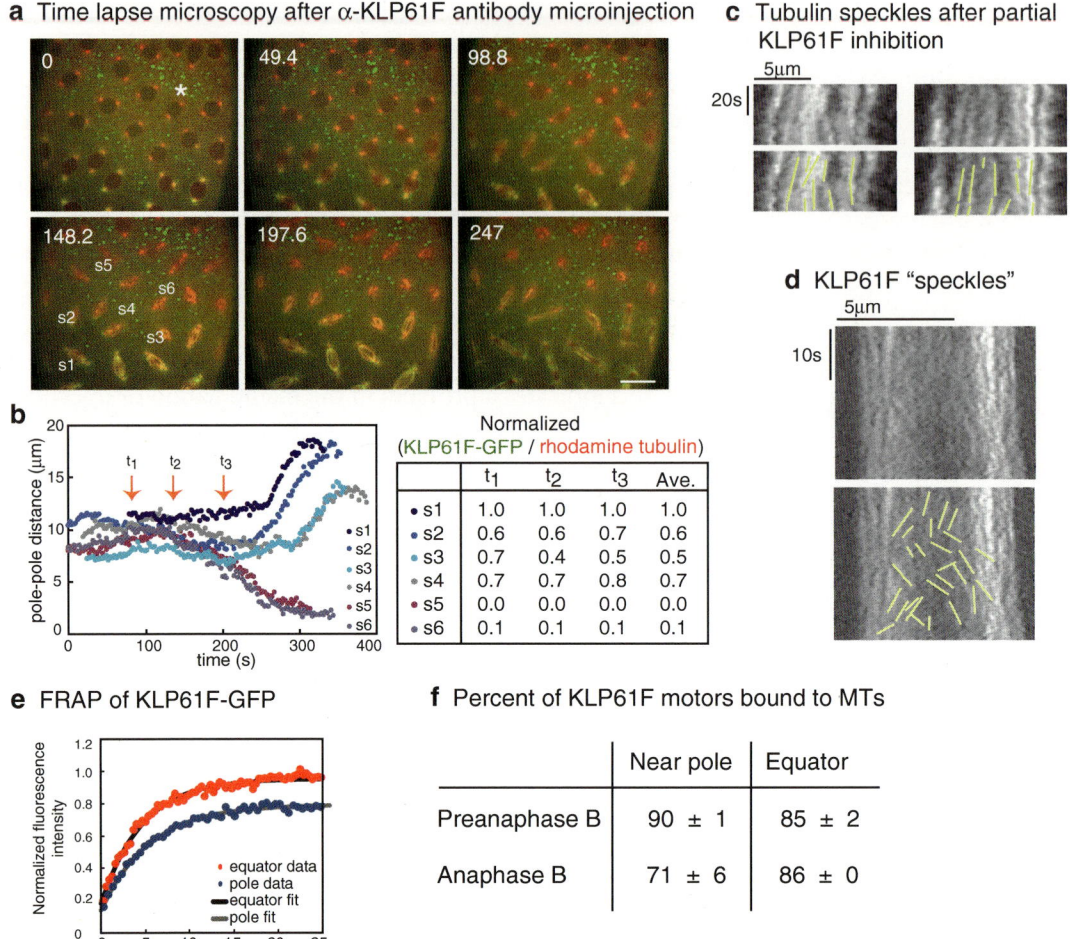

a Time lapse microscopy after α-KLP61F antibody microinjection

c Tubulin speckles after partial KLP61F inhibition

d KLP61F "speckles"

b

Normalized
(KLP61F-GFP / rhodamine tubulin)

	t_1	t_2	t_3	Ave.
● s1	1.0	1.0	1.0	1.0
● s2	0.6	0.6	0.7	0.6
● s3	0.7	0.4	0.5	0.5
● s4	0.7	0.7	0.8	0.7
● s5	0.0	0.0	0.0	0.0
● s6	0.1	0.1	0.1	0.1

e FRAP of KLP61F-GFP

f Percent of KLP61F motors bound to MTs

	Near pole	Equator
Preanaphase B	90 ± 1	85 ± 2
Anaphase B	71 ± 6	86 ± 0

Fig. 7 Function of the kinesin-5 KLP61F. (**a**) Microinjection of anti-KLP61F antibody: Images from a time-lapse movie of an embryo expressing KLP61F-GFP (*green*) injected with rhodamine tubulin (*red*) and anti-KLP61F. Time in each frame is given in seconds from NEB. Bar, 10 μm. The injection site was close to the top of the embryo (*asterisk*), KLP61F-GFP forms immunoprecipitates, and most KLP61F-GFP is depleted from the spindles. These spindles collapse, as seen at 247 s. Toward the bottom of the embryo, KLP61F-GFP is still present on the spindles, and consequently these spindles assemble, though they may exhibit defects. (**b**) Graph of pole–pole distance as a function of time (*left*) and quantification of KLP61F remaining on these spindles (*right*). The normalized ratio of KLP61F-GFP to rhodamine tubulin is used to compare the amount of motor remaining on each spindle at different time points. (**c**) Poleward flux requires the function of KLP61F. In embryos injected with anti-KLP61F antibody, kymographs obtained from spindles that did not collapse but maintained a steady length show that tubulin speckles do not flux toward the pole, indicating that the function of KLP61F is required for flux. (**d**) Kymographs of KLP61F-GFP time-lapse movie show KLP61F speckles, which have shorter runs than tubulin. In the equator region of the spindle, many speckles appear immobile consistent with KLP61F sliding antiparallel ipMTs apart while staying stationary. (**e**) FRAP analysis of KLP61F-GFP during anaphase B. (**f**) Modeling of the FRAP data shows that KP61F is partitioned between freely diffusing and microtubule-bound states, with most motors bound to MTs at any given time [7] (**a–c** adapted from [13], **d** and **f** adapted from [7])

that KLP61F has rapid dynamics in the spindle [7]. We also developed a force-balance model to test the qualitative model proposed for anaphase B spindle elongation mechanism [8]. This showed quantitatively that our model was feasible and a very robust mechanism for spindle elongation in which the rate is dependent only on a few parameters (Fig. 6a). It made predictions about which parameters are important [8], and how changes in these parameters should impact the outcome.

Acknowledgements

This protocol is currently used in our laboratory and has been refined over the years by many people including Drs. David Sharp, Mijung Kwon, Patrizia Sommi, and Dhanya Cheerambathur. We thank Dr. Bill Sullivan (UCSC), who provided us with excellent advice on the manipulation and microinjection of early *Drosophila* embryos when our work in this system was being initiated. We thank all members of the Scholey laboratory. Our work on mitosis in *Drosophila* is supported by NIH grant GM55507.

References

1. Brust-Mascher I, Scholey JM (2007) Mitotic spindle dynamics in *Drosophila*. Int Rev Cytol 259:139–172

2. Duffy JB (2002) GAL4 system in *Drosophila*: a fly geneticist's Swiss army knife. Genesis 34: 1–15

3. Buster DW, Zhang D, Sharp DJ (2007) Poleward tubulin flux in spindles: regulation and function in mitotic cells. Mol Biol Cell 18:3094–3104

4. Mitchison TJ (1989) Polewards microtubule flux in the mitotic spindle: evidence from photoactivation of fluorescence. J Cell Biol 109: 637–652

5. Waterman-Storer C, Desai A, Salmon ED (1999) Fluorescent speckle microscopy of spindle microtubule assembly and motility in living cells. Methods Cell Biol 61:155–173

6. Waterman-Storer CM, Desai A, Bulinski JC et al (1998) Fluorescent speckle microscopy, a method to visualize the dynamics of protein assemblies in living cells. Curr Biol 8:1227–1230

7. Cheerambathur DK, Brust-Mascher I, Civelekoglu-Scholey G et al (2008) Dynamic partitioning of mitotic kinesin-5 cross-linkers between microtubule-bound and freely diffusing states. J Cell Biol 182:429–436

8. Brust-Mascher I, Civelekoglu-Scholey G, Kwon M et al (2004) Model for anaphase B:

role of three mitotic motors in a switch from poleward flux to spindle elongation. Proc Natl Acad Sci U S A 101:15938–15943

9. Civelekoglu-Scholey G, Sharp DJ, Mogilner A et al (2006) Model of chromosome motility in *Drosophila* embryos: adaptation of a general mechanism for rapid mitosis. Biophys J 90: 3966–3982

10. Civelekoglu-Scholey G, Tao L, Brust-Mascher I et al (2010) Prometaphase spindle maintenance by an antagonistic motor-dependent force balance made robust by a disassembling lamin-B envelope. J Cell Biol 188:49–68

11. Cytrynbaum EN, Sommi P, Brust-Mascher I et al (2005) Early spindle assembly in *Drosophila* embryos: role of a force balance involving cytoskeletal dynamics and nuclear mechanics. Mol Biol Cell 16:4967–4981

12. Brust-Mascher I, Scholey JM (2009) Microinjection techniques for studying mitosis in the *Drosophila melanogaster* syncytial embryo. J Vis Exp (31): 1382

13. Brust-Mascher I, Sommi P, Cheerambathur DK et al (2009) Kinesin-5-dependent poleward flux and spindle length control in *Drosophila* embryo mitosis. Mol Biol Cell 20: 1749–1762

14. Axelrod D, Koppel DE, Schlessinger J et al (1976) Mobility measurement by analysis of

fluorescence photobleaching recovery kinetics. Biophys J 16:1055–1069

15. Cheerambathur DK, Civelekoglu-Scholey G, Brust-Mascher I et al (2007) Quantitative analysis of an anaphase B switch: predicted role for a microtubule catastrophe gradient. J Cell Biol 177:995–1004

16. Phair RD, Misteli T (2000) High mobility of proteins in the mammalian cell nucleus. Nature 404:604–609

17. Lele T, Wagner SR, Nickerson JA et al (2006) Methods for measuring rates of protein binding to insoluble scaffolds in living cells: histone H1-chromatin interactions. J Cell Biochem 99:1334–1342

18. Salmon ED, Leslie RJ, Saxton WM et al (1984) Spindle microtubule dynamics in sea urchin embryos: analysis using a fluorescein-labeled tubulin and measurements of fluorescence redistribution after laser photobleaching. J Cell Biol 99:2165–2174

19. Verde F, Dogterom M, Stelzer E et al (1992) Control of microtubule dynamics and length by cyclin A- and cyclin B-dependent kinases in *Xenopus* egg extracts. J Cell Biol 118:1097–1108

20. Sharp DJ, McDonald KL, Brown HM et al (1999) The bipolar kinesin, KLP61F, crosslinks microtubules within interpolar microtubule bundles of *Drosophila* embryonic mitotic spindles. J Cell Biol 144:125–138

21. Maiato H, Hergert PJ, Moutinho-Pereira S et al (2006) The ultrastructure of the kinetochore and kinetochore fiber in *Drosophila* somatic cells. Chromosoma 115:469–480

22. Hays TS, Salmon ED (1990) Poleward force at the kinetochore in metaphase depends on the number of kinetochore microtubules. J Cell Biol 110:391–404

23. Toba S, Watanabe TM, Yamaguchi-Okimoto L et al (2006) Overlapping hand-over-hand mechanism of single molecular motility of cytoplasmic dynein. Proc Natl Acad Sci U S A 103:5741–5745

24. Valentine MT, Fordyce PM, Krzysiak TC et al (2006) Individual dimers of the mitotic kinesin motor Eg5 step processively and support substantial loads in vitro. Nat Cell Biol 8: 470–476

25. Marshall WF, Marko JF, Agard DA et al (2001) Chromosome elasticity and mitotic polar ejection force measured in living *Drosophila* embryos by four-dimensional microscopy-based motion analysis. Curr Biol 11:569–578

26. Wollman R, Civelekoglu-Scholey G, Scholey JM et al (2008) Reverse engineering of force integration during mitosis in the *Drosophila* embryo. Mol Syst Biol 4:195

27. Civelekoglu-Scholey G, Scholey JM (2010) Mitotic force generators and chromosome segregation. Cell Mol Life Sci 67:2231–2250

28. Kwon M, Morales-Mulia S, Brust-Mascher I et al (2004) The chromokinesin, KLP3A, drives mitotic spindle pole separation during prometaphase and anaphase and facilitates chromatid motility. Mol Biol Cell 15:219–233

29. Sharp DJ, Yu KR, Sisson JC et al (1999) Antagonistic microtubule-sliding motors position mitotic centrosomes in *Drosophila* early embryos. Nat Cell Biol 1:51–54

30. Brust-Mascher I, Scholey JM (2002) Microtubule flux and sliding in mitotic spindles of *Drosophila* embryos. Mol Biol Cell 13:3967–3975

31. Rogers SL, Rogers GC, Sharp DJ et al (2002) *Drosophila* EB1 is important for proper assembly, dynamics, and positioning of the mitotic spindle. J Cell Biol 158:873–884

32. Rusan NM, Tulu US, Fagerstrom C et al (2002) Reorganization of the microtubule array in prophase/prometaphase requires cytoplasmic dynein-dependent microtubule transport. J Cell Biol 158:997–1003

33. Schnitzer MJ, Visscher K, Block SM (2000) Force production by single kinesin motors. Nat Cell Biol 2:718–723

34. Tao L, Mogilner A, Civelekoglu-Scholey G et al (2006) A homotetrameric kinesin-5, KLP61F, bundles microtubules and antagonizes Ncd in motility assays. Curr Biol 16:2293–2302

35. Carter NJ, Cross RA (2005) Mechanics of the kinesin step. Nature 435:308–312

36. Krzysiak TC, Grabe M, Gilbert SP (2008) Getting in sync with dimeric Eg5. Initiation and regulation of the processive run. J Biol Chem 283:2078–2087

Chapter 2

Rapid Measurement of Mitotic Spindle Orientation in Cultured Mammalian Cells

Justin Decarreau, Jonathan Driver, Charles Asbury, and Linda Wordeman

Abstract

Factors that influence the orientation of the mitotic spindle are important for the maintenance of stem cell populations and in cancer development. However, screening for these factors requires rapid quantification of alterations of the angle of the mitotic spindle in cultured cell lines. Here we describe a method to image mitotic cells and rapidly score the angle of the mitotic spindle using a simple MATLAB application to analyze a stack of Z-images.

Key words Mitotic spindle, Microtubules, MATLAB, DeltaVision

1 Introduction

Correct orientation of the mitotic spindle in dividing cells is crucial for asymmetric cell division during development as well as for the maintenance of stem cell populations. However, it is often simplest to screen for gene products involved in this process by assaying for quantitative changes in the orientation of mitotic spindles in cultured cell lines. Cultured transformed cell lines are often the system of choice for these measurements because they are easier to grow reproducibly than polarized cell sheets and easier to image than cells in tissues. As spindle orientation may exhibit baseline randomness in cultured cells, enough cells must be assayed to determine whether a particular gene product or treatment is influencing spindle orientation to a significant extent. Cell lines also have the advantage that they can be transfected with proteins to rescue spindle orientation and confirm siRNA depletions. Cultured cells also lack the intrinsic apical and basolateral polarization found in some epithelial model systems for cell polarization which can be an advantage if treatments lead to changes in cell motility or cell adhesion which could complicate the spindle orientation assays. Thus, even though cultured transformed cell lines may not

David J. Sharp (ed.), *Mitosis: Methods and Protocols*, Methods in Molecular Biology, vol. 1136,
DOI 10.1007/978-1-4939-0329-0_2, © Springer Science+Business Media New York 2014

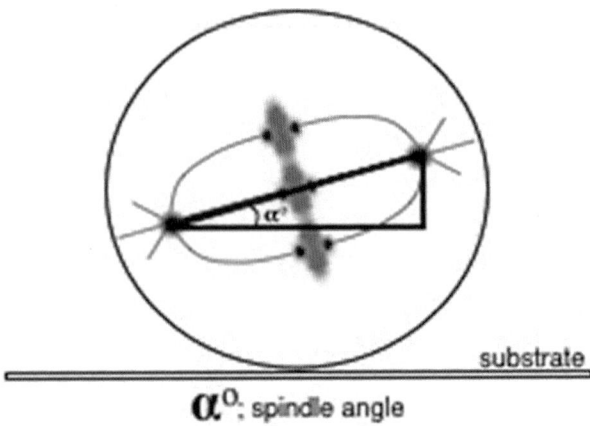

substrate

α^{o}: spindle angle

Fig. 1 Schematic showing measurement of the mitotic spindle angle relative to the growth substrate. Alpha represents the calculated spindle angle derived from the measured distance between centrosomes and the number of vertical slices between centrosome focal planes. Spindle angles between 0° and 10° are considered normal

functionally rely on oriented spindles, they still exhibit an orientation relative to the coverslip that is consistent enough to use to screen for the machinery and regulators of spindle orientation.

The mitotic spindles of cultured HeLa cells tend to naturally orient to an angle of between 5° and 15° relative to the substrate (Fig. 1). Experimental treatments and loss of the machinery for mitotic spindle orientation can lead to an increase of 30–50° in the angle of the mitotic spindle relative to the substrate [1, 2]. Alternate methods for scoring the orientation of the mitotic spindle and cleavage plane have successfully employed patterned substrates [3, 4] or in situ imaging relative to a reference orientation [5] to quantify mitotic spindle orientation. These can be quite effective but are challenging either from the standpoint of cost (for the production of patterned substrates) or with respect to imaging within tissue. Both of these challenges limit the speed and number of cells that can be effectively scored. In contrast, evaluating the orientation of the spindle relative to the cultured substrate requires no special equipment, utilizes standard cell culture methods, and can be applied to common cell lines.

2   Materials

2.1   Cell Culture

2.1.1   Cell Culture Reagents

1. Acid washed No. 1.5 glass coverslips (12 mm round) (Electron Microscopy Sciences): Acid wash by incubating in 1 N HCl solution for 2 h, swirling occasionally. Decant acid solution and wash extensively with dH₂O five times. Wash an additional five times with 95 % EtOH. Coverslips should be stored in 95 % EtOH and flamed prior to use.

2. 10 cm^2 Falcon polystyrene tissue culture plates.

3. 24-well Falcon polystyrene tissue culture plates.

4. Fetal bovine serum (FBS) hydroclone, 40 nm filtered. Stored in 50 mL aliquots at –20 °C.

5. Pen-Strep (Gibco) 100× working solution.

6. L-glutamine.

7. MEM-alpha (Gibco): Mix one 10 g package (makes 1 L total media) with 850 mL dH$_2$O, stir to mix. Add 2.2 g NaHCO$_3$ and stir to mix and then adjust to desired pH. Bring volume up to 900 mL with dH$_2$O. Add 100 mL FBS (10 % final concentration) and filter sterilize using vacuum-driven filtration system such as 0.22 μm Millipore Durapore bottles. Pen-Strep and L-glutamine can be added prior to filtration if desired.

8. 0.05 % Trypsin/EDTA for HeLa cells; 0.25 % Trypsin/EDTA for Hct116 cells (Gibco).

9. Sterile phosphate buffered saline (PBS): 0.137 M NaCl, 2.68 mM KCl, 8 mM Na$_2$HPO$_4$·7H$_2$O, 1.47 mM KH$_2$PO$_4$. PBS is sterilized by autoclaving.

10. Cultured cell lines HeLa, Hct 116, hTERT RPE-1 (*see* **Note 1**).

2.1.2 Cell Culture Equipment

1. Incubator set at 37 °C and 5 % CO$_2$ for cell maintenance.

2. Water bath set to 37 °C for warming media.

3. Tissue culture hood.

4. Microscope.

5. Hemocytometer.

2.2 Fixation and Labeling

2.2.1 Fixation Reagents

1. 16 % paraformaldehyde solution (Electron Microscopy Sciences).

2. Methanol at –20 °C.

3. Donkey serum (Jackson ImmunoResearch): Stored at –20 °C at 60 mg/mL in ddH$_2$O.

4. Tris-buffered saline (TBS): 50 mM Tris, 150 mM NaCl. Adjust pH to 7.6 (*see* **Note 2**).

5. Fixative: 4 % paraformaldehyde diluted into –20 °C methanol (*see* **Note 3**).

6. Blocking buffer: 20 % donkey serum in TBS.

7. Antibody dilution buffer (AbDil): 1× TBS, 1 % BSA (Jackson ImmunoResearch), 0.1 % Triton-X 100, 0.01 % azide.

8. Mouse monoclonal alpha-tubulin (DM1a) antibodies (Sigma) used at 1:500. Diluted in AbDil.

9. Mouse monoclonal gamma-tubulin antibodies (Sigma): used at a concentration of 1:1,000. Diluted in AbDil.

10. Donkey anti-mouse secondary antibodies with TRITC or FITC labels.

2.2.2 *Fixation Equipment*

1. Thomas cover glass staining outfit (Thomas Scientific).
2. Parafilm.
3. 15 cm² Petri dish.
4. Kimwipes.
5. Beaker (200 mL).
6. Dumont negative action style tweezers. Style N4 (Electron Microscopy Sciences).

2.3   Imaging

2.3.1 *Imaging Reagents*

1. Vectashield mounting medium with DAPI (Vector Labs).
2. Slides.

2.3.2 *Imaging Equipment*

1. DeltaVision deconvolution microscope.

2.4   Image Analysis

2.4.1   Image Analysis Software Introduction

To generate meaningful statistics that describe the distribution of spindle angles adopted by cells under a certain set of conditions, large numbers of images (~100) must be analyzed per condition. Doing this manually can be extremely tedious. A better alternative is to write software that automates the identification of spindle poles in batches of image Z-stacks with only occasional direction from the user. Here we describe such software written in MATLAB.

2.4.2   Spindle Pole Location Algorithm

To measure spindle angles, both spindle poles must be located. Labeling gamma tubulin makes this easy since the two poles are often the brightest objects in the image stack, but this is not *always* true. Our algorithm selects the two brightest objects by default but also provides other candidate poles from which the user can manually select in the event that the simplest assumption is not true.

In the first step, the algorithm creates a maximum intensity projection of the Z-stack. Doing so makes the localization less computationally expensive and makes it much easier to display the results for inspection by the user. The algorithm then integrates the maximum Z intensities over a sliding 5×5 (about $500 \, nm \times 500 \, nm$) pixel window in X and Y. Locations where those integrated intensities are greater than any other integrated intensity within 15 pixels are candidate poles. The two candidate poles with the brightest integrated intensities are selected by the algorithm as the true poles.

```
% 'spindle' is a 3D double array storing the image
stack.
% the images are 256x256.
flatspindle = max(spindle,[],3);

intensitysum = zeros(252,252);
for i = 1+2:256-2
for j = 1+2:256-2
        roi = flatspindle(i-2:i+2,j-2:j+2);
        intensitysum(i,j) = sum(roi(:));
```

```
            end
       end
       candidatepoles = zeros(1,4);
       m = 0;
       for i = 1+15:252-15
       for j = 1+15:252-15
                 roi = intensitysum(i-15:i+15,j-15: j+15)
                 if intensitysum(i,j) == max(roi(:))
                       m = m + 1;
                       % [Y,X,(Z placeholder),Intensity]
       data is stored.
                       candidatepoles(m,:) = [i,j,0,
       intensitysum(i,j);
                 end
          end
       end
       [scores,I] = sort(candidatepoles(:,4));
       finalpoles = I(end-1:end);
```

Once the final pole objects have been identified in the 2D maximum intensity projection, the objects can be located in Z by finding the maximum integrated intensity within a $5 \times 5 \times 3$ vertically sliding cube centered where the poles were found.

```
pmax = 0;
for l = 1:size(candidatepoles,1)
     for k = 2(size(spindle,3)/3)-1
roi = spindle(candidatepoles(l,1)-
2:candidatepoles(l,1)+2,candidatepoles(l,2)-
2: candidatepoles(l,2)+2,k-1:k+1);
sroi = sum(roi(:));
if sroi > pmax
       candidatepoles(l,3) = k;
       pmax = sroi;
     end
   end
end
```

The result is an array, candidatepoles, that stores all of the location information for the pole-like objects (Y,X,Z,Intensity) and a vector, finalpoles, that contains the indices of the two candidate poles that the program will accept for now as the true spindle poles. Once the user has vetted this selection, the position information can be exported in spreadsheet form using a function like xlswrite. The spindle length and angle are easy to calculate at this point.

```
   % [x1,y1,z1] and [x2,y2,z2] are the positions of
poles 1 and 2
   % pixelsize is the size of a pixel in microns
   % zscanwidth is the distance between image slices
in microns
```

```
vect = [pixelsize*(x1-x2),pixelsize*(y1-y2),
zscanwidth*(z1-z2)]';
    L = sqrt(vect(1)^2 + vect(2)^2);
    spindlelength = sqrt(vect'*vect);
    spindleangle = (180/pi)*atan(abs(vect(3))/L);
```

2.4.3 The GUI

In our experience, the algorithm described above correctly identifies the spindle poles about 99 % of the time. Nevertheless, most users will want to be able to correct erroneous results in a manner that is not painstaking. This is the purpose of the graphical user interface (GUI). We cannot describe the ~1,100 lines of code in great detail here, but we can give an overview of its construction.

Creating a GUI

A GUI is simply a customized MATLAB figure with user interface controls. To create one programmatically, you need to write an m-file with instructions for the design of the interface and the functions that execute when buttons are pressed. To combine all of this information into one file, create a master function of the same name as the file that calls the appropriate subfunction. The code below creates a small part of the interface and should be instructive as to how the rest is generated. The MathWorks website has excellent tutorials available for GUI construction.

```
%% -- SpindleGUI.m --

%% This is the master function carrying the
SpindleGUI name.
%% The 'action' argument is the name of the sub-
function called.
    function SpindleGUI(action)

if nargin < 1
    %% if no 'action' is specified, the initializa-
tion function
    %% runs
    InitializeSpindleGUI
else
    feval(action)
end
end
%% This is a partial look at the initialization
function
    function InitializeSpindleGUI
    %% The SpindleGUI figure declaration. The figure
handle is stored %% in 'SF'.
    SF = figure('Name','SpindleGUI'),...

    'NumberTitle','off',...
    'Position',[50,50,1225,800],...
    'Resize','off',...
    'MenuBar','none');
```

```
%% Axes are placed on the SpindleGUI with this
declaration.
   The %% axes handle is stored in the 'ud'
structure.
   ud.Axes = axes('Parent',SF,…
      'Units','Pixels',…
      'Position',[450,50,700,700],…
      'Box','on',…
      'XTick',[],…
      'YTick',[]);
%% This button will pull up a file dialog box for
the user to
%% select the file to run the algorithm on. Notice
the Callback
%% property stores a SpindleGUI command to be
executed by the
%% master function. The subfunction will be written
below this
%% initialization function.
   ud.FileButton = uicontrol('Style','Push',…
      'Position',[50,500,100,100],…
      'String','File',…
      'FontSize',24…
      'Callback','SpindleGUI(''FileButtonCallb
ack");');

%% Define a default filepath to retrieve image stack
data.
   ud.FilePath = 'My Documents\Data\';
%% Create a blank image object to store projection
images later.
   ud.SpindleImage = imagesc(zeros(256,256)); color-
map gray;
   set(ud.Axes,'XTick',[],'YTick',[]);
%% The 'ud' structure is written to the Spindle
figure userdata
%% property so that it can be recalled by other
functions later.
   set(SF,'UserData',ud);
   end
%% This is the function that executes when the file
button is
%% pressed. The user will first select a file from
a dialog box.
%% Then the max Z projection will be calculated
and placed on the
%% axes, and the pole locating algorithm will be
executed.

function FileButtonCallback
```

```
%% Recall the figure's user data.
ud = get(gcbf,'UserData');

%% The code for the dialog box. The conditional
handles the case
%% that the user clicks 'cancel' on the dialog box.
[filename,pathname] = uigetfile([ud.FilePath,' *.
tif']);
if filename == 0
    return
end
%% Store the new filepath as the GUI default.
ud.FilePath = pathname;
%% Redefine filename to include the full file path.
filename = [pathname,filename];
%% Query the file for information about its length.
Our files are
%% RGB, but gamma tubulin is only in the green
channel. We will
%% discard the other two.
info = imfinfo(filename);
L = length(info);
ud.NZScans = L/3;
spindle = zeros(256,256,L/3);

%% Store all of the green channel images in the
spindle array.
j = 0;
for i = 2:3:L
    j = j + 1;
    spindle(:,:,j) = imread(filename,i);
end

%% Convert the spindle array to type double for
computation.
spindle = double(spindle);

%% Execute the algorithm reproduced in the ear-
lier sections of %% this chapter.
```

3   Methods

3.1   General Cell Culture

Each of the above human cell lines can be routinely cultured in MEM-alpha media plus 10 % FBS supplemented with Pen-Strep and L-glutamine.

3.1.1   Splitting and Plating Cells for Analysis

1. Remove culture medium from cells plated in 10 cm^2 dish and rinse with 10 mL sterile PBS.

2. Add 1 mL Trypsin/EDTA and incubate at 37 °C 10 min to allow cells to come off the dish.

3. Add 4 mL MEM-Alpha to inactivate trypsin and triturate cells.

4. Add a sterile 12 mm² coverslip to each well of a 24-well plate.

5. Add 0.5 mL of media to each well.

6. Remove 9 μL of cell suspension and count cells in a hemocytometer.

7. Add 25,000–35,000 cells/cm² to each well containing a coverslip (*see* **Note 4**).

8. Culture in a CO_2 incubator for a maximum of 48 h.

3.2 Fixation and Labeling for Fluorescence Microscopy

3.2.1 Fixation

1. Place Thomas cover glass staining outfit in a 200 mL glass beaker.

2. Add 100 mL of fixative to the beaker.

3. Using negative action tweezers, pull coverslips out of 37 °C media and place directly in the fixative, settling them into the rack maintaining a consistent orientation. Incubate in fixative for 10 min.

4. Transfer rack to a 200 mL beaker containing 80 mL TBS. Wash twice in TBS for 5 min each.

3.2.2 Immunofluorescence Labeling

1. Prepare a chamber for incubating the coverslips by putting a square piece of parafilm on the bottom of a 15 cm² Petri dish. Tear a Kimwipe in half, twist each half into a strip and place around the inside edge of the dish. Moisten the Kimwipe strips with distilled water.

2. Using tweezers, place the coverslips cell side up on the parafilm in the 15 cm² dish prepared in **step 1**.

3. Add 20 μL of 20 % blocking buffer to each coverslip.

4. Cover chamber with Petri dish lid and incubate for 45–60 min at room temperature or a temperature equivalent to that which you will use for antibodies.

5. Return coverslips to the Thomas cover glass staining outfit and wash twice more in 80 mL TBS for 5 min each.

6. Remove coverslips one at a time and return them to the 15 cm² dish. Wick off excess TBS with a Kimwipe.

7. Add 6 μL each of gamma-tubulin antibody to the coverslips. Incubate for 1 h at room temperature.

8. Repeat **step 5**.

9. Repeat **step 6**.

10. Add 6 μL of the secondary antibody with the fluorophore of your choice (we use FITC) to each coverslip. Incubate for 1 h at room temperature.

11. Repeat **step 5** with three washes instead of two.

12. Repeat **steps 6–11** with the alpha-tubulin primary antibody and TRITC-labeled secondary antibody.

13. Place a small drop (~10 µL) of Vectashield on a glass slide (one drop for each coverslip). For imaging with a DeltaVision deconvolution microscope, do not mount more than three coverslips per slide.

4 Notes

1. HeLa (ATCC® CCL-2™), Hct 116 (ATCC® CCL-247™), or hTERT RPE-1 (ATCC® CRL-4000™) cell lines are available from the American Type Culture Collection (ATCC). HeLa cells require Biosafety Level 2 cell culture conditions because the cells contain human papillomavirus. Hct 116 human colorectal cancer cells and hTERT RPE-1 telomerase immortalized human retinal pigmented epithelial cells are both cultured at Biosafety Level 1. hTERT RPE-1 cells require a license agreement for commercial customer uses. hTERT RPE-1 cells are distributed for research purposes only. A signed addendum to the ATCC Material Transfer Agreement must be sent to ATCC in advance of shipment.

2. If TBS is to be used exclusively with fixed cells, then 0.05 % sodium azide can be added as a preservative and the TBS stored at room temperature.

3. This fixative can be stored at –20 °C.

4. At this point, cells can be transfected with DNA constructs or siRNA based on individual protocols to screen for effects on the spindle orientation machinery.

Acknowledgement

This work was supported by NIH grant GM69429 to L.W.

References

1. Thaiparambil JT, Eggers CM, Marcus AI (2012) AMPK regulates mitotic spindle orientation through phosphorylation of myosin regulatory light chain. Mol Cell Biol 32(16):3203–3217

2. Toyoshima F et al (2007) PtdIns(3,4,5)P3 regulates spindle orientation in adherent cells. Dev Cell 13(6):796–811

3. Fink J et al (2011) External forces control mitotic spindle positioning. Nat Cell Biol 13(7):771–778

4. Kiyomitsu T, Cheeseman IM (2012) Chromosome- and spindle-pole-derived signals generate an intrinsic code for spindle position and orientation. Nat Cell Biol 14(3):311–317

5. Godin JD et al (2010) Huntingtin is required for mitotic spindle orientation and mammalian neurogenesis. Neuron 67(3):392–406

Chapter 3

Automated Segmentation of the First Mitotic Spindle in Differential Interference Contrast Microcopy Images of *C. elegans* Embryos

Reza Farhadifar and Daniel Needleman

Abstract

Differential interference contrast (DIC) microscopy is a non-fluorescent microscopy technique that is commonly used to visualize the first mitotic spindle in *C. elegans* embryos. DIC movies are easy to acquire and provide data with high spatial and temporal resolution, allowing detailed investigations of the dynamics of the spindle—which elongates, oscillates, and is positioned asymmetrically. Despite the immense amount of information such movies provide, they are normally only used to draw qualitative conclusion based on manual inspection. We have developed an algorithm to automatically segment the mitotic spindle in DIC movies of *C. elegans* embryos, determine the position of centrosomes, quantify the morphology and motions of the spindle, and track these features over time. This method should be widely useful for studying the first mitotic spindle in *C. elegans*.

Key words DIC microscopy, *C. elegans*, Mitotic spindle, Active contour method, Centrosome

1 Introduction

Automated analysis of DIC movies is challenging because of the complex nature of the images. Fluorescence microscopy of spindle-associated proteins, such as gamma-tubulin::GFP [1], provides high-contrast images that are more readily amenable to automated image analysis, but requires genetic manipulation that could produce unintended perturbations, is laborious if large numbers of different genetic backgrounds need to be studied, and requires the use of expensive confocal microscopes to obtain high resolution images. Thus, DIC is still widely used to study the first mitotic cell division in *C. elegans* embryos, and it is highly desirable to extract quantitative data from the obtained movies in a robust fashion.

In this chapter, we introduce an algorithm, based on active contour methods [2], to segment spindles in DIC images of *C. elegans* embryos and determine the position of the centrosomes—the

David J. Sharp (ed.), *Mitosis: Methods and Protocols*, Methods in Molecular Biology, vol. 1136,
DOI 10.1007/978-1-4939-0329-0_3, © Springer Science+Business Media New York 2014

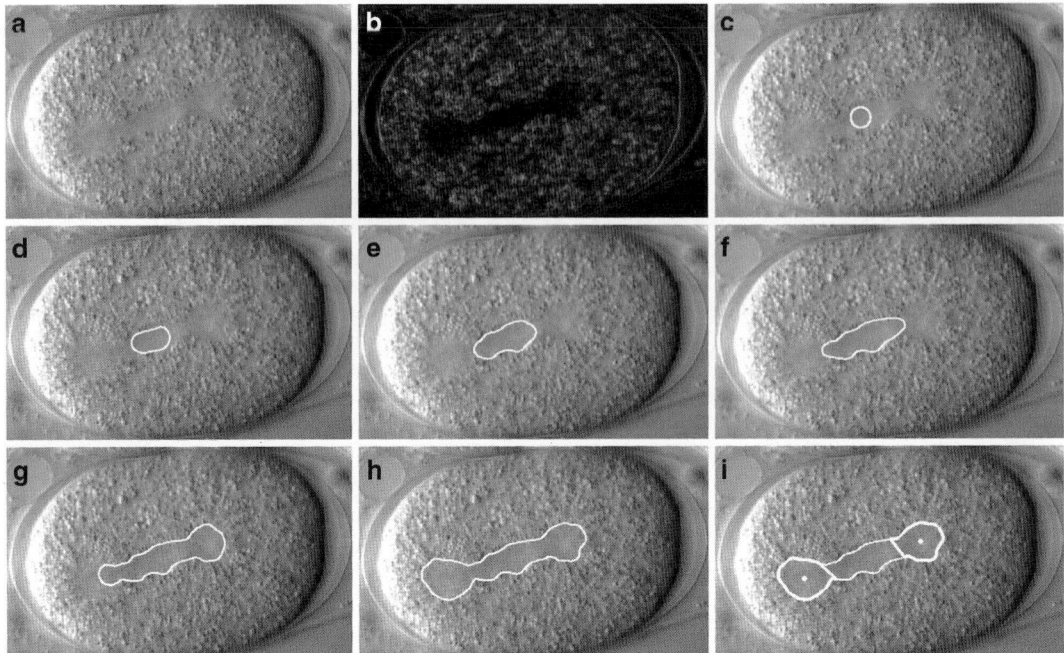

Fig. 1 Segmentation of the first *C. elegans* mitotic spindle in a DIC microscopy image: (**a**) Raw image. (**b**) Smoothed Gaussian derivative of the raw image in (**a**). (**c**–**h**): Time evolution of the active contour starting from an arbitrary shape in (**c**). (**i**): Centrosome positions for segmented spindle

organizing centers that define the bipolar structure of the spindle. In this algorithm, we use a deformable contour and define internal and external forces such that the contour maintains a smooth shape and evolves from an initial configuration to find the shape of the spindle. These forces are derived from a potential function, which has only two free parameters.

2   Methods

1. Obtain a time-lapse DIC microscopy movie of the first mitotic division of a *C. elegans* embryo as described in [3] (*see* **Note 1**).

2. Select a frame of the movie to initiate the image-processing steps (*see* Fig. 1a). The pixel intensity at position (i,j) of this image is denoted by $I(i,j)$.

3. Calculate the smoothed gradients of the image along the x and y directions

$$G_x(x,y) = \sum_{i,j} I(i,j) \frac{i-x}{2\pi\sigma^4} \exp\left(-\frac{(x-i)^2 + (y-j)^2}{2\sigma^2}\right)$$

$$G_y(x,y) = \sum_{i,j} I(i,j) \frac{j-y}{2\pi\sigma^4} \exp\left(-\frac{(x-i)^2 + (y-j)^2}{2\sigma^2}\right)$$

Here, σ is the width of the smoothing Gaussian filter (*see* **Note 2**).

4. Calculate the gradient image $G(x,y) = \sqrt{G_x^2(x,y) + G_y^2(x,y)}$ (*see* Fig. 1b).

5. Calculate the normalized gradient image $N(x,y) = G(x,y)/\max(G)$, where $\max(G)$ is the maximum value of $G(x,y)$.

6. Define an arbitrary closed curve $\Gamma(s) = (x(s), y(s))$ (s is the arc length) inside the spindle region as an initial guess for spindle shape (*see* **Note 3** and Fig. 1c).

7. The final shape of the spindle is determined by a curve $\Gamma_m(s)$ that minimizes the energy functional $E[\Gamma(s)]$

$$E\left[\Gamma(s)\right] = \oint_{\Gamma[s]} \frac{\alpha}{2}|\Gamma_{ss}|^2 \, \mathrm{d}s + \iint_R \left(N(x,y) - \beta\right) \mathrm{d}x\mathrm{d}y,$$

where $\Gamma_{ss} = \partial^2\Gamma/\partial s^2$. The first term is a bending energy that keeps the contour smooth. The second integral is over the area enclosed by the curve $\Gamma(s)$. A value of β is chosen such that the second term makes the curve grow if it is inside the spindle region and shrink if it is outside the spindle region. This energy can be minimized by iteratively updating the contour using a gradient decent method (*see* **Notes 4** and **5** and Fig. 1c–h).

8. To dissect a centrosomal region from the minimal curve $\Gamma_m(s)$, first rotate and center the contour along the x-axis such that the center of the spindle is located at $x=0$. For the region of the contour that is located in $x>0$, connect the inflection points or the negative curvature points on the two sides of this region. The centrosome position is chosen to be the geometric center of the dissected area (*see* Fig. 1i). Repeat this procedure for the region in $x<0$.

9. To segment the spindle for the next frame of the movie, choose $\Gamma_m(s)$ of the current frame as initial curve for the minimization. Then iterate until all images in the movie have been segmented (*see* Fig. 2a–f).

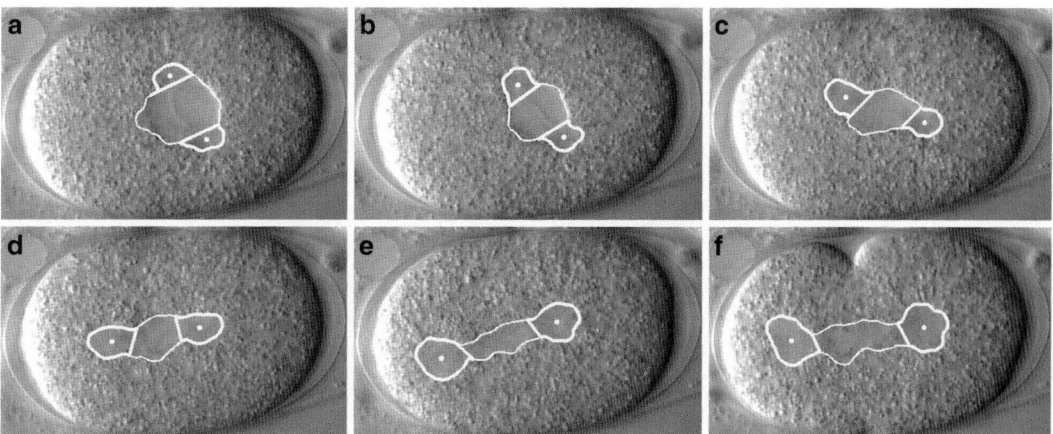

Fig. 2 Tracking spindle for a time-lapse movie: (**a**)–(**f**) segmented spindle and centrosomes for sequential frames of a time-lapse movie

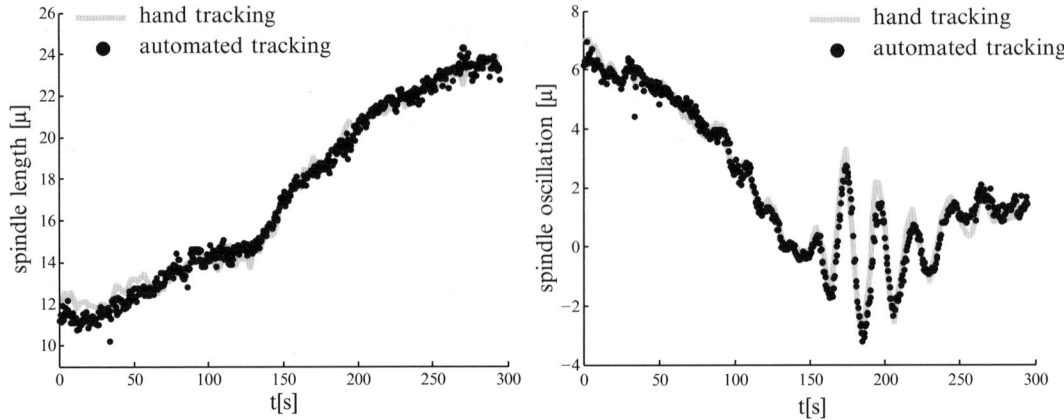

Fig. 3 Spindle length and spindle oscillation as a function of time for embryo shown in Fig. 2. *Gray solid line* is manual tracking

10. Now the positions of the centrosomes have been calculated for each image. The length of the spindle (i.e., the distance between centrosomes) can be plotted as a function of time, allowing spindle elongation to be studied in detail. The oscillations of the spindle can be studied by plotting the Y position of the centrosome as a function of time. We confirmed that the presented automated tracking procedure gives the same results as manually tracking the spindle length and spindle oscillation by hand (Fig. 3a, b).

3   Notes

1. The spindle should appear as a connected smooth region inside the cell surrounded by granule lipid particles.

2. The value of σ should be large enough such that spindle appears as a smooth black region in the derivative image (*see* Fig. 1b).

3. Although this method is robust to variation in shape of the initial curve, to ensure efficient minimization, choose the initial curve not too different from the spindle shape.

4. The first term in $E(\Gamma(s))$ ensures smoothness of the curve by preventing formation of sharp corners with high curvatures. The second term is the sum of pixel intensity of the normalized derivative image inside the region enclosed by the curve $\Gamma(s)$. A balance between the second term and the third term, which is a constant multiplied by the area of this region, determines the final shape of the spindle.

5. One method to minimize the energy functional $E[\Gamma(s)]$ is the steepest descent method. In this method, we assume that the

contour is a function of both arc length and time, i.e., $\Gamma = \Gamma(s,t)$. The steady solution of the equation

$$\Gamma_t\left(s,t\right) = -\frac{\delta E\left[\Gamma\right]}{\delta \Gamma} = -\alpha \Gamma_{ssss}\left(s,t\right) - \left(N\left(\Gamma\left(s,t\right)\right) - \beta\right)\vec{n},$$

where $\vec{n} = \left(y_{s_,} - x_s\right)$ is the normal vector to the curve $\Gamma(s,t)$, gives us the minimal curve.

Acknowledgment

This work was supported by grant RGP0034/2010 from the Human Frontiers Science Program.

References

1. Jaensch S, Decker M, Hyman AA, Myers E (2010) Automated tracking and analysis of centrosomes in early *Caenorhabditis elegans* embryos. Bioinformatics 26:13–20

2. Kass M, Witkin A, Terzopoulos D (1987) Snakes—active contour models. Int J Comput Vis 1:321–331

3. Sulston JE, Schierenberg E, White JG, Thomson JN (1983) The embryonic-cell lineage of the nematode *Caenorhabditis elegans*. Dev Biol 100:64–119

Chapter 4

Imaging the Mitotic Spindle by Spinning Disk Microscopy in Tobacco Suspension Cultured Cells

Takashi Murata and Tobias I. Baskin

Abstract

Plants are valuable systems for analyzing the acentriolar mitotic spindle. We have developed methods for imaging the mitotic spindle in living tobacco (*Nicotiana tabacum*) suspension culture cells expressing GFP-α-tubulin. The methods allow the spindle to be observed in living cells at high spatial and temporal resolution and rely on water immersion objectives, spinning disk optics, and high-sensitivity cameras. Here, we describe these methods and provide step-by-step protocols for certain key steps. We also describe a method for application and removal of inhibitors.

1 Introduction

1.1 Mitotic Spindle Development in Plant Cells

Plant cells serve as important systems for analyzing acentriolar microtubule organization. In mitosis, plant cells naturally form mitotic spindles in the absence of centrosomes [1]. Microtubules of the mitotic spindle originate on the nuclear envelope in prophase [2]. They gradually organize to form a prophase spindle, with focused poles, one on each side of the nucleus. Presumably, microtubule rearrangement around the nucleus is self-organized by motor protein-dependent interactions. After nuclear envelope breakdown, microtubules increase in number and are organized to form the metaphase spindle. In this step, kinesin motors are involved, as in animal spindles [3, 4].

Also unique to plant cells is the mechanism of division site determination. The division site is predetermined by a hoop of microtubules called the preprophase band. The preprophase band affects the distributions of actin and at least one kinesin and disappears before metaphase [5]. At cytokinesis, the nascent cell wall (called the cell plate), which divides the daughter cells, orients during its development by interaction with sites in the cell cortex conditioned by the preprophase band [6].

David J. Sharp (ed.), *Mitosis: Methods and Protocols*, Methods in Molecular Biology, vol. 1136, DOI 10.1007/978-1-4939-0329-0_4, © Springer Science+Business Media New York 2014

Here, we describe methods for imaging spindle development in living plant cells. We include methods for applying microtubule inhibitors, which are useful for analyzing microtubule dynamics.

1.2 Tobacco BY-2 Cells as a Material for Mitotic Spindle Studies

For live imaging spindle development in plants, there are several choices of material. Perhaps the most obvious choice would be root meristem cells of *Arabidopsis thaliana*. Community resources for this species are outstanding, including stable lines expressing various cytoskeletal proteins tagged with fluorescent proteins and well-characterized mutants in motor proteins and others associated with the cytoskeleton. Among organs, the root meristem is relatively suitable for imaging insofar as it lacks chlorophyll and has a dense population of dividing cells. However, in the root of this species, dividing cells are difficult to synchronize. Additionally, the cells have small spindles (~5 μm long) and are surrounded for the most part by nondividing tissue (peripheral root cap), features that hinder imaging at high resolution.

An emerging model plant for cellular and developmental research is the moss, *Physcomitrella patens*. This organism makes cellular filaments called protonema, which are excellent material for imaging living cells, and *P. patens* is being taken advantage of by researchers interested in cell division and cytoskeletal organization (e.g., [7]). Protocols for culture and microscopy of *P. patens* cells are available online (http://www.nibb.ac.jp/evodevo/titleE.html).

Tobacco BY-2 cells are also excellent material for analyzing cell division. The cells are not surrounded with the neighboring cells, have large spindles, and are readily synchronized [8]. They are easily transformed by Agrobacterium [9], and stable transformants can be obtained within 1.5 months. Numerous cell lines are available with fluorescent markers. One disadvantage of this material is the lack of a sequenced genome. Nevertheless, these cells are used by many researchers to study the mitotic spindle [10–12]. Here, for this material, we describe methods for high-resolution imaging of spindle development.

1.3 Choice of Optics

We use a spinning disk confocal system, comprising a Yokogawa spinning confocal unit, a chilled CCD or CMOS camera, and an inverted microscope. Because mitosis in tobacco BY-2 cells is photosensitive, we minimize excitation light. For doing so, a system with a chilled CCD or CMOS camera is superior to galvano-mirror confocal systems. Image acquisition by the galvano-mirror confocal system is possible, but the signal-to-noise ratio or resolution of the image will generally suffer under laser power that has been reduced sufficiently to minimize phototoxicity.

Plant cells do not lie flat on a glass surface; therefore, the spindle is at an appreciable distance from the coverslip. In the case of tobacco BY-2 cells, the distance is usually more than 20 μm.

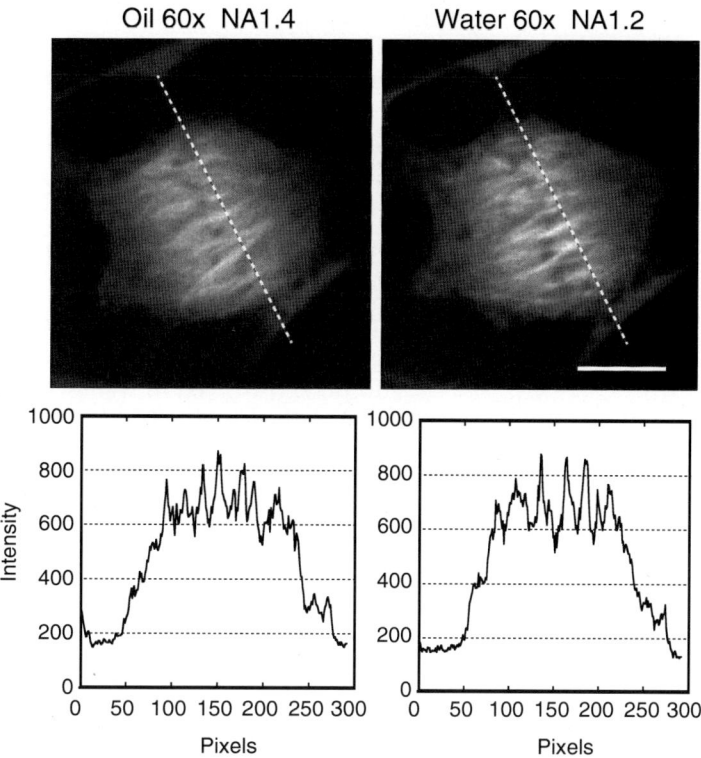

Fig. 1 Comparison between an oil immersion and a water immersion objective lens. A metaphase tobacco BY-2 cell expressing GFP-α-tubulin was observed on an inverted microscope (Olympus IX81) equipped with spinning disk optics (Yokogawa CSU21), and images taken with a CMOS camera (Hamamatsu Orca-Flash 4.0) through either an oil immersion (Olympus PlanAPO 60× 1.40 NA) or a water immersion (Olympus UPLSAPO 60× 1.20 NA) lens. Pixel intensities along the *dotted lines* are shown below. Approximately 3 min elapsed between images. Conditions of image acquisition and image processing are identical in the two images. Scale bar = 10 μm

For such distances, an oil immersion objective lens with a high numerical aperture (NA) does not focus well because of refractive index mismatch between immersion oil and culture medium or cytoplasm [13]. Therefore, water- or silicone-fluid immersion lenses are needed, in which refractive index of the immersion medium is close to that of culture medium (n = approx. 1.33) or cytoplasm (n = approx. 1.37). Despite its lower NA, a water immersion lens improves image quality compared to a similar quality oil immersion lens (Fig. 1).

1.4   Tobacco BY-2 Cells Expressing GFP-α-Tubulin

Tobacco (*Nicotiana tabacum*) BY-2 cells expressing GFP-α-tubulin [14]: Obtained from RIKEN Bioresource Center, Japan (http://www.brc.riken.jp/lab/epd/Eng/catalog/pcellc.shtml).

1.4.1   Culture Medium

Modified LS medium [15]: Murashige and Skoog basal salt mixture supplemented with 3 % sucrose, 0.2 g/L KH_2PO_4, 0.1 g/L myo-inositol, 1 mg/L thiamine-HCl, 0.2 mg/L 2,4-dichloroacetic acid. The pH is adjusted to 5.5–5.7 with KOH. The medium is sterilized at 120 °C for 15 min.

1.4.2 Poly-L-Lysine Treatment of Glass-Bottom Dishes

(a) Hydrophilic treatment of the surface of a glass-bottom dish. Put the dish in a glow discharge plasma for 30–60 s. Exposure to the plasma effectively removes hydrocarbons and other contaminants from the glass surface and might also give the glass a net negative charge.

(b) Immediately after plasma cleaning, add a few μL of 1 mg/mL high molecular weight poly-L-lysine (mol. wt. 150,000–300,000) dissolved in water onto the surface of glass. Remove excess solution. Air-dry.

(c) Wash the glass surface with distilled water at least six times, and then air-dry.

(d) The poly-L-lysine-coated dish should be used within 1–2 days.

1.4.3 Anti-microtubule Drugs

Propyzamide-containing medium: Modified LS medium containing 5 or 10 μg/mL propyzamide. Dissolve propyzamide at 100 mM in dimethyl sulfoxide. The stock solution is stable at –20 °C. Dilute the stock solution 10- or 20-fold with dimethyl sulfoxide and then with culture medium at 1:1,000 dilution.

1.5 Microscope Equipment

Imaging is performed on an inverted fluorescence microscope, such as Olympus IX81 with a 60× 1.2 NA water immersion lens. A spinning disk confocal unit (e.g., Yokogawa CSU-X1) equipped with appropriate filters for fluorescent proteins used is attached to the inverted microscope. A chilled CCD (e.g., CoolSNAP HQ, Roper) or a scientific CMOS (e.g., Orca-Flash 4.0, Hamamatsu) is used for image acquisition. The microscope, shutters of the spinning disk unit, and the camera are driven by control software, such as Metamorph (Universal Imaging) or iQ (Andor). The microscope is set in a dark room, in which the temperature is stable (*see* **Note 1**). Change in temperature during microscopy results in the focal plane shifting. Temperature on the microscope stage is between 25 and 26 °C, to match the conditions of the cells during culture.

Removal of medium without moving the culture dish is difficult by hand. Instead, we use a metal capillary connected to a peristaltic pump (Fig. 2).

2 Methods

2.1 Preparation of Cells

1. Tobacco BY-2 cells expressing GFP-α-tubulin are grown in the dark in 300 mL flasks on a rotary shaker at 120 rpm at 25–26 °C. Flasks are covered tightly enough to minimize contamination but loosely enough for gas exchange. Cells are subcultured at 1-week intervals by adding 2 mL of culture to

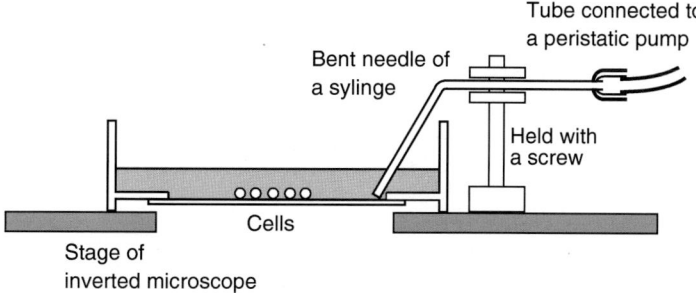

Fig. 2 Medium-removal device. A bent stainless-steel capillary is fixed on the microscope stage with a bolt and screws. The culture dish is fixed with dental wax (not shown). The culture medium is removed by the peristaltic pump. Fresh medium is manually applied with a pipette

80 mL of fresh medium. Upon subculturing to a fresh medium, the cells start to divide. The cells divide for 3–4 days (log phase), and then division activity gradually ceases and elongation becomes dominant (stationary phase). In log phase, cells tend to grow as a cluster, and conditions for microscopy are suboptimal.

2. Transfer 4 mL of a 6- or 7-day-old suspension to 40 mL of fresh culture medium in a 200 mL flask and culture as usual for about 20–24 h prior to observation. At this time, there is high frequency of division and good conditions for microscopy. Division frequency notably decreases after 24 h.

2.2 Attachment of Cells onto the Poly-L-Lysine-Coated Dishes

1. Mix 0.5 mL of the 1-day-old culture with 5 mL of 3 % sucrose in water in a 50 mL Falcon tube. To avoid shear damage, approximately 5 mm is cut off from the pointed end of all plastic micropipette tips used for cell transfer.

2. Take 0.5 mL of the mixture and pipette vigorously onto the surface of the poly-lysine-coated culture dish. Repeat three to four times to attach sufficient cells onto the glass surface.

3. Remove the excess solution and add 0.5 mL of fresh medium. The cells are ready to observe (*see* **Note 2**).

2.3 Time-Lapse Observation of Spindle Development

1. Choose the cell with GFP fluorescence around the nucleus. Along with bright perinuclear fluorescence, a good sign for a cell entering mitosis is the presence of a preprophase band, a band of microtubules at the cell cortex encircling the cell at the position of the nucleus (Fig. 3, time zero).

2. Set exposure time and time interval between images. Minimize light intensity and exposure time, because progression of mitosis is photosensitive. Time interval is dependent on purpose of experiments (*see* **Note 3**).

3. Record images for 2–3 h (Fig. 3). If the cell is not damaged, breakdown of the nuclear envelope occurs within 30–40 min.

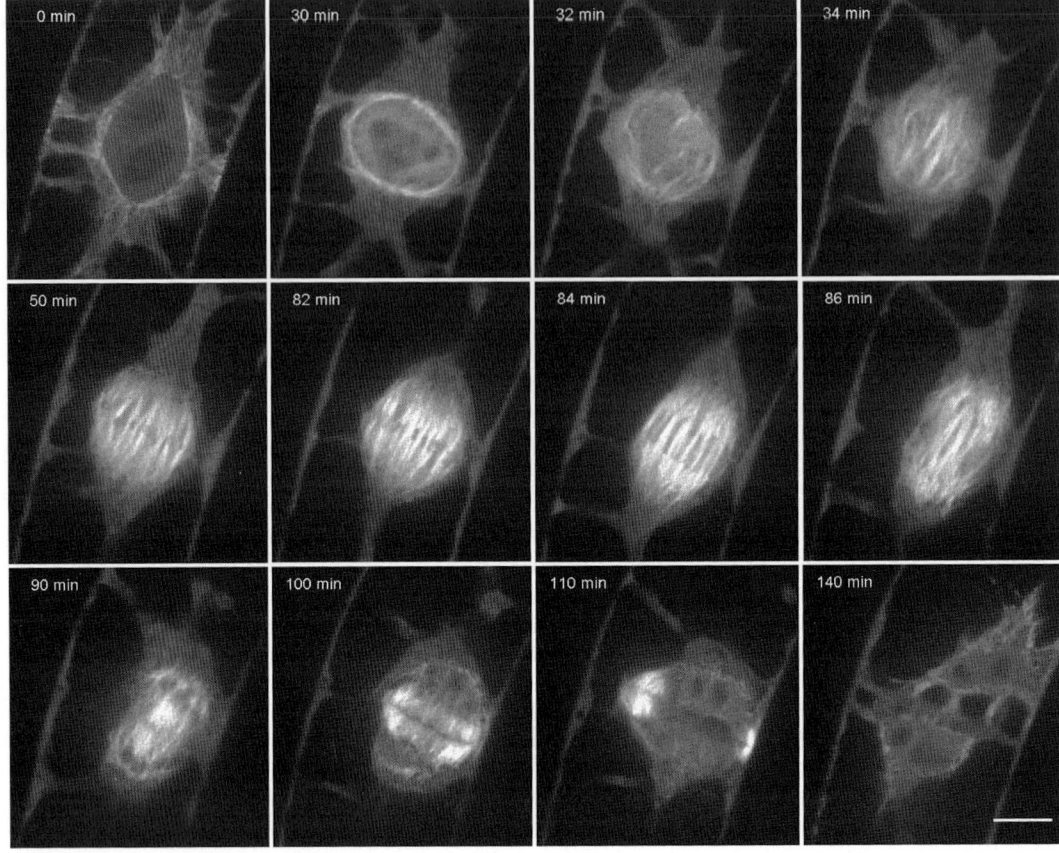

Fig. 3 Time-lapse observation of spindle development in a tobacco BY-2 cell expressing GFP-α-tubulin. Imaging setup as for Fig. 1 except for use of silicone-fluid immersion lens (UPLSAPO 60XS, Olympus). Time after the start of observation is indicated in each panel. 30–34 min: nuclear envelope breakdown; 82–86 min: anaphase; 90–140 min: cytokinesis. Scale bar = 10 μm

2.4 Analyses of Chromosome Segregation

1. Choose a cell with a metaphase spindle. The chromosomes are distinguished as dark regions within the bright spindle.

2. Set exposure time and time interval between images. Minimize light intensity and exposure time, because onset of anaphase is very photosensitive (*see* **Note 4**).

3. Record images for 20–30 min. If the cell is not damaged, it will progress through anaphase (Fig. 4).

4. Open the images as a stack in ImageJ (http://imagej.nih.gov/ij/). A kymograph of the image can be made by using the Reslice command, found in Stacks sub-menu within the Image menu.

5. The velocity of chromosome movement and spindle elongation can be analyzed, as reported by Hayashi et al. [12].

a. Progression of anaphase

b. Kymograph

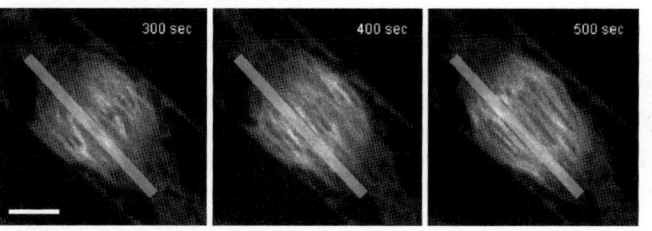

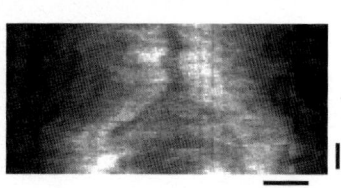

Fig. 4 Examples of analyzing chromosome segregation. (**a**) Several frames from a time-lapse observation of anaphase and (**b**) the corresponding kymograph, made from the region marked with the *thin gray rectangle*. Scale bar in (**a**) = 10 μm

2.5 Observation of Spindle Reformation After Depolymerization of Microtubules

1. Place the needle of the medium-removal device in the observation dish.

2. Choose a cell at metaphase. Set exposure time and time interval as in Subheading 2.4. Start image acquisition.

3. Remove the culture medium by the pump. Without letting cells become dry, add 0.5 mL of propyzamide-containing medium by hand. Repeat. With practice, the two exchanges of medium can be done in less than 10 s.

4. After adequate time (e.g., 10 min) of incubation with propyzamide-containing medium, remove the medium and add 0.5 mL of drug-free medium. Repeat the medium exchange five times.

5. Continue image acquisition. Microtubules recover soon after medium exchange (Fig. 5). Note that microtubules recover around the chromosomes, as reported by Falconer et al. [16], and a typical looking spindle eventually forms.

3 Notes

1. If the temperature in the microscope room fluctuates, the focal plane will move during image acquisition. In such a case, focus correction software (e.g., Perfect Focus by Nikon) can help. Alternatively, the focus can be corrected manually, although doing so requires considerable skill.

2. If the cells die soon after attachment, the concentration of poly-l-lysine might be too high or the culture dishes have been washed insufficiently. Try more washes or diluting the poly-lysine concentration.

3. For recording from the start of nuclear envelope breakdown to the end of cytokinesis, more than 1 h is needed. During this time, the water immersion lens might dry out. For prolonged observation, use of silicone-fluid immersion lens (UPLSAPO

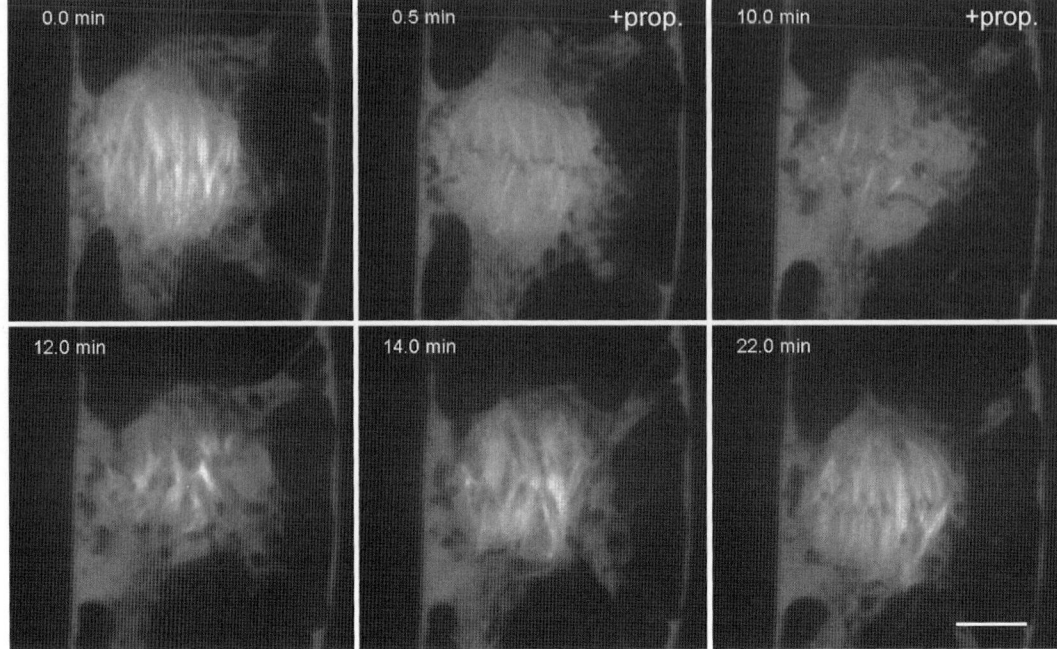

Fig. 5 Disruption and recovery of spindle microtubules by propyzamide treatment. At 0 min, 10 μM propyzamide is applied and removed at 10 min. Microtubules recover around the chromosomes and a typical looking spindle eventually forms. Image setup as for Fig. 1. Scale bar = 10 μm

60XS, Olympus) combined with silicone-fluid as immersion medium is helpful. Use of inert immersion media such as Immersol W 2010 (Zeiss) for water immersion lens might work.

4. Because of this sensitivity, we bin images by 2× during acquisition to reduce excitation light and increase signal intensity, although this decreases resolution.

Acknowledgements

This work was supported by Invitation Fellowship Programs for Research in Japan (no. L-08551) to T.B. and Grant-in-Aid for Scientific Research (B) (no. 21370026) to T.M. from the Japanese Society for Promotion of Science.

References

1. Baskin TI, Cande WZ (1990) The structure and function of the mitotic spindle in flowering plants. Annu Rev Plant Physiol Plant Mol Biol 41:277–315

2. De Mey J, Lambert AM et al (1982) Visualization of microtubules in interphase and mitotic plant cells of *Haemanthus* endosperm with the immuno-gold staining method. Proc Natl Acad Sci U S A 79:1898–1902

3. Bannigan A, Scheible WR et al (2007) A conserved role for kinesin-5 in plant mitosis. J Cell Sci 120:2819–2827

4. Ambrose JC, Li WX et al (2005) A minus-end-directed kinesin with plus-end tracking protein activity is involved in spindle morphogenesis. Mol Biol Cell 16:1584–1592

5. Rasmussen CG, Humphries JA et al (2011) Determination of symmetric and asymmetric division planes in plant cells. Annu Rev Plant Biol 62:387–409

6. Mineyuki Y (1999) The preprophase band of microtubules: its function as a cytokinetic apparatus in higher plants. Int Rev Cytol 187:1–49

7. Hiwatashi Y, Obara M et al (2008) Kinesins are indispensable for interdigitation of phragmoplast microtubules in the moss *Physcomitrella patens*. Plant Cell 20:3094–3106

8. Kumagai-Sano F, Hayashi T et al (2006) Cell cycle synchronization of tobacco BY-2 cells. Nat Protoc 1:2621–2627

9. An GH (1985) High-efficiency transformation of cultured tobacco cells. Plant Physiol 79:568–570

10. Ambrose JC, Cyr R (2007) The kinesin ATK5 functions in early spindle assembly in *Arabidopsis*. Plant Cell 19:226–236

11. Dhonukshe P, Vischer N et al (2006) Contribution of microtubule growth polarity and flux to spindle assembly and functioning in plant cells. J Cell Sci 119:3193–3205

12. Hayashi T, Sano T et al (2007) Contribution of anaphase B to chromosome separation in higher plant cells estimated by image processing. Plant Cell Physiol 48:1509–1513

13. Gibson SF, Lanni F (1991) Experimental test of an analytical model of aberration in an oil-immersion objective lens used in 3-dimensional light-microscopy. J Opt Soc Am A 8:1601–1613

14. Kumagai F, Yoneda A et al (2001) Fate of nascent microtubules organized at the M/G1 interface, as visualized by synchronized tobacco BY-2 cells stably expressing GFP-tubulin: time-sequence observations of the reorganization of cortical microtubules in living plant cells. Plant Cell Physiol 42:723–732

15. Nagata T, Nemoto Y et al (1992) Tobacco BY-2 cell-line as the Hela-cell in the cell biology of higher-plants. Int Rev Cytol 132:1–30

16. Falconer MM, Donaldson G et al (1988) MTOCs in higher-plant cells—an immunofluorescent study of microtubule assembly sites following depolymerization by APM. Protoplasma 144:46–55

Chapter 5

Image-Based Computational Tracking and Analysis of Spindle Protein Dynamics

Ge Yang

Abstract

The spindle is a highly complex and dynamic molecular machine that is assembled during cell division for accurate segregation of replicated chromosomes. Successful completion of cell division relies on the right spindle proteins to be at the right place at the right time to serve their functions. Quantitative characterization and analysis of spatiotemporal behaviors of spindle proteins are therefore essential to understanding related cell division mechanisms. The main goal of this chapter is to introduce basic concepts and methods for computational tracking and analysis of spindle protein spatiotemporal dynamics that is visualized and recorded in fluorescence microscopy images. An emphasis is placed on providing practical and useful information on related software tools. Examples are used to demonstrate applications of related computational methods and software tools.

Key words Cell division, Spindle, Spindle protein, Protein dynamics, Spatiotemporal dynamics, Fluorescence microscopy, Tracking, Image analysis, Computer vision, Computational biology

1 Introduction

1.1 Background

The spindle is a microtubule-based molecular machine that is assembled during cell division for accurate segregation of replicated chromosomes [1]. Its structure is highly complex and dynamic yet exquisitely organized [2]. Successful completion of cell division relies on precise spatiotemporal coordination of the spindle proteins involved. Spatiotemporal behaviors of spindle proteins therefore contain critical information about related cell division mechanisms.

Spatiotemporal dynamics of spindle proteins can be visualized and recorded effectively using fluorescence microscopy techniques [3–6]. Indeed, advances in fluorescence microscopy techniques in areas such as single-molecule imaging, super-resolution imaging, and fluorescence probe development have made it possible to visualize spindle protein dynamics with remarkable performance, as measured by metrics such as resolution, sensitivity, specificity, and multiplicity. Recorded images of spindle protein dynamics are

David J. Sharp (ed.), *Mitosis: Methods and Protocols*, Methods in Molecular Biology, vol. 1136, DOI 10.1007/978-1-4939-0329-0_5, © Springer Science+Business Media New York 2014

often examined and interpreted directly by investigators based on their knowledge and expertise. However, quantification of spindle protein behaviors requires quantitative image analysis. Simplified image analysis often can be performed manually to obtain approximate measurements of spindle protein dynamics. In this case, images are processed directly by investigators, usually with the support of software for human–computer interaction.

Alternatively, images can be processed by computer software under investigator supervision, a process henceforth referred to simply as computational image analysis. Depending on the level of supervision required, computational image analysis can be partially or fully automated. With proper quality control, computational image analysis can offer significant advantages over manual image analysis in key performance metrics such as accuracy, precision, resolution, sensitivity, efficiency, and consistency [7, 8]. It is particularly well suited for obtaining comprehensive measurements of spindle protein dynamics required for quantitative modeling [9, 10]. These performance advantages will be further discussed later in this chapter.

1.2 Image-Based Computational Tracking and Analysis of Spindle Protein Dynamics

To obtain high-quality comprehensive measurements is an overall goal of using computer software to process images of spindle protein dynamics. At the core of the image processing is the recovery of trajectories of spindle protein spatiotemporal behaviors using *image feature tracking* techniques, which will be introduced in Section 3. Based on recovered trajectories, spindle protein dynamics can be quantitatively characterized and analyzed for understanding the underlying molecular mechanisms. These two processes, namely, the computational tracking and analysis of spindle protein dynamics based on recorded images, are the main subject of this chapter. The main goal is to introduce related basic concepts and methods, with an emphasis on providing practical and useful information on related software tools.

The general procedure for image-based computational tracking and analysis of spindle protein dynamics is outlined in Fig. 1. Recorded time-lapse images are usually preprocessed prior to computational analysis. For example, image noise is often suppressed using *image filtering* techniques [8, 11, 12]. Spindles may drift in their positions during live imaging due to external forces. To characterize protein dynamics within spindles, such drift often must be corrected to prevent analysis artifacts by using *image registration* techniques [13, 14]. After image preprocessing, objects within the images, often referred to as *image features*, are identified using *image feature detection* techniques [15]. The identified image features correspond to different spindle structural components. For example, fluorescently labeled spindle proteins that are diffraction limited in size and well separated in space are detected as *particles*. Dynamic spatiotemporal behaviors of detected image

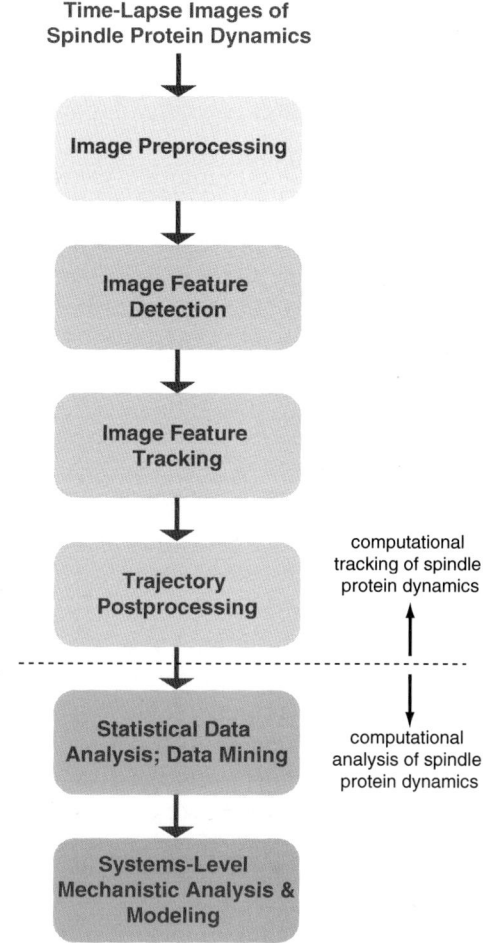

Fig. 1 General procedure for computational tracking and analysis of spindle protein dynamics

features are then followed using *image feature tracking* techniques [8, 16, 17]. For example, single particle tracking techniques [18, 19] can be used to follow spatiotemporal behaviors of detected particles of spindle proteins with high resolution, accuracy, and precision.

Spatiotemporal behaviors of spindle proteins followed by image feature tracking are usually recorded as trajectories [8, 16, 17]. In the example of a particle within a sequence of 2D time-lapse images, its trajectory can be represented by a time series such as $\{(x_1^k, y_1^k; I_1^k), (x_2^k, y_2^k; I_2^k), \ldots, (x_j^k, y_j^k; I_j^k), \ldots, (x_N^k, y_N^k; I_N^k)\}$ where (x_j^k, y_j^k) and I_j^k denote the Cartesian coordinate and intensity, respectively, of this kth numbered particle in the jth image frame and N denotes the total number of image frames, i.e., trajectory length. The recovered raw trajectories may need to be further processed. For example, erroneous trajectories sometimes can be manually corrected or removed. Overall, the main goal of computational tracking of spindle protein dynamics is to recover spatiotemporal trajectories.

After successful completion of computational tracking of spindle protein dynamics, recovered trajectories are analyzed to obtain quantitative measurements of spindle protein dynamics. Such measurements may include diffusion coefficient or velocity of spindle protein movement, spatial density or pattern of spindle protein distribution, etc. Since a substantial amount of quantitative measurement data is often generated at this step, statistical analysis techniques are essential to ensure rigorous data analysis. Furthermore, data mining techniques can be used to search for valuable but hidden information [20]. Quantitative measurements of spindle protein dynamics can be further analyzed for systems-level mechanistic studies and quantitative modeling. Overall, the main goal of computational analysis of spindle protein dynamics is to obtain quantitative measurements and mechanistic knowledge.

1.3   A Simple Classification of Images of Spindle Protein Dynamics

Different experimental methods and imaging modalities can produce very different images of spindle protein dynamics, which in turn may require different computational image analysis techniques [21]. In general, however, images of spindle protein dynamics can be classified into two categories. Images in the first category are henceforth referred to as *single particle images*. Primary features of these images are *particles*, which are images of fluorescently labeled spindle proteins that are diffraction limited in size and well separated in space. Under the Rayleigh criterion, the diffraction limit for fluorescence light microscopy is ~200 nm [22]. Images in the second category are henceforth referred to as *continuous region images*. Primary features of these images are *regions*, which are images of fluorescently labeled spindle proteins that cannot be resolved spatially.

Figure 2 illustrates differences between these two types of images using fluorescence imaging of single as well as spindle microtubules as an example. Single microtubules assembled in vitro can be visualized by adding fluorescently labeled tubulin since labeled tubulin dimers can be randomly incorporated during polymerization (Fig. 2a). By tuning the fractions of labeled tubulin relative to the total tubulin pool, single particle images (e.g., Fig. 2a upper panel) or continuous region images (e.g., Fig. 2a lower two panels) can be produced. Similarly, single particle images (e.g., Fig. 2b upper panel) or continuous region images (e.g., Fig. 2b lower panel) can be produced for spindle microtubules assembled in vitro. Inspired by the speckled appearance of single particle images, the method of imaging cytoskeleton dynamics using low fractions of labeled cytoskeletal filament subunits is often referred to as fluorescent speckle microscopy (FSM) [23]. Spindle microtubules assembled in cells can be visualized by expressing tubulin

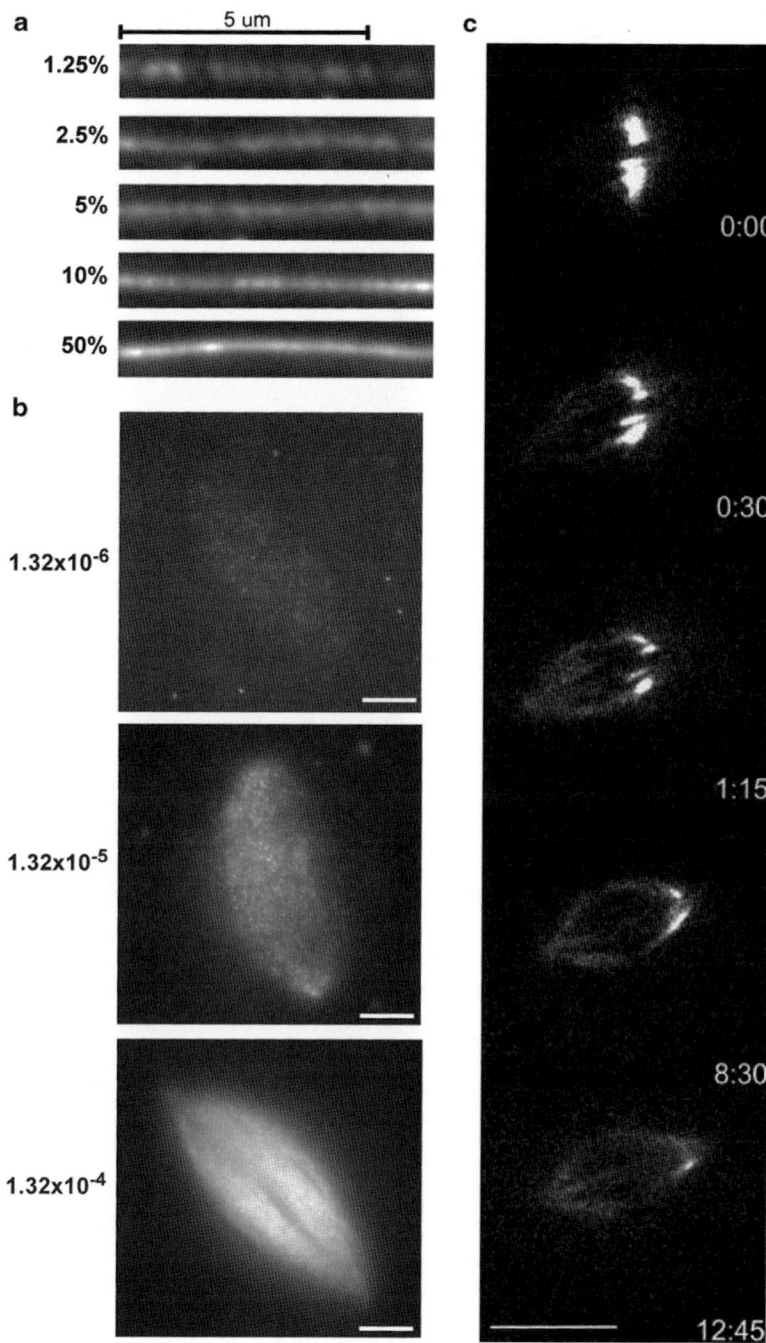

Fig. 2 Single particle and continuous region images of single and spindle micro-tubules (**a**) Images of single microtubules under different fractions (*left* of each image) of labeled tubulin (Reprinted with permission from ref. 23. Copyright 2006). (**b**) Images of *Xenopus* egg extract spindles under different fractions (*left* of each image) of labeled tubulin. Bars: 10 μm. (**c**) Poleward movement (flux) of spindle microtubules in a PtK1 cell, visualized by photoactivated tubulin-GFP (Reprinted with permission from ref. 24. Copyright 2006). Bar: 10 μm

labeled with genetically encoded fluorescent proteins [4, 24]. This usually produces continuous region images (Fig. 2c).

For computational tracking and analysis of spindle protein dynamics, single particle images are preferred for their higher spatiotemporal resolutions. This is because individual particles serve as sensitive and local spindle dynamics reporters, whose activities can be followed with high spatiotemporal resolution using single particle tracking [18, 19]. For example, single particle tracking has made it possible to analyze movement of individual spindle microtubules [25]. However, to produce single particle images is often technically more challenging. As an example, for FSM of spindle microtubule in cells, labeled tubulin needs to be delivered by microinjection, and imaging conditions must be optimized for reliable collection of dim speckle signals [23]. Furthermore, if particles are too sparsely distributed, spindle protein dynamics may not be fully represented by dynamics of individual particles.

In comparison, continuous region images provide lower spatiotemporal resolutions because of their low contrasts. For example, individual microtubules cannot be resolved from continuous region images (*see*, e.g., Fig. 2b lower panel) due to low spatial resolution. However, continuous region images are often more convenient to produce. Furthermore, specialized imaging techniques such as photoactivation (PA) (Fig. 2c) and fluorescence recovery after photobleaching (FRAP) can be used to directly generate image contrast [26]. Computational image analysis techniques are available for both single particle images and continuous region images. These techniques will be introduced in Section 3.

1.4 Chapter Outline

A wide variety of techniques are available for computational tracking and analysis of spindle protein dynamics. The focus of this chapter is on basic concepts and methods, with an emphasis on providing practical and useful information on related software tools. Information on more advanced techniques can be found from the references. The rest of the chapter is organized as follows. Section 2 briefly discusses the computer equipment and software tools required. Section 3 introduces methods for computational tracking and analysis of spindle protein dynamics, respectively, and uses a detailed example to demonstrate applications of introduced computational methods and software. Section 4 concludes with a discussion of several important practical issues.

2 Materials

Materials required for computational tracking and analysis of spindle protein dynamics include computers as well as general-purpose and specialized software tools.

2.1 Computers

The computing power required for image-based computational tracking and analysis of spindle protein dynamics can vary significantly depending on specific tasks. However, substantial CPU processing power, system memory, and storage capacity are usually required. Consequently, workstations are often preferred for their performance, reliability, flexibility, and expandability. Today, entry-level workstations are available at relatively low costs, making it feasible to start with low to medium range system configurations and upgrade when necessary. Well-configured personal computers can also provide substantial computing power and are cost-effective. However, they are generally not designed for intensive scientific computing and are therefore not recommended. When large volumes of image data need to be processed or when concurrent usage by multiple users needs to be supported, computing servers or computer clusters may become necessary but generally will carry higher system and maintenance costs. In addition to computing equipment, high-capacity storage equipment such as storage servers are often required for keeping large volumes of image data and related analysis results. Currently, Linux and Windows are widely adopted operating systems on workstations and computing servers. Mac OS is widely used on personal computers but less so on workstations and computing servers. Linux is also widely adopted on computer clusters. Different software for computational tracking and analysis of spindle protein dynamics may have different support for these operating systems.

2.2 Software Tools

Both general-purpose and specialized software tools are available for computational tracking and analysis of spindle protein dynamics. General-purpose software tools are briefly introduced here. Specialized software tools are introduced in Section 3. Usually a combination of these tools is needed.

General-purpose software tools for biological image analysis are reviewed in [27]. Some of them are commercially available, e.g., Imaris (Bitplane Scientific Software). Others are free open source, e.g., ImageJ [28], ICY [29], and Fiji [30]. Many open-source software tools are developed with an open architecture so that new functions can be added as plug-ins.

Although general-purpose software provides many useful functions such as image visualization, feature detection, and feature tracking, they are not designed to address specific questions on mitosis. Instead, specialized software tools often need to be developed. A commonly used scientific computing language and software environment for developing specialized biological image analysis software is MATLAB (MathWorks). Its function is extended by a large number of toolboxes. A commonly used computer language and software environment for statistical data analysis is R. In addition to MATLAB and R, several general-purpose computer languages, such as C/C++, Java, Python, and Perl, are often used. However, compared to specialized computer languages such as MATLAB and R, these general-purpose languages are usually less efficient in supporting rapid prototyping of software tools for specific biological applications. Further discussion on software selection can be found in **Note 1**.

3   Methods

The procedure for computational tracking of spindle protein dynamics consists of four steps: image preprocessing, image feature detection, image feature tracking, and trajectory postprocessing (Fig. 1). This section first introduces computational methods and software tools for single particle images. Methods and tools for continuous region images are more limited due to resolution limitations and are introduced briefly. Then, methods for computational analysis of spindle protein dynamics are described. Lastly, an example is used to demonstrate applications of the introduced computational tracking and analysis methods. Unless noted otherwise, introduced software tools can be downloaded from a web page maintained by the author's research group: http://ccdl.compbio.cmu.edu/software.html. Independent of what specific software tools are adopted, it is essential to conduct quality control and to avoid artifacts in image data analysis. This is discussed in detail in **Note 2**.

3.1   Methods for Tracking Spindle Protein Dynamics in Single Particle Images

A MATLAB software package has been developed by the author's research group for analyzing single particle images of spindle protein dynamics. It consists of three modules: *corrAligner* for spindle drift correction, *ptDetector* for particle feature detection, and *LAPtrack* for trajectory recovery and analysis. The package has been used in several studies on spindle protein dynamics [24, 25, 31–35]. Detailed instructions for each module are provided in the documentation included in the software package. Here the introduction focuses on related basic concepts and methods as well as practical considerations. Other related software tools are also introduced.

3.1.1   Image Preprocessing

Live imaging is required for visualizing and recording spindle protein dynamics. However, a commonly encountered problem is that spindles drift in their positions during imaging due to external forces. An example is shown in Fig. 3a in which a *Xenopus* egg extract spindle underwent both translational and rotational drift movement. Spindle drift must be corrected for computational tracking of spindle protein dynamics since otherwise it will cause biases and artifacts in quantification. Correction of spindle drift is usually completed using image registration techniques [13, 14] (Fig. 3b).

The software module *corrAligner* (*corr*elation *aligner*) is developed for spindle drift correction based on maximization of image correlation [13] and has been used in several studies of spindle protein dynamics [24, 25, 31–35]. Its strategy is to start with

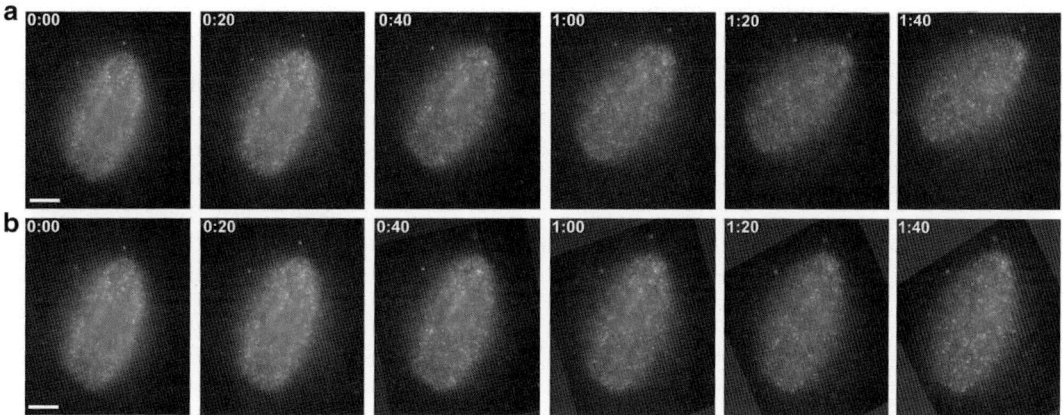

Fig. 3 Drift correction for a *Xenopus* egg extract spindle using image registration. (**a**) Original sequence of images. (**b**) Sequence of images after drift correction. Note that background intensity of images in (**b**) has been adjusted to avoid distortion of image contrast. Bars: 10 μm

aligning the second frame of the time-lapse image sequence to the first frame. Then the third frame is aligned to the *aligned* second frame and so on. This process is repeated until the entire image sequence is processed. Spindles often undergo significant shape changes during imaging. The strategy of *corrAligner* provides an effective solution since shape changes between two consecutive frames are usually small. Figure 3b shows the result of using *corr Aligner* to remove spindle drift shown in Fig. 3a.

To align a pair of images, the computational algorithm underlying *corrAligner* computes the optimal translation and rotation of the source image such that, after it is transformed by computed translation and rotation, the result image is maximally correlated with the target image [13]. This process is more reliable when images to be aligned contain rich texture. When images from multiple channels need to be aligned simultaneously, images from the channel with the richest texture should be selected for image registration. After the optimal translation and rotation are computed for images from this channel, they can be applied to images of other channels. Multiple channel image alignment is supported in *corrAligner*.

Several additional drift correction software tools are available. Microscope stage drift can cause drift of the entire image field of view. An ImageJ plug-in *ImageStabilizer*, based on the Lucas–Kanade algorithm for template tracking [36], is developed for correcting this type of drift. Another MATLAB software *ptAligner*, based on tracking correlated movement of automatically selected corner features [37], is developed for drift correction when the number of image features is low. Image registration is an important image processing technique and has many applications in biological image analysis beyond drift correction. Additional image registration tools are available in several general-purpose biological image analysis packages [27].

A key factor to be considered in selecting image registration software is the type of image transformation allowed. For computation tracking of spindle protein dynamics, it is often necessary to limit the allowed image transformation to translation and rotation, which are referred to collectively as rigid body transformation as it preserves distance and angle [14]. Some image registration software allows image warping, i.e., nonrigid transformation. These tools are often developed for specialized medical imaging applications and are not recommended for spindle drift correction because the warping operation may generate artificial motion. Only rigid body transformation is allowed in *corrAligner*. In addition to drift correction, another common image preprocessing operation is noise suppression. It is often integrated into the next step, namely, particle feature detection.

3.1.2 Particle Feature Detection

Following image preprocessing, particles within preprocessed images are detected (Fig. 4). The *ptDetector* (*point detector*) software module is developed as a general-purpose tool for localizing 2D particle features in biological images. Two detection resolutions are available. Particles can be detected at a pixel-level resolution based on statistically testing whether local maxima of pixel intensities are significantly higher than their local background intensities

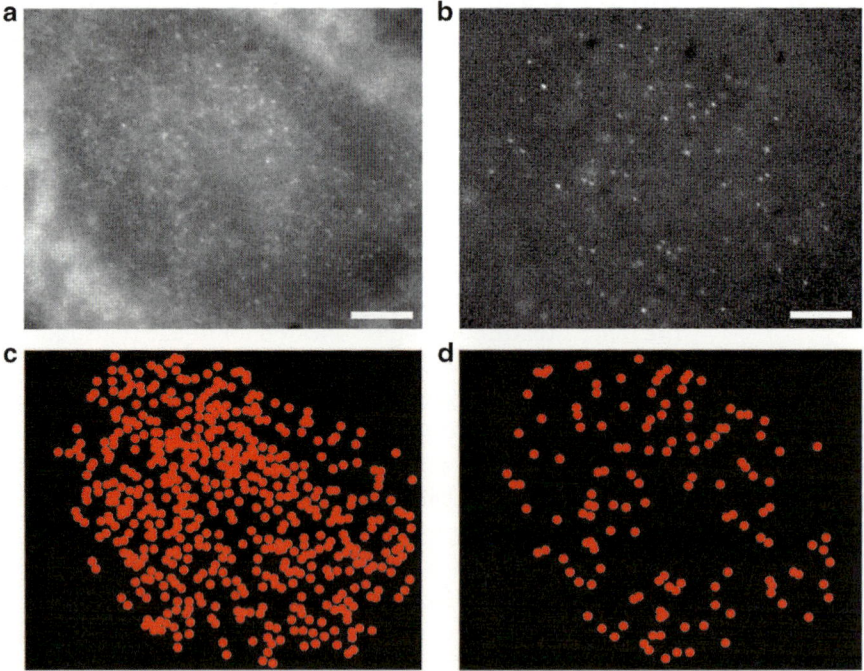

Fig. 4 Particle feature detection. (**a**) A control *Xenopus* egg extract spindle, imaged with Alexa-488-labeled tubulin at a concentration of ~2 nM. (**b**) Same spindle as in (**a**), imaged nearly simultaneously with X-rhodamine-labeled tubulin at a concentration of ~0.033 nM. (**c**, **d**) Particles detected using *ptDetector* software. (**a–d**) Because the effective pixel size is 160 nm, only pixel resolution detection is possible. Bars: 10 μm. Reprinted with permission from ref. 25. Copyright 2007

[38]. After pixel-level resolution detection, localization of particles can be further improved to a sub-pixel-level resolution by fitting their images using point spread function (PSF) of the microscope [39, 40]. Under a high-image signal-to-noise ratio (e.g., >10), single particles can be localized up to 1 nm in resolution [40]. In practice, however, the signal-to-noise ratio of fluorescence images of spindles is often much lower such that single particles can be localized to tens of nanometers in resolution. Detailed introduction to the computational method underlying *ptDetector* can be found in [38, 41].

Sub-pixel resolution detection is required for many subsequent trajectory analyses, such as calculation of particle diffusion coefficients. To achieve sub-pixel-level resolution, proper imaging conditions must be satisfied since otherwise the PSF fitting may give worse results than pixel resolution detection. In general, the effective pixel size should be less than one third of the radius of the Airy disk r, which is defined as $r=0.61\lambda/\text{NA}$ where λ is the emission wavelength and NA is the numerical aperture of the microscope objective lens. Since the radius of the Airy disk is ~200 nm for fluorescence microscopy, the effective pixel size should be ~70 nm or smaller. When the image signal-to-noise ratio is high, the effective pixel size can be slightly larger.

The *ptDetector* software allows user to control the level of selectivity in particle detection. For detection of dim particles, a low level of selectivity is required. However, this will result in an increase in the number of false-positive detections. Conversely, to reduce the number of false-positive detections, a high level of selectivity is required. However, this will result in the loss of dim particles. For computational tracking of spindle protein dynamics, a low selectivity level is preferred to avoid loss of dim particles. Since false-positive particles are generated by image noise and disappear quickly, their trajectories will be short. The high number of false-positive particles can be effectively excluded based on their trajectories by setting a threshold on the minimum trajectory length.

Particle detection is an important image analysis technique and is provided as a function in multiple general-purpose software packages for biological image analysis [27]. Furthermore, PSF fitting-based sub-pixel resolution particle detection is a standard technique in localization-based super-resolution imaging [42]. Recent advances in this area have produced many excellent software packages. For example, computational techniques have been developed for significant acceleration of particle detection through parallel computing [43].

3.1.3 Particle Feature Tracking

Spatiotemporal behaviors of detected particles are followed using image feature tracking techniques (Fig. 5). The goal is to establish correspondence between detected particles within different image frames, i.e., at different time points, so that complete spatiotemporal trajectories of particle can be recovered. This process is often

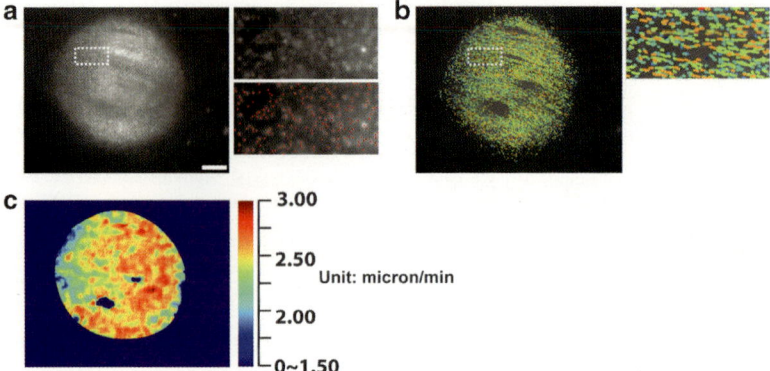

Fig. 5 Single particle tracking of spindle microtubule dynamics. (**a**) *Left panel*: an X-rhodamine tubulin speckle image of a *Xenopus* egg extract spindle. The *dotted rectangular* region is magnified in the *right upper panel*. *Right lower panel*: tubulin speckles detected by software are marked in *red* and overlaid onto the magnified image. Scale bar: 10 µm. (**b**) *Left panel*: speckle trajectories recovered using single particle tracking. *Right panel*: trajectories within the *dotted rectangular* region. *Warmer colors* indicate faster speckle velocities. (**c**) Global distribution of speckle movement velocities with the spindle. (**a–c**) Reprinted with permission from ref. 31. Copyright 2008

referred to as *single particle tracking*. It has been used in many studies on spindle protein dynamics.

The *LAPtrack* (*l*inear *a*ssignment *p*rogramming *track*ing) software module is developed as a general-purpose tool for single particle tracking. It recovers complete particle trajectories through identifying correspondence between particles in each pair of consecutive frames. It can be configured in several ways. For example, it can be configured to minimize the total distance between corresponding particles from each pair of consecutive frames. This is also referred to as global nearest neighbor tracking [44]. Alternatively, it can be configured to minimize the total variation in particle movement velocity and direction between each pair of consecutive frames [45]. Additional customization of *LAPtrack* is developed specifically for tracking bidirectional spindle microtubule flux [31]. In this case, users can interactively specify the approximate direction of microtubule flux [31].

A key parameter for *LAPtrack* is the search radius, which sets the upper limit of allowed particle movement between two consecutive frames. It is usually estimated based on kymographs. First, maximum velocities of particle movement within different manually selected regions of a spindle are estimated from their kymographs. Then, the overall maximum velocity among all analyzed regions is used to calculate the search radius, often multiplied by a safety factor to account for the fact that only a limited number of spindle regions are manually sampled.

Because of its many important applications in biological image analysis, single particle tracking has been extensively studied and is provided as a function in multiple general-purpose biological image analysis software packages, such as ICY [29]. Early approaches tried to solve the problem using optimal graph assignment [45] or statistical data association [46]. These approaches often perform poorly in resolving complex movement of large numbers of particles, which is common in biological applications. This is mainly due to their high computational cost and limitations in handling events such as particle appearance/disappearance as well as merging/splitting. Recently, significant progress has been made towards overcoming these limitations. To effectively resolve events such as particle appearance/disappearance and merging/splitting, algorithms based on the framework of multiple hypothesis testing have been developed [44, 47, 48]. These algorithms can follow complex particle movement events effectively but may not be suitable for tracking large numbers (e.g., >1,000) of particles due to their high computational costs. For all these algorithms, the specific cellular processes under investigation often impose constraints on particle movement. Such constraints can be integrated in different single tracking algorithms for improved performance. One example is the use of microtubule flux direction information in tubulin speckle tracking [31]. Comprehensive reviews of single particle tracking techniques are provided in [16, 17].

In single particle tracking, each particle is represented by its Cartesian coordinates. This general representation makes it possible to use the same technique to follow objects that are not particles. For example, region features that are well separated can be represented by their centroid coordinates and followed using single particle tracking. As another example, microtubule plus-end tracking proteins such as EB1 can be followed using single particle tracking for analyzing spindle microtubule dynamics [49, 50].

3.1.4 Trajectory Postprocessing

Recovered particle trajectories often need to be further processed. An important application of trajectory postprocessing is to correct or remove erroneous trajectories and to recover missing trajectories. These operations are performed manually by users with software support for human–computer interaction. For 2D images, a particularly useful strategy is to overlay software recovered trajectories of particles from 2D images onto their corresponding kymographs for inspection and correction [41]. However, these operations are only feasible when the numbers of particles and images are small. Manual analysis and correction of 2D trajectories are supported in the *LAPtrack* software module.

Another important application of trajectory postprocessing is to link recovered trajectory fragments into complete trajectories. This process is a part of several feature tracking techniques that are designed to first recover fragments of trajectories, often referred to as tracklets, and then connect them into full trajectories in postprocessing [51, 52].

3.2 Methods for Tracking Spindle Protein Dynamics in Continuous Region Images

Continuous region images provide lower spatiotemporal resolutions but are more convenient to generate experimentally. The procedure for tracking spindle protein dynamics in continuous region images is similar to the procedure for single particle images. For example, the same drift correction software tools can be used for continuous region images. There are, however, several important differences. First, there is no explicit feature detection for continuous region images. Instead, tracking is performed on automatically selected patches of pixels, often referred to as templates [11, 12]. Second, a different set of computational methods are required for tracking image templates [11, 12, 53, 54]. Third, the tracking of these templates reports vector fields, i.e., instantaneous directions, of spindle protein movement instead of complete trajectories. Further analysis can be performed on recovered vector fields. For example, their divergence can be used to infer local assembly and disassembly of cellular components if mass conservation is assumed [55]. Overall, however, readouts from continuous region images are more limited compared to single particle images.

3.3 Methods for Computational Analysis of Spindle Protein Dynamics

A wide variety of computational analysis of local and global spindle protein dynamics can be performed based on recovered trajectories. Here the focus is on protein spindle protein movement.

3.3.1 Computational Analysis of Local Spindle Protein Dynamics

From recovered particle trajectories, simple quantitative descriptors can be calculated to characterize local spindle protein dynamics at the single particle level. In the case of a particle undergoing directed motion, with its trajectory represented by $\{(x_1^k, y_1^k; I_1^k), (x_2^k, y_2^k; I_2^k), \ldots, (x_j^k, y_j^k; I_j^k), \ldots, (x_N^k, y_N^k; I_N^k)\}$, its behavior can be characterized using it average velocity, defined as

$$v^k = \frac{1}{(N-1) \cdot T} \sum_{i=2}^{N} \sqrt{\left(x_i^k - x_{i-1}^k\right)^2 + \left(y_i^k - y_{i-1}^k\right)^2}$$ where T is the time

interval between two consecutive frames. In the case of a particle undergoing diffusive motion, its behavior can be characterized using the relation of its mean-square displacement (MSD) with respect to time [56, 57]. A particle undergoing pure diffusion satisfies the linear relation of $\mathrm{MSD}(t) = 2Dt$ where D is its diffusion coefficient. For a particle undergoing diffusion superimposed with directed movement or local flow, its MSD follows $\mathrm{MSD}(t) = 2Dt + v^2 t^2$ where v is the movement or flow velocity. For a particle undergoing locally confined diffusion, its MSD follows $\mathrm{MSD}(t) = \mathrm{MSD}(\infty) [1 - e^{-t/\tau}]$ where τ is a constant representing constraint. This type of single particle analysis has been used broadly in computational analysis of spindle protein dynamics [31, 33, 34].

The computational analysis described above assumes implicitly that the particle has the same behavior in different regions. This may not be the case in the heterogeneous spindle structure. Heterogeneous behaviors can be described using hidden Markov model so that transitions between multiple states can be quantified and modeled [58, 59]. For particles undergoing directed movement but switch between different velocity modes, statistical clustering analysis can be used to identify their velocity modes [60, 61], and probabilities of switching between these modes can be estimated using hidden Markov models.

3.3.2 Computational Analysis of Global Spindle Protein Dynamics

Spindle protein dynamics can be characterized at the global level by analyzing behaviors of particle populations within selected regions of interest. An important class of analysis is to examine the spatial patterns of particles. A variety of statistical analysis tools have been developed for this purpose [62, 63]. Another important class of analysis is to detect spatial heterogeneity. This often requires a partition of selected regions of interest into smaller units and characterizes spatiotemporal dynamics within each unit [64]. A simple application of this approach in analyzing spatial heterogeneity of spindle microtubule flux is described in [31]. Overall, quantitative studies of global spindle protein dynamics remain limited. Please *see* **Notes 3, 4,** and **5** for further discussions.

3.4 Example: Computational Tracking and Analysis of Kinsin-5 Dynamics In Vitro and In Vivo

The mitotic spindle is a molecule machine that is assembled from dynamic microtubules and molecular motors. Molecular motor kinesin-5 is an essential spindle protein that serves important roles in spindle assembly, organization, and function [1, 2]. Understanding spatiotemporal dynamics of kinesin-5 is important to understanding molecular mechanisms of spindle assembly and chromosome segregation.

Image-based computational tracking and analysis have been used for high-resolution studies of kinesin-5 dynamics in vitro [33]. Movement of single kinesin-5 with and without the C-terminal tail domain along microtubules can be visualized using kymographs (Fig. 6a, b). For high-resolution analysis of dynamic behavior of kinesin-5, single particle tracking is required to recover their complete trajectories [33]. From the recovered trajectories, behavior properties such as mean velocity, diffusion constant, and microtubule association time under different experimental conditions can be calculated (Fig. 6c, d). Kinesin-5 without the C-terminal tail domain exhibits increased movement velocity and diffusion constants and decreased microtubule association time under low salt condition, indicating that the tail domain promotes kinesin-5 processivity and microtubule association [33].

Image-based computational tracking and analysis have also been used for high-resolution analysis of kinesin-5 dynamics in vivo in cultured mammalian cells [34]. Early studies of kinesin-5 in *Xenopus* egg extract spindles suggested that spindle kinesin-5 was largely stationary [65]. This conclusion was revised by a later study using photoactivation and fluorescence recovery after photobleaching techniques, which revealed active transport of spindle kinesin-5 [66]. However, the continuous region images used in

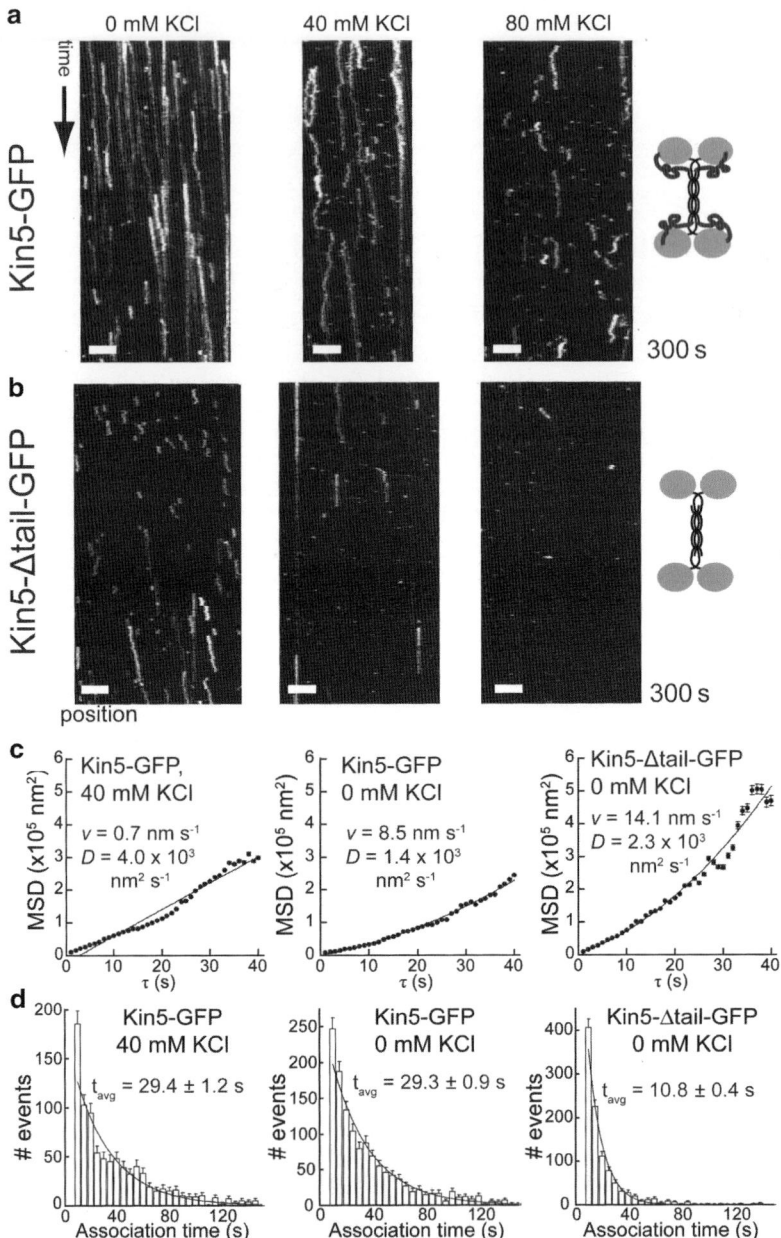

Fig. 6 Image-based analysis of kinesin-5 dynamics in vitro. (**a**, **b**) Kymographs of movement of kinesin-5-GFP and kinesin-5-Δtail-GFP under increasing salt concentrations. Bars: 3 μm. (**c**) Mean velocity and diffusion coefficient under different conditions. (**d**) Microtubule association time under different conditions. (**c**, **d**) Are computed from kinesin-5 trajectories recovered by single particle tracking. Reprinted with permission from ref. 33. Copyright 2011

the study did not have sufficient resolution to examine whether kinesin-5 dynamics varies in different spindle regions. This question was addressed in a recent study that combines high-resolution imaging with single particle tracking of kinesin-5 [34] (Fig. 7a–d).

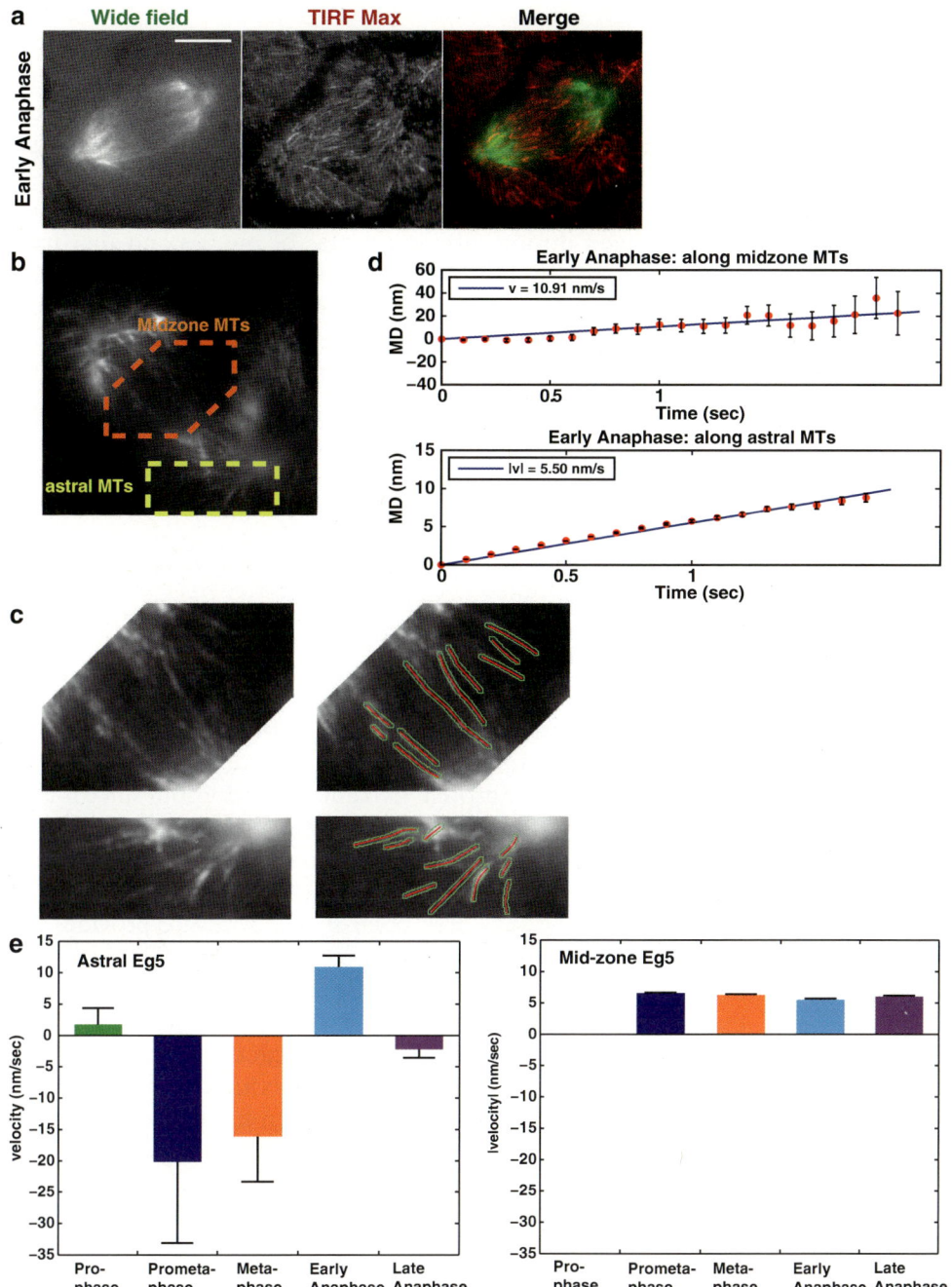

Fig. 7 Image-based analysis of kinesin-5 motility in mammalian cells. (**a**) Kinesin-5 distribution in an early anaphase spindle of LLC-Pk1 cells visualized by fluorescence microscopy. *Left panel*: kinesin-5 visualized using wide field imaging, which produces a low-contrast continuous region image. *Middle panel*: kinesin-5 visualized using TIRF, which produces a single particle image. *Right panel*: overlay. TIRF imaging provides significantly improved image contrast. Scale bar: 10 μm. (**b**) Identified regions of midzone and astral microtubules. (**c**) Detected microtubules within the identified regions in (**b**). (**d**) Velocities of kinesin-5 along midzone and astral microtubules are determined using linear regression of kinesin-5 particle mean displacement (MD), calculated from individual trajectories. (**e**) Kinesin-5 velocities on astral and midzone microtubules, respectively, during different stages of mitosis. Reprinted with permission from ref. 34. Copyright 2012

It found that kinesin-5 movement varies spatially between midzone microtubules and astral microtubules (Fig. 7d) and that kinesin-5 movement changes directions at different stages of cell division on astral microtubules but not on interzonal microtubules (Fig. 7e). Overall, the study revealed that spindle kinesin-5 is dynamically reorganized during different stages of mitosis and that its spatial distribution is tightly regulated in synchronization with the progression of mitosis. Together, these in vitro and in vivo studies of kinesin-5 dynamics provide a representative example of using image-based computational tracking and analysis for understanding complex spatiotemporal behaviors of an essential spindle protein.

4   Notes

1. A wide variety of commercial and free open-source software tools are available for biological image analysis [27]. Commercial software generally offers better usability, reliability, documentation, and technical support than free open-source software. However, to be financially viable, commercial software usually must support a broad range of applications and is therefore not designed to address specific applications such as spindle protein dynamics. In addition, since it is proprietary, modification and customization are often not feasible. Commercial software can be costly, especially when the number of users is limited.

   In many ways, free open-source software provides a necessary complement to commercial software. They are provided with source code to facilitate modification and customization and free of charge to increase accessibility. Many are developed by investigators for specific biological applications. However, free open-source software also has its limitations. Since resources available to its developers are often limited, documentation is often limited or lacking, and technical support may not be available. Overall, to combine commercial software with free open-source software is often a practical solution. Research and development in biological image analysis are advancing rapidly and will make more choices of software tools available.

2. Proper quality control is crucial to computational tracking and analysis of spindle protein dynamics, especially considering that a wide variety of computational analysis techniques must be integrated (Fig. 1). The basic requirements are to ensure accuracy and precision in data analysis and to avoid analysis artifacts.

   Several strategies are frequently adopted to ensure accuracy and precision in data analysis. First, most software tools provide visualization of computational image analysis results so that quality of the results can be directly inspected. For example, a useful approach to check correctness of recovered particle trajectories in 2D images is to overlay them onto kymographs and look for discrepancy [41, 47]. Second, investigator can develop experiments with known outcomes to test the performance of computational image analysis. Third, general quality control techniques are available for validating and benchmarking of computational image analysis techniques [67]. Specialized techniques have also been developed for specific biological applications such as particle localization [39]. These techniques can be used for benchmarking accuracy and precision of computational tracking and analysis of spindle protein dynamics.

Analysis artifacts are false outcomes generated by the data analysis process rather than the actual biological process. Rigorous data analysis is essential to preventing analysis artifacts. However, analysis artifacts can be difficult to detect as they may be caused by many factors, such as erroneous imaging settings or inherent limitations of adopted data analysis techniques. Although there are no guaranteed solutions to prevent analysis artifacts, effective quality control practices can be adopted, such as comparing results under different image settings, using different analysis methods, and under different controlled experimental conditions.

3. Manual image analysis is often convenient to perform and allows investigators to quickly obtain approximate results. This is particularly useful when the image data set is small. For example, single particles can be tracked manually using kymographs. Importantly, manual image analysis allows investigators to utilize their specialized knowledge and expertise in image analysis. This is difficult to implement in computational image analysis. However, manual image analysis has its limitations. First, quality of manual image analysis varies depending on individual investigators. Manual image analysis often is inferior in metrics such as accuracy, precision, resolution, sensitivity, and consistency. Furthermore, when the image data set is large, manual data analysis is no longer feasible. In such cases, manual data analysis at best can be performed on a small fraction of the data. This runs the risk of losing important information.

4. As is the case for many areas, developing biological image analysis techniques requires specialized training and expertise. Although there are practical barriers, interdisciplinary collaboration between experimental biologists and image analysis practitioners is a practical solution.

5. Image-based computational tracking and analysis are particularly well suited for obtaining measurements of spindle protein dynamics for quantitative modeling. So far, however, the potential remains to be fully utilized. Current applications of computational tracking and analysis of spindle protein dynamics have

focused primarily on reporting ensemble measurements. To understand at the systems-level mechanisms of cell division, computational tracking and analysis of spindle protein dynamics should be closely integrated with experimental investigation and quantitative modeling.

Acknowledgements

I am grateful to my colleagues for valuable discussions and inputs. I would like to thank Gaudenz Danuser, Ted Salmon, Tarun Kapoor, Yixian Zheng, and Patricia Wadsworth for sharing their insights into the biology of mitosis. I would also like to thank Yu-li Wang and Jelena Kovacevic for their support. This work is supported in part by NSF grants MCB-1052660 and DBI-1052925 and NSF Faculty Early Career Award DBI-1149494.

References

1. Wittmann T, Hyman A, Desai A (2001) The spindle: a dynamic assembly of microtubules and motors. Nat Cell Biol 3(1):E28–E34

2. Walczak CE, Heald R (2008) Mechanisms of mitotic spindle assembly and function. Int Rev Cytol 265:111–158

3. Rieder CL, Khodjakov A (2003) Mitosis through the microscope: advances in seeing inside live dividing cells. Science 300(5616):91–96

4. Ferenz NP et al (2010) Imaging protein dynamics in live mitotic cells. Methods 51(2):193–196

5. Maddox PS et al (2012) Imaging the mitotic spindle. In: Michael Conn P (ed) Methods in enzymology. Academic, San Diego, pp 81–103

6. Toya M et al (2012) Imaging of mitotic spindle dynamics in *Caenorhabditis elegans* embryos. In: Cassimeris L, Tran P (eds) Methods in cell biology. Academic, San Diego, pp 359–372

7. Danuser G (2011) Computer vision in cell biology. Cell 147(5):973–978

8. Dorn JF, Danuser G, Yang G (2008) Computational processing and analysis of dynamic fluorescence image data. In: Sullivan KF (ed) Methods in cell biology. Academic, San Diego, pp 497–538

9. Mogilner A et al (2006) Modeling mitosis. Trends Cell Biol 16(2):88–96

10. Mogilner A, Craig E (2010) Towards a quantitative understanding of mitotic spindle assembly and mechanics. J Cell Sci 123(20):3435–3445

11. Sonka M, Hlavac V, Boyle R (2007) Image processing, analysis, and machine vision, 3rd edn. Thomson Engineering, Toronto

12. Szeliski R (2010) Computer vision: algorithms and applications. Springer, New York

13. Zitova B, Flusser J (2003) Image registration methods: a survey. Image Vision Comput 21(11):977–1000

14. Maintz JBA, Viergever MA (1998) A survey of medical image registration. Med Image Anal 2(1):1–36

15. Nixon M, Aguado A (2008) Feature extraction & image processing, 2nd edn. Academic, Oxford

16. Meijering E et al (2009) Tracking in cell and developmental biology. Semin Cell Dev Biol 20(8):894–902

17. Meijering E, Dzyubachyk O, Smal I (2012) Methods for cell and particle tracking. Methods Enzymol 504:183–200

18. Saxton MJ (2008) Single-particle tracking: connecting the dots. Nat Methods 5(8):671–672

19. Saxton MJ, Jacobson K (1997) Single-particle tracking: applications to membrane dynamics. Annu Rev Biophys Biomol Struct 26(1):373–399

20. Witten IH, Frank E, Hall MA (2011) Data mining: practical machine learning tools and techniques, 3rd edn. Morgan Kaufmann, Burlington, MA

21. Yang G (2013) Bioimage informatics for understanding spatiotemporal dynamics of cellular processes. Wiley Interdiscip Rev Syst Biol Med 5(3):367–380

22. Lichtman JW, Conchello JA (2005) Fluorescence microscopy. Nat Methods 2:910–919

23. Danuser G, Waterman-Storer CM (2006) Quantitative fluorescent speckle microscopy of cytoskeleton dynamics. Annu Rev Biophys Biomol Struct 35(1):361–387

24. Cameron LA et al (2006) Kinesin-5 independent poleward flux of kinetochore microtubules in PtK1 cells. J Cell Biol 173(2): 173–179

25. Yang G et al (2007) Architectural dynamics of the meiotic spindle revealed by single-fluorophore imaging. Nat Cell Biol 9(11): 1233–1242

26. Lippincott-Schwartz J, Snapp E, Kenworthy A (2001) Studying protein dynamics in living cells. Nat Rev Mol Cell Biol 2(6):444–456

27. Eliceiri KW et al (2012) Biological imaging software tools. Nat Methods 9(7):697–710

28. Schneider CA, Rasband WS, Eliceiri KW (2012) NIH image to ImageJ: 25 years of image analysis. Nat Methods 9(7):671–675

29. de Chaumont F et al (2012) Icy: an open bio-image informatics platform for extended reproducible research. Nat Methods 9(7): 690–696

30. Schindelin J et al (2012) Fiji: an open-source platform for biological-image analysis. Nat Methods 9(7):676–682

31. Yang G et al (2008) Regional variation of microtubule flux reveals microtubule organization in the metaphase meiotic spindle. J Cell Biol 182(4):631–639

32. Houghtaling BR et al (2009) Op18 reveals the contribution of nonkinetochore microtubules to the dynamic organization of the vertebrate meiotic spindle. Proc Natl Acad Sci U S A 106(36):15338–15343

33. Weinger JS et al (2011) A nonmotor microtubule binding site in kinesin-5 is required for filament crosslinking and sliding. Curr Biol 21(2):154–160

34. Gable A et al (2012) Dynamic reorganization of Eg5 in the mammalian spindle throughout mitosis requires dynein and TPX2. Mol Biol Cell 23:1254–1266

35. Goodman B et al (2010) Lamin B counteracts the kinesin Eg5 to restrain spindle pole separation during spindle assembly. J Biol Chem 285(45):35238–35244

36. Baker S, Matthews I (2004) Lucas-Kanade 20 years on: a unifying framework. Int J Comput Vision 56(3):221–255

37. Qiu M, Yang G (2013) Drift correction for fluorescence live cell imaging through correlated motion identification. In: Proceedings of 2013 IEEE international symposium on bio-

38. Ponti A et al (2003) Computational analysis of F-Actin turnover in cortical actin meshworks using fluorescent speckle microscopy. Biophys J 84(5):3336–3352

39. Cheezum MK, Walker WF, Guilford WH (2001) Quantitative comparison of algorithms for tracking single fluorescent particles. Biophys J 81(4):2378–2388

40. Yildiz A, Selvin PR (2005) Fluorescence imaging with one nanometer accuracy: application to molecular motors. Acc Chem Res 38(7):574–582

41. Qiu M, Lee H-C, Yang G (2012) Nanometer resolution tracking and modeling of bidirectional axonal cargo transport. In: Proceedings of 2012 IEEE international symposium on biomedical imaging (ISBI). Barcelona, Spain, pp 992–995

42. Huang B, Bates M, Zhuang X (2009) Super-resolution fluorescence microscopy. Annu Rev Biochem 78(1):993–1016

43. Smith CS et al (2010) Fast, single-molecule localization that achieves theoretically minimum uncertainty. Nat Methods 7(5): 373–375

44. Blackman S, Popoli R (1999) Design and analysis of modern tracking systems. Artech House, Norwood

45. Veenman CJ, Reinders MJT, Backer E (2001) Resolving motion correspondence for densely moving points. IEEE Trans Patt Anal Mach Intel 23(1):54–72

46. Cox I (1993) A review of statistical data association techniques for motion correspondence. Int J Comput Vision 10(1):53–66

47. Mukherjee A et al (2011) Automated kymograph analysis for profiling axonal transport of secretory granules. Med Image Anal 15(3): 354–367

48. Padfield D, Rittscher J, Roysam B (2011) Coupled minimum-cost flow cell tracking for high-throughput quantitative analysis. Med Image Anal 15(4):650–668

49. Applegate KT et al (2011) plusTipTracker: quantitative image analysis software for the measurement of microtubule dynamics. J Struct Biol 176(2):168–184

50. Matov A et al (2010) Analysis of microtubule dynamic instability using a plus-end growth marker. Nat Methods 7(9):761–768

51. Jaqaman K (2008) Robust single-particle tracking in live-cell time-lapse sequences. Nat Methods 5:695–702

52. Li K et al (2008) Cell population tracking and lineage construction with spatiotemporal context. Med Image Anal 12(5):546–566

medical imaging (ISBI). San Francisco, CA, pp 452–455

53. Ji L, Danuser G (2005) Tracking quasi-stationary flow of weak fluorescent signals by adaptive multi-frame correlation. J Microsc 220(3):150–167

54. Ye M, Haralick RM, Shapiro LG (2003) Estimating piecewise-smooth optical flow with global matching and graduated optimization. IEEE Trans Patt Anal Mach Intel 25(12): 1625–1630

55. Wilson CA et al (2010) Myosin II contributes to cell-scale actin network treadmilling through network disassembly. Nature 465(7296):373–377

56. Qian H, Sheetz MP, Elson EL (1991) Single particle tracking: analysis of diffusion and flow in two-dimensional systems. Biophys J 60(4): 910–921

57. Saxton MJ (2007) Modeling 2D and 3D diffusion. Methods Mol Biol 400:295–321

58. Das R, Cairo CW, Coombs D (2009) A hidden Markov model for single particle tracks quantifies dynamic interactions between LFA-1 and the actin cytoskeleton. PLoS Comput Biol 5(11):e1000556

59. Rabiner LR (1989) A tutorial on hidden Markov models and selected applications in speech recognition. Proc IEEE 77(2):257–286

60. Reis GF et al (2012) Molecular motor function in axonal transport in vivo probed by genetic and computational analysis in *Drosophila*. Mol Biol Cell 23(9):1700

61. Fraley C, Raftery AE (2002) Model-based clustering, discriminant analysis and density estimation. J Am Stat Assoc 97:611–631

62. Diggle PJ (2007) Spatio-temporal point processes: methods and applications. In: Finkenstad B, Held L, Isham V (eds) Statistical methods for spatio-temporal systems. Chapman & Hall, Boca Raton, FL, pp 1–45

63. Illian J et al (2008) Statistical analysis and modeling of spatial point patterns. Wiley-Interscience, Chichester

64. Cressie N, Wikle CK (2011) Statistics for spatio-temporal data. Wiley-Interscience, Hoboken, NJ

65. Kapoor TM, Mitchison TJ (2001) Eg5 is static in bipolar spindles relative to tubulin. J Cell Biol 154(6):1125–1134

66. Uteng M et al (2008) Poleward transport of Eg5 by dynein/dynactin in *Xenopus laevis* egg extract spindles. J Cell Biol 182(4):715–726

67. Christensen HI, Philips PJ (eds) (2002) Empirical evaluation methods in computer vision. World Scientific Publishing, Singapore

Part II

Analysis and Manipulation of the Microtubule Cytoskeleton in Cells and In Vitro

Chapter 6

The Use of Cultured *Drosophila* Cells for Studying the Microtubule Cytoskeleton

Jonathan Nye, Daniel W. Buster, and Gregory C. Rogers

Abstract

Cultured *Drosophila* cell lines have been developed into a powerful tool for studying a wide variety of cellular processes. Their ability to be easily and cheaply cultured as well as their susceptibility to protein knockdown via double-stranded RNA-mediated interference (RNAi) has made them the model system of choice for many researchers in the fields of cell biology and functional genomics. Here we describe basic techniques for gene knockdown, transgene expression, preparation for fluorescence microscopy, and centrosome enrichment using cultured *Drosophila* cells with an emphasis on studying the microtubule cytoskeleton.

Key words Cell culture, Centrosomes, *Drosophila*, dsRNA-mediated interference, Kc cells, RNAi, S2 cells, Schneider cells

1 Introduction

Since the early 1900s when Thomas Morgan was performing his groundbreaking work on the role of chromosomes in heredity that eventually won him a Nobel Prize, *Drosophila melanogaster* has proven to be one of the most productive and widely used model systems for the study of a broad range of biological processes. More recently, the use of cultured *Drosophila* cell lines has increased in popularity due to their ability to be cheaply and easily maintained while exploiting the extensive knowledge gained from over a century of fly research. There are over 100 different cell lines available from the Drosophila Genomics Research Center (DGRC), a nonprofit repository of *Drosophila* cell lines and DNA clones. The most commonly used cell lines include the S2 and Kc cells: these cell lines, derived from spontaneously immortalized cells arising in cultures of disrupted fly embryos, are conveniently maintained at room temperature (23–28 °C), in ambient atmosphere, using inexpensive media.

Interestingly, many cultured *Drosophila* cells have the unique ability to take up dsRNA through receptor-mediated endocytosis,

David J. Sharp (ed.), *Mitosis: Methods and Protocols*, Methods in Molecular Biology, vol. 1136,
DOI 10.1007/978-1-4939-0329-0_6, © Springer Science+Business Media New York 2014

meaning that selective knockdown of targeted proteins can be accomplished by simply adding long double-stranded (ds)RNA oligomers directly to the media and "soaking" the cells [1, 2]. This eliminates the need for the often costly and toxic short interfering (si)RNA transfection methods used for mammalian cell cultures and provides a more rapid and cost-effective way to study the effects of gene silencing. Typically, a protein of interest can be depleted from RNAi-treated cells within 3–7 days, although the length of time required will vary depending on the turnover rate of the target protein and the possible effects of target protein knockdown on the overall health of the cells. These characteristics—combined with a fully sequenced and well-annotated genome—have made cultured *Drosophila* cells ideal for the experimental manipulation and analysis of cellular processes. For example, RNAi-based high-throughput genome-wide screens in S2 cells are responsible for significant advances in our understanding of mitotic phenomena, including the identification of new proteins necessary for proper mitotic spindle assembly and function as well as centrosome assembly/duplication [3–5].

In this chapter, we will outline techniques that take advantage of the useful traits of this extraordinary model system including gene knockdown, transfection of exogenous genes, and fluorescence microscopy of live and fixed cells. Though our focus is on the use of cultured fly cells for studying the microtubule cytoskeleton, mitosis, and centrosome biology, this system and the protocols described here can be exploited to examine numerous aspects of cell biology. Importantly, most cultured *Drosophila* cells, like many cell types in the fruit fly, display an unconventional cycle of centrosome function compared to mammalian cells (Fig. 1). Centrosomes do not exist during most of the cell cycle in these cells [6]. Instead, interphase microtubule nucleation is apparently random and independent of centrioles and γ-tubulin [7]. Most cultured cells do contain centrioles though, but these do not nucleate microtubule growth until mitotic entry when they undergo a maturation process and recruit pericentriolar material (PCM). Therefore, bona fide centrosomes only exist during mitosis to assist in mitotic spindle assembly. Centrosomes shed their PCM upon mitotic exit, leaving only denuded centrioles (Fig. 2). Notably, within a culture there exist cells with supernumerary centrioles and cells that completely lack centrioles, but the presence of too many or few centrioles has surprisingly little consequence on overall cell health or proliferation save for a subtle increase in the duration of cell division. This is due to centrosome-independent mechanisms of spindle assembly [8], as well as the robust ability of spindles to cluster excess centrosomes and achieve bipolarity (Fig. 3) [10]. Thus, cultured *Drosophila* cells present an extraordinary system to gain molecular insight into the important processes of microtubule nucleation and spindle assembly in addition to new mechanisms evolved to cope with centrosome amplification, a phenomenon that is frequently observed in human cancer cells.

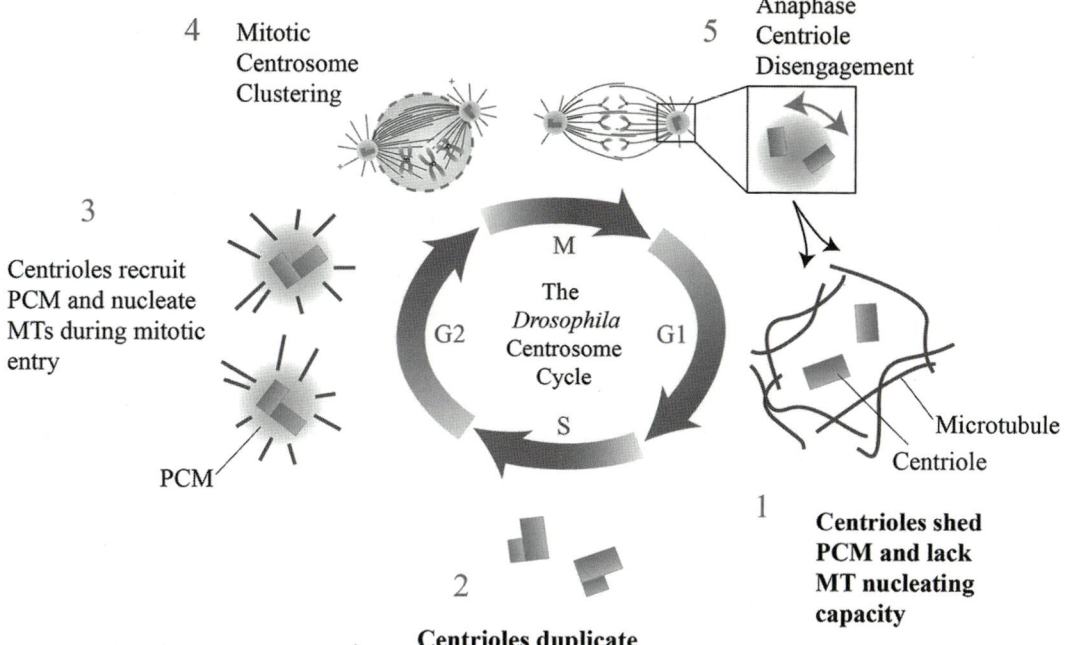

Fig. 1 Illustration of the centrosome cycle in cultured *Drosophila* cells. (*1*) Throughout interphase, microtubule (MT) nucleation is apparently random and occurs by a mechanism which is independent of both centrioles and γ-tubulin. Centrioles do not recruit pericentriolar material (PCM) or nucleate/organize MT arrays and exist as singlets (at the light microscope level) during G1. (*2*) During late G1 or S phase, procentrioles assemble on parent centrioles. (*3*) During G2, procentrioles elongate and continue to do so into mitosis. Prior to mitosis, centrosomes assemble as mitotic kinases promote PCM recruitment to centrioles and nucleate MT growth. (*4*) After nuclear envelope breakdown, excess centrosomes cluster to achieve spindle bipolarity. (*5*) In late anaphase, mother–daughter centriole pairs separate (disengage). Lastly, centrosomes fragment after mitotic exit and release centriole singlets into the cytoplasm that eventually shed their PCM

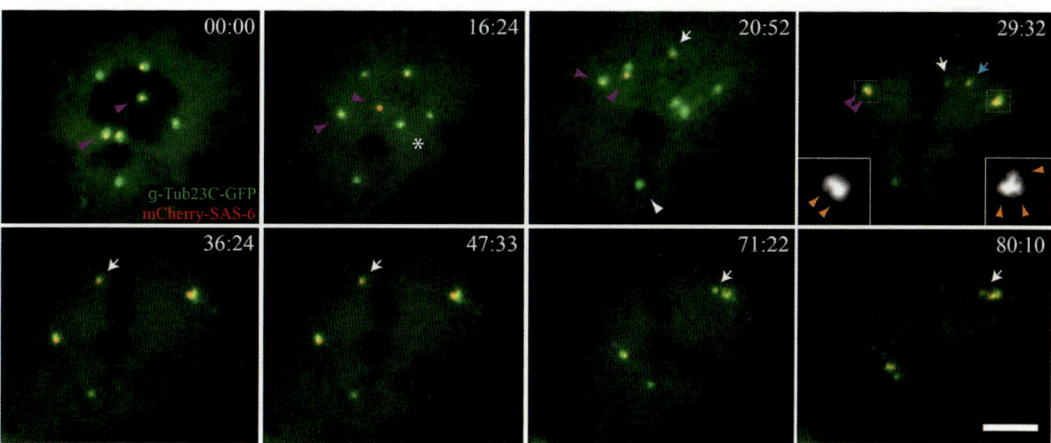

Fig. 2 Centrosome clustering suppresses multipolar spindle formation. Time series of bona fide centrosomes labeled with the centriole marker mCherry-SAS-6 (*red, insets* with *orange arrowheads*) and PCM marker γ-Tub23C-GFP (*green*) in a mitotic cell beginning with eight centrosomes (*asterisk* marks the site of a centrosome positioned above the focal plane). A population of cells in an S2 culture will contain excess centrioles that generate supernumerary mitotic centrosomes. Centrosomes cluster to achieve spindle bipolarity (two centrosomes that cluster at the left pole are marked with *purple arrowheads* in the *upper panel*). Some centrosomes cluster by utilizing a novel behavior of motility by moving along the metaphase spindle periphery (*white* and *blue arrows*). One remote centrosome (*white arrowhead*) is delayed in spindle incorporation. Scale, 5 μm

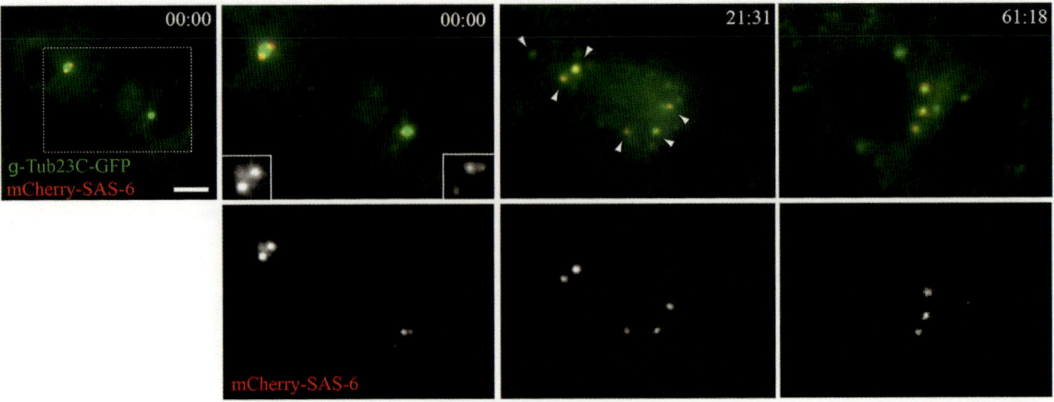

Fig. 3 Centriole disengagement is followed by centrosome fragmentation as S2 cells exit mitosis. Using time-lapse fluorescence microscopy, bona fide centrosomes were visualized in live stable cells expressing the PCM marker γ-Tub23C-GFP (*green*) and the centriole marker mCherry-SAS-6 (*red, insets*). Time series of a cell exiting mitosis. The use of concanavalin-A to flatten cells prevents the successful completion of cytokinesis. Nevertheless, cell cycle progression continues as nuclear envelopes reform and rotate to position their attached centrosomes in the middle of the cell; each centrosome appears as a large perinuclear γ-tubulin spot containing three disengaged centrioles. Although the expected number of centrioles in each γ-tubulin spot should be two at this final stage of mitosis, for reasons unknown, some cells contain more than expected number. Within 30 min, each centrosome fragmented, releasing the centrioles into the cytoplasm. These then move throughout the cell. As centrosomes fragment, centrioles retained some γ-tubulin. Presumably they eventually shed all of their γ-tubulin because interphase centrioles do not co-localize with γ-tubulin or nucleate microtubule growth during interphase [7]. Similar PCM shedding has been observed in live larval dividing neuroblasts [9]. Scale, 5 μm

2   Materials

2.1   Maintaining Drosophila Cell Lines

1. Tissue culture flasks: Tissue culture grade plastic Petri dishes or culture flasks are suitable. For general maintenance, we use "T25" (25 cm^2 culture surface area) culture flasks with plug-seal caps containing 5–10 ml of medium. For transfected cells, we use 6-well culture plates, with 1–2 ml of medium per well.

2. Medium: There are a variety of media that can be used for *Drosophila* cell culture. A number of serum-free media are available such as SF-900 II, Insectagro Sf9, SFX-Insect, and Insect-XPRESS, which are cheap alternatives to others that require supplementation with heat-inactivated fetal calf serum (FCS) such as Schneider's medium or Shields and Sang's M3 medium. Obviously, the specific cell line dictates the medium needed, but some cell lines can be maintained in more than one medium (e.g., SF-900 II or Schneider's/10 % FCS in the case of S2 cells). The specific experimental procedure may also be a factor to consider. For example, Schneider's/10 %

FCS medium may be a better option for live-cell microscopy experiments since it exhibits less autofluorescence than SF900 II. Also, some serum-free media induce a higher basal level of expression of those transgenes controlled by the inducible metallothionein promoter, compared to Schneider's/10 % FCS medium. Finally, antibiotics may be added to cultures but are not necessary if a researcher's sterile technique is adequate. Antibiotics are used at the same concentrations for mammalian cell culture: 50–100 U/ml penicillin G and 50–100 μg/ml streptomycin. We use a commercial cocktail of antibiotics and fungicide (penicillin-streptomycin-amphotericin B).

3. Incubator: An incubator is not needed unless the ambient temperature fluctuates significantly from room temperature (~25 °C). If capped culture flasks are used, then it is usually unnecessary to use a humidified chamber.

2.2 Design and Production of dsRNA for RNAi

2.2.1 Production of DNA Template

1. Taq DNA polymerase.
2. DNA: (1) cDNA, (2) genomic DNA, or (3) aliquot from a previous PCR reaction.
3. 10× PCR Buffer: 200 mM Tris–HCl (pH 8.4), 500 mM KCl, 15 mM $MgCl_2$. (Vary the $MgCl_2$ concentration as needed.)
4. 10× dNTP mix: 1.25 mM each nucleotide (5 mM total) in nuclease-free H_2O (pH 7.5); sodium salts of dATP, dCTP, dGTP, and dTTP.
5. Primers (see below).

2.2.2 Purification of DNA Template

1. Phenol/chloroform/isoamyl alcohol (25:24:1, by volume).
2. Chloroform/isoamyl alcohol (24:1, by volume).

2.2.3 In Vitro Transcription Reaction

1. 5× Transcription Reaction Buffer: 400 mM HEPES (pH to 7.5 with KOH), 120 mM $MgCl_2$, 200 mM DTT, 10 mM spermidine. Store at –20 °C. Optionally, 50 mg/ml PEG-8000 can be added to increase yield.
2. rNTP Solution: mixture containing 62.5 mM of each nucleotide (sodium salts of ATP, CTP, GTP, and UTP) in RNase-free water; pH 8.0; stored at –20 °C. Since sodium, potassium, and ammonium reportedly inhibit T7 RNA polymerase, then using the lithium or Tris salts of the nucleotides could increase the reaction speed/yield.
3. Pyrophosphatase: 0.1 U/μl.
4. T7 RNA polymerase: 80 U/μl.
5. RNase-free DNase: 1 U/μl.
6. Nuclease-free H_2O.

2.3 Nucleofection of Cultured Drosophila Cells

1. Nucleofection Solution: 5 mM KCl, 15 mM MgCl$_2$, 60 mM Na$_2$PO$_4$, 60 mM NaH$_2$PO$_4$, 50 mM D-mannitol. Adjust pH to 7.2, if necessary. Sterile filter and aliquot for use. Store at 4 °C (*see* **Note 1**).

2. Electroporation cuvettes: 2 mm gap size, sterile, disposable electroporation cuvettes. We reuse cuvettes that have been *extensively* washed and then sterilized with UV light.

3. Transfer pipettes: sterile, fine tip, small bulb, disposable transfer pipettes (Samco Scientific).

4. Nucleofection Device: marketed by Lonza; the Nucleofector 2b device is currently the most basic model.

2.4 Preparation of Glass-Bottom Dishes

1. 35 × 10 mm plastic Petri dishes: Dishes do not need to be sterile or tissue culture grade. Verify that the dishes will fit on the stage of your inverted microscope.

2. Cover glass: Either round or square microscope-quality cover glass can be used. Verify that the cover glass is sufficiently large to cover the ¾ in. hole that you will make in the Petri dish, but not so large that it extends beyond the edge of the dish when it is glued in place. We use number 1.5, 22 × 22 mm^2 cover glass.

3. Drill bit: ¾ in. spade bit.

4. Grinding bit: Conical-shaped coarse grinding bit. Ours is ¾ in. diameter.

5. Electric drill: Variable-speed electric drill, preferably with a lock-on button.

6. Adhesive: Sylgard 184 Silicone Elastomer Kit (Dow Corning). This kit contains two components: the base, and the catalyst that cures the adhesive after base and catalyst are mixed in a 10:1 weight ratio.

7. Drilling surface: The surface used for drilling should not offer much resistance to the bit but should be firm enough to support the weight of the drill. We use a thick-walled styrofoam box whose surface has been covered with a layer or two of wide-diameter shipping tape (e.g., Scotch packaging tape) to prevent bits of styrofoam from being dislodged while drilling.

8. Device for gripping the dish: Pliers or vise grips with short pieces of Tygon tubing slipped over the jaws provide a good way to grip the dish while drilling. We use Irwin GV6 pliers.

2.5 Immunofluorescence and Live-Cell Imaging of Cultured Cells

1. Concanavalin-A: 0.5 mg/ml in distilled H$_2$O. If to be used for imaging live cells, then sterile filter the solution and aliquot. Store at –20 °C.

2. Fixation buffer: Many different fixation buffers are mentioned in the literature; we describe two buffers below. The first is preferred for visualizing the microtubules of fixed cells; the

second is often adequate for generally preserving epitopes and is a starting point when developing a new fixation protocol. Including a mild detergent (e.g., 1 mg/ml saponin) may improve fixation.

    (a) PME: 100 mM PIPES (pH to 6.2 with KOH), 2.5 mM $MgCl_2$, 0.5 mM EDTA, 5 mM EGTA. The composition of BRB80, another common microtubule-preserving buffer, is similar: 80 mM PIPES (pH 6.8), 1 mM $MgCl_2$, 1 mM EGTA.

    (b) PBS: 6.13 mM NaH_2PO_4, 3.87 mM Na_2HPO_4, 4 mM KCl, 130 mM NaCl, pH 7.0.

3. Fixative: Optimal fixation conditions are empirically determined. Some guides to selecting the appropriate fix method are mentioned further below.

    (a) Aldehyde fix: Commonly, a solution of 10 % formaldehyde in buffer is used for fixation. However, mixtures of formaldehyde and glutaraldehyde are often used (typically 2.5–5 % formaldehyde, 0.1–0.5 % glutaraldehyde), but then a subsequent wash with $NaBH_4$ (0.5 % in PBS, 5 min.) is needed to reduce residual glutaraldehyde to prevent autofluorescence. Make aldehyde solutions fresh. Buffers with free amines (e.g., Tris) should not be used for aldehyde fixation.

    (b) Methanol fix: Cold (–20 °C), anhydrous, premium-grade methanol is preferred; traces of water can be removed by adding drying beads (3 Å pore size molecular sieves) to the methanol storage container.

    (c) Methanol/formaldehyde fix: 90 % methanol/10 % formaldehyde (by volume), 5 mM sodium bicarbonate (pH 9.0), –70 °C, 10 min. This fix is not optimal for microtubule preservation but has been used to preserve a variety of epitopes.

4. PBS/Triton X-100: 0.1 % (by volume) Triton X-100 in PBS.

5. Blocking buffer: 5 % (by volume) normal goat serum in PBS/Triton X-100. Sterile filter (or add 0.02 % [w/v] sodium azide) and store at 4 °C.

6. Mounting medium: 90 % (by volume) glycerol, 10 % PBS (10×), 0.1 M propyl gallate. Rock overnight at room temperature to completely dissolve propyl gallate. Protect from light and store at –20 °C.

7. Commonly used stains and antibodies: For microtubule immunostaining, the anti-α-tubulin mouse monoclonal, DM1A (use at 2,000× dilution), is excellent. Fluorophore conjugates (e.g., FITC) of DM1A are also available but must be used at low (~5-fold) dilution. For centrosome immunostaining: anti-γ-tubulin

mouse monoclonal, GTU-88. Centriole immunostaining: polyclonal anti-D-PLP [11]. Chromosomes: Hoechst 33258. Specifically mitotic chromosomes: anti-phospho-Ser10-histone H3 (available as rabbit polyclonal and mouse monoclonal).

2.6 Purification of Centrosomes

1. PBS: 6 mM NaH_2PO_4, 4 mM Na_2HPO_4, 4 mM KCl, 130 mM NaCl, pH 7.0 (*see* **Note 2**). (Need ~20 ml/prep.)

    (a) 0.1× PBS/8 % sucrose: 8 % (w/v) ultrapure sucrose in 0.1× PBS, 0.05 % NaN_3. (NaN_3 is needed only to preserve solutions if stored for extended periods.) (Need ~15 ml/prep.)

2. 8 % sucrose: 8 % (w/v) ultrapure sucrose in double-distilled H_2O, 0.05 % NaN_3. (Need ~20 ml/prep.)

3. Tris buffer: 1 mM Tris–HCl, pH 8.0, 8 mM β-mercaptoethanol (β-ME; add fresh).

4. Tris/Igepal CA-630: 0.5 % (v/v) Igepal CA-630 in Tris buffer; if necessary, warm to 30 °C to dissolve detergent; 8 mM β-ME (add fresh). (Need ~50 ml/prep.)

5. PE/Igepal CA-630: 0.1 % (v/v) Igepal CA-630 in 1× PE (see below), 0.05 % NaN_3.

6. Ficoll cushion: 20 % (w/w) Ficoll 400 (average M_W = 400,000) in PE/Igepal CA-630. [To 10 g of Ficoll, add PE/Igepal CA-630 until reaching 50 g. Stir at room temperature until Ficoll dissolves. Just before use, add 8 mM β-ME.] (Need 3 ml/prep.)

7. 50× PE: 500 mM PIPES (pH to 7.2 with KOH), 50 mM EDTA, 0.05 % NaN_3, 400 mM β-ME (add β-ME fresh). (Need <1 ml/prep.)

8. PE/TX-100: 0.1 % (v/v) TX-100 in 1× PE, 0.05 % NaN_3. (Need ~45 ml/prep.)

9. 20 % sucrose gradient solution: 20 % (w/w) ultrapure sucrose in PE/TX-100 solution. [To 20 g of sucrose, add PE/TX-100 until mass is 100 g. Just before use, add 8 mM β-ME.] (Need ~18 ml/prep.)

10. 70 % sucrose gradient solution: 70 % (w/w) ultrapure sucrose in PE/TX-100 solution. [To 70 g of sucrose, add PE/TX-100 until mass is 100 g. Just before use, add 8 mM β-ME.] (Need ~20 ml/prep.)

11. 20–70 % sucrose gradient: To make a 32 ml gradient in an ultracentrifuge tube (we use Beckmann tube #344058; capacity of these tubes is ~38 ml), follow the recipe in Table 1. For each layer, mix the indicated amounts of 20 and 70 % solutions, then gently add the mixture to the top of the gradient. When all layers have been added, let stand 1 h, room temperature, without agitation. Then move to 4 °C long enough to chill the gradient.

Table 1
Recipe for a 5-layer step sucrose gradient

	(Bottom)				(Top)
Solution	Layer 1	Layer 2	Layer 3	Layer 4	Layer 5
20 %	0 ml	1.56	3.13	4.69	6.25 ml
70 %	7 ml	4.69	3.13	1.56	0 ml
Final %	70	57.5	45	32.5	20
Volume[a]	7 ml	6.25	6.25	6.25	6.25 ml

[a]Total volume of gradient = 32 ml

3   Methods

3.1   Maintaining Drosophila Cell Lines

1. Choice of cell line: A number of immortalized *Drosophila* cell lines are available commercially and from public sources like DGRC (https://dgrc.cgb.indiana.edu/), ATCC (http://www.atcc.org/), and the ECACC (http://www.hpacultures.org.uk/collections/ecacc.jsp). The choice of cell line will largely be determined by their experimental purpose. Perhaps the most commonly used cell lines are S2, S2R+, and Kc, but the unique characteristics of some cell lines are especially useful for studying specific cellular processes. As examples, D17-c3 cells are a model system for studying cell migration [12] and BG2-c6 cells for studying neurogenesis [13]. Table 2 has a short list of experimentally useful cultured *Drosophila* cells. These protocols have been optimized for S2 and Kc cell lines; however, they should be useful guides for culturing other *Drosophila* cell lines.

2. Cultured *Drosophila* cells are incubated at room temperature (~25 °C) with atmospheric CO_2 levels in non-ventilated tissue culture-treated flasks or Petri dishes. The frequency of passage naturally depends on the cell line's doubling time (which in turn is affected by culture temperature and medium) (Table 2). In the case of S2 cells, passaging is required approximately every 4–5 days (*see* **Note 3**).

3. Dislodge the cells using a method appropriate for the specific cell line (see below). Then disperse any clumps of cells using gentle trituration (*see* **Note 4**).

   For loosely adherent lines like S2 or Kc cells, dislodging cells is accomplished by pipetting medium onto cells to release them from the bottom of the flask.

   For more tightly adherent lines like S2R+, Kc_{167}, or D17-c3 cells, several techniques can be employed to dislodge cells: (1)

Table 2
Characteristics of some common and some specialized *Drosophila* cell lines

Cell line	Origin	Ploidy (primarily)	Commonly used medium	Doubling time (h)[a]	Original reference
S2	Late embryo	Tetraploid (often >4n)	Serum-free (SF-900 II)	24	[14]
S2R+	From Wg receptor expressing S2	(Unknown)	Schneider's, 10 % FCS[b]	39	[15]
Kc	Late embryo	Diploid	Schneider's, 10 % FCS[b]	18	[16]
Kc$_{167}$	From Kc at passage 167	Tetraploid	Serum-free (SFX)	24	[17]
D17-c3	Larval imaginal disc (haltere)	(Unknown; likely diploid)	Schneider's, 10 % FCS[b], 10 μg/ml insulin	70	[18]
BG2-c6	Larval CNS	Diploid	M3, BPYE[c], 10 % FCS, 10 μg/ml insulin	29	[13]

[a]Times are approximate and vary with growth conditions (temperature and medium). Most of these reported doubling times were determined for cells grown at 25 °C
[b]FCS is traditionally heat inactivated (56 °C for 30 min with frequent mixing); however, heat inactivation may be unnecessary [Invitrogen. 1995. Expressions. 2(2):11]
[c]BPYE *b*acto*p*eptone (2.5 g/L) and *y*eastolate (1 g/L)

Fully confluent cultures may dislodge as a sheet when medium is pipetted onto them. (2) Scrape the cells from the surface. (3) Cells can be released using commercial cell disassociation solutions which can contain proteases (e.g., trypsin/EDTA or ICT's Accutase) or can be enzyme-free. When using trypsin, cells must first be washed to remove FCS, then trypsinized, and finally treated with medium/FCS to inactivate the trypsin.

4. Remove a portion of the cell suspension and transfer to a new flask at a dilution ratio in the range of 1:3 to 1:5 (cell suspension/fresh media). S2 cells are most adherent when transferred to new tissue culture-treated plastic; however, once confluent they can begin to grow as large clumps in suspension.

3.2 Design and Production of dsRNA for RNAi

These protocols can serve as a less expensive alternative to other commercially available dsRNA synthesis kits and can be very useful if large amounts of dsRNA are needed.

3.2.1 Primer Design

In order to produce dsRNA using in vitro transcription, a dsDNA template containing T7 promoter sequences (at the 5′ end of each strand) must first be made by PCR amplification. Primers should be designed to amplify a ~500 bp region of the gene encoding the target protein, although dsRNAs of 150–3,000 bp have been shown to work. The chosen template sequence can span one or more exons of the target gene, but a stringent requirement is that the sequence should be free of any ≥19 bp stretches present in the cDNA sequence for any other protein. Be aware that a single protein can have isoforms that may vary by sequence and that the template sequence must be chosen appropriately. "Snapdragon" is a simple primer design program, available through the Drosophila RNAi Screening Center, which will identify primer sequences appropriate to generate long dsRNAs while also minimizing off-target effects (http://www.flyrnai.org/snapdragon_doc1.html). While this method targets exon sequences for knockdown, it may be advantageous to target the 5′ or 3′ UTR of a gene. dsRNA generated from a gene's UTRs is used in gene replacement experiments to knock down endogenous protein (because mRNA contains the UTRs and so is depleted by RNAi) while sparing the exogenous replacement protein (because the transgene lacks the UTRs). Finally, after selecting the gene-specific template sequences for both primers, the T7 promoter sequence (5′-TAATACGACTCACTATAGGG) must to be added to the 5′ end of each primer to allow in vitro transcription by T7 RNA polymerase. Custom synthesis of primers is commercially available from many vendors.

3.2.2 Production of the DNA Template

1. PCR amplification: DNA template for in vitro transcription reactions is generated using standard PCR procedures. We first perform a small-scale PCR reaction with new primers and confirm by agarose gel that a band of expected size has been amplified. Then five identical 100 μl PCR reactions are performed using low-cost Taq polymerase (we sacrifice perfect fidelity for economy), many reaction cycles (~35), and an initial template that is determined by availability: (1) 50 ng (per 100 μl reaction) of cloned EST cDNA obtained from a *Drosophila* cDNA library (e.g., DGRC, https://dgrc.cgb.indiana.edu/index.html), (2) 1 μg (per reaction) of S2 cell genomic DNA (be aware that intron sequences are present in the amplified product if the target sequence spans multiple exons), or least preferably (3) 0.5–2 μl (per reaction) of a previous PCR reaction which amplified the same target sequence.

2. Template Purification

   1. Combine all of the PCR reactions in a single RNase-free 1.7 ml plastic microcentrifuge tube. Add an equal volume phenol/chloroform/isoamyl alcohol and mix well by hand. Centrifuge for 5 min at $20,000 \times g$, room temperature.

   2. Carefully and slowly remove the upper aqueous phase while avoiding the lower organic phase and interface, and transfer to a new 1.7 ml tube. Add an equal volume of chloroform/isoamyl alcohol and mix well. Centrifuge as before.

   3. Carefully remove the upper aqueous phase and transfer to a new tube. Add 1/10 the volume 3 M sodium acetate, mix, and then add 2 volumes cold (–20 °C) 100 % ethanol and mix. Incubate at –80 °C for 30 min. (Use only an explosion-proof freezer.)

4. Centrifuge for 10–20 min 20,000 × g at 4 °C.

5. A white pellet should be visible at this point. Discard the supernatant. Add 1 ml of 70 % ethanol. (The pellet will not dissolve in this wash but may dislodge or fragment.) Centrifuge for 5 min, 20,000 × g, at 4 °C.

6. Remove the supernatant as completely as possible. Then add 50 μl of nuclease-free H_2O to dissolve the pellet.

7. Analyze by agarose gel to confirm the quality of product; determine its concentration by measuring the OD_{260} of a diluted aliquot (μg/ml concentration = OD_{260} × 50 × dilution). Store the stock solution at −20 °C.

3.2.3 *dsRNA Synthesis*

Thaw all stock solutions to room temperature; keep only the enzymes on ice. Assemble the in vitro transcription reaction at room temperature to avoid precipitating spermidine in the reaction buffer.

	Volume (μl)	Final quantity
Nuclease-free H_2O	54	(To bring total reaction volume to 100 μl)
5× Reaction buffer	20	(1×)
rNTP solution	12	7.5 mM each nucleotide
DNA template (1 mg/ml)	10	10 μg
Pyrophosphatase	3	0.3 U
T7 RNA polymerase	1	80 U

1. Carefully mix the components (avoid shearing the DNA) and incubate at 37 °C for 24–72 h (yield often improves with longer incubation).

2. Hydrolyze the DNA template by adding 1 μl of RNase-free DNase to the solution. Mix thoroughly but not excessively. Incubate at 37 °C for 1–2 h.

3. Quantitate the dsRNA by gel densitometry. Standard TAE-based agarose gels are adequate, presumably because dsRNA is less prone to form intramolecular secondary structures than ssRNA and so denaturing gels are unnecessary. To calibrate the mass of dsRNA from band intensity, one of the following can be run on the same gel: (1) a commercial RNA ladder containing RNA oligos of known concentration; (2) a previously quantitated dsRNA sample (preferably of a size similar to the newly synthesized dsRNA); and (3) a commercial DNA ladder with DNA fragments of known concentration; however, the fluorescence increase of ethidium bromide when

bound to dsDNA is about $1.2\times$ greater than when bound to dsRNA, so this method will slightly underestimate the dsRNA concentration.

4. If desired, dilute the dsRNA with RNase-free water. Aliquot into autoclaved, RNase-free microcentrifuge tubes and store at -20 °C. The aliquots will be sterile and ready to use if reasonable care has been taken to keep the in vitro transcription reaction free from contamination.

Expected result: When analyzed on an agarose gel, an aliquot of the synthesized dsRNA should yield a single, robust band with the same apparent size as the DNA template. Any smearing can often be reduced by diluting the gel sample. Lack of product can result if the polymerase is inhibited, for example, by NaCl (>30 mM) or pyrophosphate (a by-product of RNA synthesis, but should be hydrolyzed if sufficient pyrophosphatase is present).

3.3 RNAi of Cultured Drosophila Cells

1. Harvest log-phase cells and transfer them to a well of a tissue culture grade 6-well plate (or 35 mm dish), using sufficient cells to reach 50 % confluency. (Low confluency may cause cells to stop proliferating or die; high cell confluency will decrease the efficiency of RNAi.) Allow the cells to adhere to the bottom of the well (~10 min for healthy S2 cells) (*see* **Note 5**).

2. Once the proper confluency has been achieved and the cells have attached to the bottom of the well, then remove the old medium and add 1 ml of fresh medium. (Usually 1 ml of medium is sufficient for a single well of a 6-well plate.)

3. Prepare dsRNA by adding 10 µg of each dsRNA to a sterile 0.5 ml microcentrifuge tube.

4. Remove a small amount of medium (~100 µl) from the well to be treated and add to the 0.5 ml microcentrifuge tube containing the dsRNA. Mix gently and return the medium/dsRNA mixture back to the well. Gently and briefly swirl the plate by hand to completely distribute the dsRNA. Length of treatment will depend on a variety of factors (*see* **Note 6**).

5. In the literature, different frequencies of dsRNA application have been used, but in our experience, treating cells every other day with new dsRNA is sufficient to deplete proteins. Therefore, repeat **steps 2–4** every other day until the treatment time course is completed. If the cells are not stored in a humidified chamber, then the sides of the plate or dish can be wrapped with Parafilm to prevent evaporation of the medium.

6. Treated cells can be removed from the well or dish and cell lysates prepared or transferred to cover glass for live-cell imaging or immunofluorescence (see below).

Cells cultured in serum-containing medium: J.E. Dixon's lab (UCSD) has reported that RNAi efficiency in S2 cells is increased 10–100-fold if dsRNA is introduced to cells in serum-free rather than serum-containing medium. Therefore, to RNAi-treat *Drosophila* cells maintained in serum-containing medium, first remove the serum-containing medium, add a mixture of serum-free medium/fresh dsRNA to the cells and incubate 30–60 min., and then add more medium and sufficient FCS to restore the required serum concentration.

Expected outcome: In 2000, RNAi treatment of cultured *Drosophila* cells was discovered to require only the addition of dsRNA to the culture medium [19]. Since then, numerous labs have utilized this technique to knock down most of the proteins encoded by the fly genome. Endogenous proteins are usually depleted in less than a week of RNAi treatment, but factors like target protein turnover rate, toxicity, and characteristics of the cell line can impact the efficiency of RNAi. In short, the details of the protocol should be optimized for each target.

3.4 Nucleofection of Cultured Drosophila Cells

This method has proven to be the most cost-effective and efficient way of transfecting S2 cells. Our protocol is optimized for the Nucleofector (Lonza) electroporation device; however, use of other devices may be possible (*see* **Note 7**).

1. Resuspend cells by pipetting media onto cells to dislodge them from the tissue culture flask (*see* **Note 4**).

2. Determine cell density by counting an aliquot of cell resuspension.

3. Centrifuge $\sim$2–5 × 10^6 cells at 1,500 × g for 2 min, room temperature. Substantial cell death may occur if too few cells are transfected (*see* **Note 8**).

4. Carefully aspirate away the medium. Gently resuspend the cell pellet in a solution containing 2 μg of plasmid DNA in sufficient transfection solution to make 100 μl total volume.

5. Transfer this DNA/cell mixture to a clean, sterile electroporation cuvette and close with cap.

6. Place cuvette in the Nucleofector device, select program G-030 (G-30 for Nucleofector I; G-030 for Nucleofector II and 2b), and press start to electroporate cells.

7. Immediately add 1 ml of fresh medium to the cell mixture in the cuvette. Transfer the cuvette contents to a well of a 6-well tissue culture plate using a sterile transfer pipette.

8. Though cells may begin expressing the transgene (e.g., a constitutively expressed GFP-tagged protein) within hours of transfection, we generally allow the transfected cells 24 h to recover before using them.

Expected result. Nucleofection will successfully transfect 70–90 % of treated cells in our experience. In addition, the health of nucleofected cells is usually good if a sufficient number of healthy cells is used and if the expressed protein is not toxic.

3.5 Preparation of Glass-Bottom Dishes

This section briefly describes the preparation of glass-bottom cell culture dishes that can be used to visualize live or immunostained cells on an inverted microscope.

3.5.1 Drilling Culture Dishes

1. Place the bottom half of the 35 mm dish on the taped surface of the styrofoam; the dish should be oriented bottom-side-down. Grip the sides of the dish with the pliers.

2. Drill a hole in the dish bottom using the ¾ in. spade bit and the electric drill.

3. After drilling a dish, the edge of the hole may be rough. The bottom surface of the dish needs to be smooth so that the coverslip that will be attached to this surface will lie flat. Remove any rough edges on the dish bottom using the coarse grinding bit and the electric drill. It is convenient to immobilize the drill and to use the lock-on button to run the grinder continuously, and then manipulate the dish by hand over the grinder to remove any roughness on the bottom surface near the drilled hole.

4. Remove most of the loose plastic debris from the dishes by briefly rinsing the dish bottoms and then allowing them to dry.

3.5.2 Attaching the Coverslip

1. Prepare the adhesive by mixing the Sylgard 184 base with the curing agent at a 10:1 weight ratio (base/catalyst). Mix thoroughly using a disposable transfer pipette, ignoring any small bubbles that appear. 8 g of the adhesive mix should be enough for 160 dishes.

2. To apply the adhesive, try cutting the tip of a 3 ml transfer pipette at an angle to create an enlarged, oblong opening. Apply the adhesive from this pipette by depositing a small ring of adhesive around the hole on the outside bottom surface of the dish.

3. Position a coverslip over the hole. The surfaces of the coverslip and dish bottom should make full contact. The adhesive will slowly spread between the surfaces. Store the dish bottom-side-up for 48 h while the adhesive cures; the curing rate can be increased by storing the dish at 37 °C.

4. If necessary, the glass-bottom dishes can be sterilized by placing them in a tissue culture hood under the sterilizing UV light for several hours. Separate the dish bottoms and lids and lay them out in the hood so that their interior surfaces face the UV light source.

3.6 Immunofluo-rescence and Live-Cell Imaging of Cultured Cells

In order to get the best results from immunofluorescence micros-copy, it is necessary to coat the surface of the glass slide with the lectin concanavalin-A (ConA) which has been found to induce extensive cell spreading and dramatically improves fixed or live-cell imaging [1].

3.6.1 Concanavalin-A Treatment

1. Add sufficient ConA solution to completely cover the upper side (i.e., lid-facing side) of the attached cover glass on the glass-bottom dish.

2. After spreading, immediately remove any excess ConA and let dry.

3. ConA-coated dishes can be sterilized by separating the dishes and their lids, placing them in a tissue culture hood so that the surfaces to be sterilized face the UV lamp, and leaving them for at least 1 h with the UV lamp on. They can then be stored at room temperature for months.

3.6.2 Fixation and Staining

Some proteins require a fast fixation and/or a harsh extraction for successful immunostaining. These include some kinetochore and centriole proteins as well as +TIP proteins, such as EB1. In these cases, methanol fixation is optimal. However, a formaldehyde fixa-tive should be used for preserving microtubules (interphase or mitosis). It is important to test different fixation conditions when immunostaining an uncharacterized protein.

1. Apply a small amount of fresh media (~150 μl) to the coverslip of the glass-bottom dish.

2. Resuspend cells by pipetting a stream of media onto the cells to dislodge them from the tissue culture-treated flask. Add the cell suspension to the coverslip in a dropwise manner and gently swirl to get an even distribution. For best results, cells should be ~50 % confluent. Cells will begin tightly adhering to the coverslip within 5–10 min. Cells will continue to spread and flatten out on the coverslip for about 1 h.

   At this point, cells can be used for live-cell imaging or pro-cessed for immunostaining as described below.

3. *Formaldehyde fixation*: (1) Rapidly remove all media from dish. (2) Briefly wash with appropriate buffer (e.g., PME to preserve microtubules) by gently adding room-temperature buffer to dish and immediately removing. (3) Gently add 2 ml of fixa-tion solution, and incubate for 12 min., room temperature.

   Methanol fixation: For this fixation, it is necessary to submerge the specimen in cold methanol. (1) Prechill 300 ml of anhy-drous methanol in a covered 1 L glass beaker in an explosion-proof −20 °C freezer. (2) Rapidly remove the medium from the dish. (3) Use long forceps to grip the dish with adhering cells and rapidly plunge it into the beaker with prechilled methanol. (4) Return the beaker to the freezer and leave for 15 min.

4. Dump the fixation solution into an appropriate waste container and permeabilize the cells by adding 2 ml of PBS/Triton X-100. Immediately dump the buffer and repeat two more times for a total of three quick washes.

5. Remove the PBS/Triton X-100 and add 2 ml of blocking buffer. Incubate cells at room temperature for 15 min.

6. Prepare primary antibody solution by diluting to the appropriate concentration in blocking buffer and add ~150 µl to the coverslip. Incubate cells at room temperature for 1 h or overnight at 4 °C in an air-tight container with a wetted paper tissue to prevent drying of the samples.

7. Remove antibody solution and wash out excess antibody by adding 2 ml of PBS/Triton X-100 for 5 min a total of three times.

8. Dilute secondary antibody in blocking buffer and add to the coverslip. Incubate for 30 min at room temperature.

9. Wash cells with 2 ml of PBS Triton/X-100 three times for 5 min. If you require DNA labeling with a Hoechst (or DAPI) dye, then dilute the dye in PBS/Triton X-100 and apply to the cells for 5 min. Remove the dye and wash two more times with PBS/Triton X-100, 5 min.

10. Dump the last buffer and add ~200 µl of mounting medium to the coverslip. This mounting medium will not harden and, if the cells need to be re-stained or labeled with a different antibody, can be removed with gentle washing in PBS/Triton X-100. Store dishes at room temperature, protected from the light.

3.6.3 Live-Cell Imaging

Live *Drosophila* cells can be imaged if plated on glass-bottom dishes, using an inverted microscope at room temperature (23–28 °C), in ambient atmosphere. No special considerations are required for imaging live fly cells, other than those that apply to all cells. The usual cautions are required to minimize photobleaching of fluorescently tagged proteins and to prevent phototoxicity.

3.7 Purification of Centrosomes

This procedure to enrich centrosomes from cultured *Drosophila* S2 or Kc cells was derived primarily from a procedure to purify centrosomes from cultured mammalian cells [11, 20] (*see* **Note 9**).

3.7.1 Cell Production and Harvesting

1. Grow up ten T150 flasks of healthy cultured cells to a high confluency.

2. Dislodge cells from flasks and gently pellet the cells. To pellet, we (1) fill four disposable 50 ml conical tubes with cell suspension and (2) centrifuge in swing-bucket rotor at $500 \times g$, 4 °C, 3 min, without braking during deceleration. (3) Discard supernatants by aspiration. (4) Then add more cell suspension to the tubes and centrifuge again. Repeat the previous two steps until all of the suspension has been centrifuged.

3. Gently resuspend all cell pellets in ~12 ml (total) cold PBS and transfer to a 15 ml conical tube. Centrifuge in swing-bucket rotor at $500 \times g$, 4 °C, 2 min, without braking.

4. Discard the supernatant and rapidly wash the pellet by gently resuspending the cell pellet in each of the wash solutions listed below and then centrifuging. For each wash, use ~12 ml of ice-cold wash solution, then pellet the cells by gently centrifuging in a swing-bucket rotor at $1,000 \times g$, 2 min., and decelerate with a weak brake; use no braking if the resulting supernatant is cloudy with cells. Aspirate off and discard each supernatant. Washes: (1) 0.1× PBS/8 % sucrose; (2) 8 % sucrose; (3) Tris buffer (remember to add fresh β-ME) (*see* **Note 10**).

3.7.2 Centrosome Enrichment

1. Lyse the cells by resuspending the washed, pelleted cells in a total of 20 ml of ice-cold Tris buffer/Igepal CA-630 in a disposable 50 ml conical tube and rocking gently at 4 °C, 10 min.

2. Add sufficient 50× PE to the lysate to make the PE a 1× final concentration. (The required volume of 50× PE will probably be ~0.4 ml.) Mix gently.

3. Transfer to a centrifuge tube (e.g., polycarbonate, 28.7 mm diameter, Sorvall) and centrifuge at $1,500 \times g$, 4 °C, 3 min. Save the supernatant and discard the small brown pellet that should be apparent.

4. Add 3 ml of 20 % Ficoll (+fresh β-ME) to a centrifuge tube. Then carefully load the supernatant on top of the Ficoll cushion. Centrifuge at $26,000 \times g$, 15 min, 4 °C. (We use Ultra-Clear centrifuge tubes [Beckman] in a swing-bucket SW28 rotor, 14,000 rpm. Fixed-angle rotors can be used, e.g., Beckman JA-20 rotor, Sorvall centrifuge tube, 14,750 rpm).

5. Carefully aspirate off (and discard) the supernatant until almost reaching the interface with the Ficoll cushion (leave about 2–3 ml above the cushion). Collect and pool the remaining material just above and at the interface until a total volume of about 3 ml has been collected. Mix the collected material with 3 ml Tris buffer/Igepal CA-630. (This decreases the Ficoll concentration to 10 % or less.) Total volume should not exceed 6 ml.

6. Layer the solution onto the 20–70 % sucrose gradient. Centrifuge for 1.5 h, $131,100 \times g$, 2 °C in a swing-bucket rotor (e.g., SW28 rotor, 27,000 rpm) (*see* **Note 11**).

7. Carefully collect 0.5 ml fractions starting from the top of gradient.

8. Fractions can be stored for the short term at 4 °C. (Most fractions will freeze if stored at −20 °C.)

9. Western blot every second or third fraction and the "input" material (i.e., the clarified lysate loaded onto the gradient). Probe with appropriate primary (such as the centriole antibody anti-D-PLP) and secondary antibodies.

Expected result: About 76 fractions (0.5 ml) will be collected from the sucrose gradient. When analyzed by SDS-PAGE, Coomassie-stained bands are usually visible in about the first 25 fractions only. When analyzed by Western blotting for a centrosomal/centriolar marker, the major peak of centrosome-containing fractions is centered at roughly fraction 50 (~60 % sucrose). Western blots of centrosome preps following this protocol can be found in [11, 21]. However, some fractions near the top and bottom of the gradient also contain the centrosome marker, indicating that either the centrosomes are not homogeneous or the centrosomes have not reached an equilibrium position within the gradient.

4   Notes

1. We have not tested solutions with a different pH. Since medium for insect cells usually has a low pH (~6.6), then using a nucleofection solution with a pH lower than 7.2 may decrease trauma of treated *Drosophila* cells.

2. The original acid/base molar ratio used by Mitchison [20] is changed to generate a pH 7.0.

3. Cell density can have an important impact on cell viability. For example, S2 cell cultures will decline if passaged at too low a density, presumably because trophic factors released by the cultured cells are too dilute when cells are plated sparsely. By limiting the dilution at each passage, the subcultured cells receive more conditioned medium and have increased viability. As an approximate guideline, S2 cells should be maintained at $1–25 \times 10^6$ cells/ml. S2 cells (especially if recently transfected) are healthier if maintained near the high end of the density range.

4. If resuspension is too vigorous or extensive, then cells can be damaged, especially if their health has been compromised by an experimental manipulation like a recent transfection. If cell debris is apparent in the culture flasks after passaging, then resuspension was too vigorous.

5. During the multiday treatment period, it may be necessary to subculture the cells to a new well or dish to maintain the optimal confluency, particularly for extended treatments and fast-growing cells. *A common mistake is to allow the cells to become over-confluent.*

6. The length of dsRNA treatment necessary to deplete the target protein depends on the protein's turnover rate. For relatively rapid-dividing cells (e.g., S2 cells), depletion can usually be achieved by treating with dsRNA for 3–7 days, while depletion in slower dividing cells (e.g., D17-c3 cells [12]) may require longer treatment times. Treatment may need to be stopped prematurely if extensive depletion of the target protein is toxic to cells, in which case the partial knockdown would be experimentally analogous to generating a hypomorphic allele. Knockdown of multiple proteins is also possible although the efficiency of depletion often decreases (presumably because the endogenous Dicer/RISC machinery becomes saturated). Therefore, the RNAi treatment regime must be tailored for each target, and protein depletion should be confirmed by Western blot, if possible. Finally, successful RNAi depends on cell health; only use cells that have been grown under optimal conditions for several passages.

7. The characteristics of the voltage pulse generated by the Nucleofector device are proprietary. We do not know if other commercial electroporators could supply the required pulse. However, a pulse method for a non-Lonza electroporator has been developed for mammalian cells, which transfects as efficiently as the Lonza nucleofector system [22], and a similar strategy would probably work for cultured insect cells.

8. Successful transfection requires healthy cells. Using cells from declining cultures or recently thawed cells will give disappointing results.

9. ~97–99 % of the cells in an asynchronous population will be in interphase of the cell cycle when centrioles do not associate with much pericentriolar material (PCM) and do not nucleate microtubule growth [7]. Thus, without modification, this protocol will purify centrioles. To enrich for centrosomes (i.e., centrioles with PCM), cells must be treated with 30 µM colchicine for 8–12 h to accumulate cells in mitosis, although this will only increase the mitotic index to 15–25 %. In addition, this protocol differs from Mitchison's [20] in that (a) the addition of reagents to depolymerize the actin cytoskeleton is not necessary and (b) a different sucrose gradient is used (20–70 % instead of 20–62.5 %).

10. Washes need to be done quickly and gently, but decelerating too rapidly causes vortexing in the tube and some cells to remain in suspension.

11. If a balance tube is needed, use another centrifuge tube containing 5 M NaCl or another solution of relatively high density.

References

1. Rogers SL, Rogers GC, Sharp DJ, Vale RD (2002) *Drosophila* EB1 is important for proper assembly, dynamics, and positioning of the mitotic spindle. J Cell Biol 158(5):873–884

2. Ulvila J, Parikka M, Kleino A, Sormunen R, Esekowitz RA, Kocks C, Rämet M (2006) Double-stranded RNA is internalized by scavenger receptor-mediated endocytosis in *Drosophila* S2 cells. J Biol Chem 281:14370–14375

3. Dobbelaere J, Josué F, Suijkerbuijk S, Baum B, Tapon N, Raff J (2008) A genome-wide RNAi screen to dissect centriole duplication and centrosome maturation in *Drosophila*. PLoS Biol 6(9):e224

4. Goshima G, Wollman R, Goodwin SS, Zhang N, Scholey JM, Vale RD, Stuurman N (2007) Genes required for mitotic spindle assembly in *Drosophila* S2 cells. Science 316:417–421

5. Kwon M, Godinho SA, Chandhok NS, Ganem NJ, Azioune A, Thery M, Pellman D (2008) Mechanisms to suppress multipolar divisions in cancer cells with extra centrosomes. Genes Dev 22:2189–2203

6. Rusan NM, Rogers GC (2009) Centrosome function: sometimes less is more. Traffic 10:472–481

7. Rogers GC, Rusan NM, Peifer M, Rogers SL (2008) A multicomponent assembly pathway contributes to the formation of acentrosomal microtubule arrays in interphase *Drosophila* cells. Mol Biol Cell 19:3163–3178

8. Wadsworth P, Khodjakov A (2004) *E pluribus unum*: towards a universal mechanism for spindle assembly. Trends Cell Biol 14: 413–419

9. Rusan RM, Peifer M (2007) A role for a novel centrosome cycle in asymmetric cell division. J Cell Biol 177:13–20

10. Goshima G, Vale RD (2003) The roles of microtubule-based motor proteins in mitosis: comprehensive RNAi analysis in the *Drosophila* S2 cell line. J Cell Biol 162:1003–1016

11. Rogers GC, Rusan NM, Roberts DM, Peifer M, Rogers SL (2009) The SCF Slimb ubiquitin ligase regulates Plk4/Sak levels to block centriole reduplication. J Cell Biol 184(2): 225–239

12. Currie JD, Rogers SL (2011) Using the *Drosophila melanogaster* D17-c3 cell culture system to study cell motility. Nat Protoc 6: 1632–1641

13. Ui K, Nishihara S, Sakuma M, Togashi S, Ueda R, Miyata Y, Miyake T (1994) Newly established cell lines from *Drosophila* larval CNS express neural specific characteristics. In Vitro Cell Dev Biol Anim 30A:209–216

14. Schneider I (1972) Cell lines derived from late embryonic stages of *Drosophila melanogaster*. J Embryol Exp Morphol 27:353–365

15. Yanagawa S, Lee JS, Ishimoto A (1998) Identification and characterization of a novel line of *Drosophila* Schneider S2 cells that respond to wingless signaling. J Biol Chem 273:32353–32359

16. Echalier G, Ohanessian A (1970) In vitro culture of *Drosophila melanogaster* embryonic cells. In Vitro 6:162–172

17. Cherbas P, Cherbas L, Lee SS, Nakanishi K (1988) 26-^{125}I-iodoponasterone A is a potent ecdysone and a sensitive radioligand for ecdysone receptors. Proc Natl Acad Sci U S A 85: 2096–2100

18. Ui K, Ueda R, Miyake T (1987) Cell lines from imaginal discs of *Drosophila melanogaster*. In Vitro Cell Dev Biol 23:707–711

19. Clemens JC, Worby CA, Simonson-Leff N, Muda M, Maehama T, Hemmings BA, Dixon JE (2000) Use of double-stranded RNA interference in *Drosophila* cell lines to dissect signal transduction pathways. Proc Natl Acad Sci U S A 97:6499–6503

20. Mitchison TJ, Kirschner MW (1986) Isolation of mammalian centrosomes. Methods Enzymol 134:261–268

21. Brownlee CW, Klebba JE, Buster DW, Rogers GC (2011) The protein phosphatase 2A regulatory subunit twins stabilizes Plk4 to induce centriole amplification. J Cell Biol 195: 231–243

22. Stroh T, Erben U, Kühl AA, Zeitz M, Siegmund B (2010) Combined pulse electroporation—a novel strategy for highly efficient transfection of human and mouse cells. PLoS ONE 5:e9488

Chapter 7

Measuring Microtubule Growth and Gliding in *Caenorhabditis elegans* Embryos

Justus Tegha-Dunghu, Eva M. Gusnowski, and Martin Srayko

Abstract

Microtubule plus-tip tracking is a powerful method to measure microtubule growth dynamics in vivo. Here we outline an approach that exploits live confocal microscopy of a GFP-tagged EB1-like protein to measure microtubule growth behavior and minus-end-directed microtubule motor activity at the cortex of *Caenorhabditis elegans* embryos. The EB1 velocity assay (EVA) provides a method to reproducibly monitor motor- and non-motor-assisted microtubule movements.

Key words Microtubule gliding, Plus-tip tracking, End-binding protein (EB1), Dynein, EB1 velocity assay, *Caenorhabditis elegans*

1 Introduction

Caenorhabditis elegans development occurs in a reproducible manner [1]. This feature has been a critical advantage for embryologists studying cell fate patterning and is equally attractive for studying the complicated morphological changes that occur to the cytoskeleton during cell division. Advancements in live-cell imaging and in vivo probes combined with RNAi methods make *C. elegans* particularly compelling for this type of analysis as even subtle deviations from wild type can be identified within a remarkably small sample size.

Microtubule motor proteins perform an array of functions, from vesicle trafficking to the nucleation, organization, and assembly of microtubule-based structures. These specialized machines help to position intracellular objects with nanometer precision. Motor movement occurs along the microtubule protofilament wall (the lattice) and is directed either towards the growing plus end of the microtubule or the relatively static minus end of the microtubule, depending on the class of motor protein. Although there are some important exceptions, kinesins tend to move towards the plus end and cytoplasmic dynein towards the minus end (reviewed in [2]).

David J. Sharp (ed.), *Mitosis: Methods and Protocols*, Methods in Molecular Biology, vol. 1136,
DOI 10.1007/978-1-4939-0329-0_7, © Springer Science+Business Media New York 2014

In vitro, motor protein activity can be measured by tracking individual motors as they move along the microtubule [3]. Another approach involves perfusing stabilized fluorescently labelled microtubules over a bed of motor proteins that have been attached to the surface of a coverslip [4]. In this case, tracking the microtubules as they glide along the surface provides a readout for the activity of the fixed motors. Both methods have provided a wealth of information about microtubule motor activity, but having access to similar assays that can be performed in a cellular context would be beneficial. Measuring motor protein activity within a living cell using microtubule movement as a readout, however, is more challenging because the microtubules in vivo are not typically static; rather they grow and shrink with dynamic instability [5–7]. Furthermore, the rate at which polymers grow and the amount of time they spend growing versus shrinking can vary widely between cell cycle stages, terminally differentiated cell types, and different species. Microtubules in vivo can also occupy multiple orientations within the 3D space of a cell, adding to the challenges associated with tracking their movement.

In *C. elegans*, dynein motors can propel free microtubules along the inner surface of the one-cell embryo [8], thus making it possible to track their movements along a relatively flat surface. We describe in detail this method that uses a combination of ectopic katanin microtubule severing to release microtubules from the centrosome and an EB1 tip-tracking probe to track microtubules as they grow and glide along the inner cell cortex. For more information on principles of imaging, we refer readers to [9] and for alternative imaging methods, to [10–13].

2 Materials

2.1 Dissection Components

1. Two 25–27 gauge, 1″–1½″ hypodermic needles (Becton Dickinson & Co.; *see* **Note 1**).

2. Plastic pasteur pipette (BrandTech Scientific, Inc.; *see* **Note 2**).

3. Standard microscope slides.

4. 22 × 22 mm coverslips (Corning No. 1½).

5. Egg buffer: 188 mM NaCl, 48 mM KCl, 2 mM $CaCl_2$, 2 mM $MgCl_2$, 25 mM Hepes; pH 7.3 [14] (*see* **Note 3**).

6. 2 % Agarose in egg buffer: Weigh out 2 g of electrophoresis grade agarose. Make volume to 100 mL with egg buffer. Heat to completely dissolve. Aliquot molten agarose into microfuge tubes while still hot (1 mL each) and let cool. Place in a microfuge rack and store at room temperature (will last at least 2 months).

7. Two-spacer slides: Stick lab tape lengthwise along two slides as shown in Fig. 1b. These act as spacers for the agarose pad and can be reused.

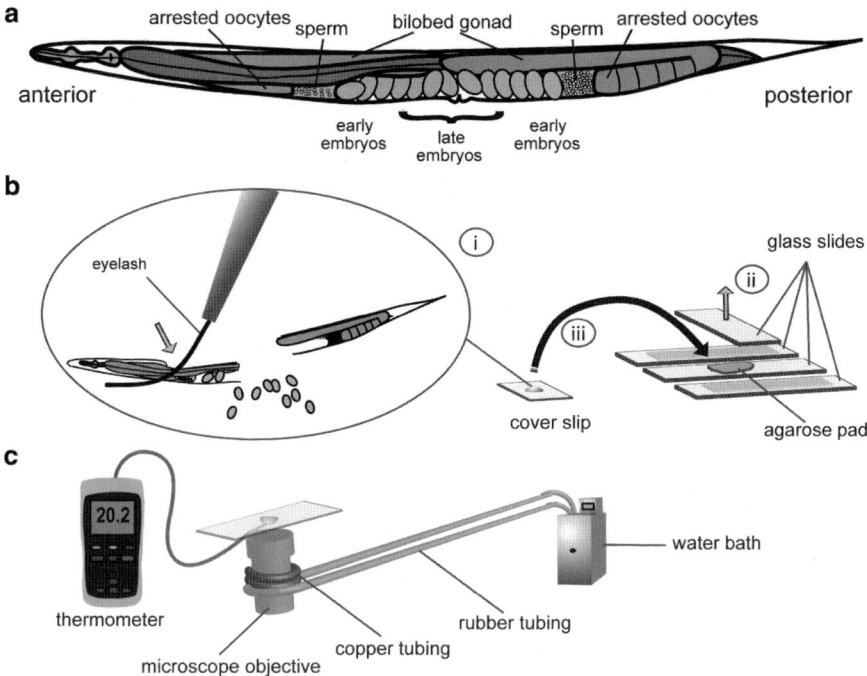

Fig. 1 (**a**) *C. elegans* adult anatomy, showing eggs within the uterus. (**b**) Embryo dissection is performed on a coverslip. To retrieve eggs from the uterus, the eyelash tool is used as shown (*i*). The glass slide is removed from the agarose pad (*ii*) and the coverslip with dissected eggs is gently inverted onto the agarose pad (*iii*). (**c**) The temperature of the objective is maintained via the copper tubing connected to the water bath. Immediately after imaging, the temperature of the immersion oil on the coverslip surface is measured with the digital thermometer probe

8. Agarose pads: Melt 2 % agarose (prepared as above) by placing the tube in a heating block (*see* **Note 4**). Place a slide between the two-spacer slides. Using a plastic pasteur pipette, transfer about 100 µL of molten agarose onto the slide (avoid bubbles; *see* **Note 5**). Quickly place a clean slide on top of the molten agarose and apply gentle pressure. Agarose should solidify between the slides within about a minute. Remove the top slide only when you are ready to use the agarose pad, to prevent dehydration.

9. Eyelash tool (*see* **Note 6**).

10. Worm pick tool (*see* **Note 7**).

11. Nematode growth media (NGM): Mix 1.7 % (w/v) Bactoagar, 50 mM NaCl, 0.25 %(w/v) peptone, 5 µg/mL cholesterol, and appropriate volume of water; autoclave and when this is hand-warm, add 1 mM $CaCl_2$, 1 mM $MgSO_4$, 25 mM KH_2PO_4 pH 6.

12. Lab timer.

2.2 C. elegans Strains Required

1. MAS37: *unc-119(ed3) abcIs3 [pie-1(promoter)-ebp-2::gfp; unc-119(+)]* (*see* **Note 8**).

2. MAS94: *unc-119(ed3) abcIs3 [pie-1(promoter)-ebp-2::gfp; unc-119(+)]; mei-1(ct46) unc-13(e1091)* (*see* **Note 9**).

Strains are available through the Caenorhabditis Genetics Center (http://www.cbs.umn.edu/CGC/).

2.3 Microscope Components

1. Inverted spinning disk confocal microscope: The microscope should be equipped with 488 nm laser and accessories for DIC optics, controlled by software that allows multiple configuration time lapse and live stream imaging.

2. Microscope temperature control capable of maintaining sample temperature of ±0.5 °C (*see* **Note 10**) and a suitable digital thermometer with a fine wire K-type probe.

3. Stereo microscope for observation and dissection of appropriate worms.

3 Methods

3.1 Worm Picking and Embryo Manipulation

1. Transfer about 30–40 L3–L4 worms (*see* **Note 11**) onto the plates using a worm pick and incubate at an appropriate temperature (usually 20 °C). The next day (24–30 h), examine the adult worms under a dissection stereo microscope to confirm that they contain embryos within their uterus (*see* **Note 12**).

2. Using a worm pick, gently transfer one gravid worm (*see* **Note 13**) into 5–8 µL of egg buffer that has been dispensed onto a coverslip, under the dissection microscope (*see* **Note 14**). Using two 25 gauge needles (*see* **Note 15**), slice the worm in half by cutting near the vulva, trying not to damage the early embryos, which reside close to the spermatheca (Fig. 1a).

3. Once all of the embryos have been released from the body cavity into the buffer (*see* **Note 16**), separate the two slides from the pre-prepared agarose pad to expose the agar surface. Trim the exposed agarose pad with a microscope slide or razor blade to remove any uneven edges.

4. Invert the coverslip with the dissected embryos onto the agarose pad. Do not apply pressure to the coverslip; the embryos will disperse and should embed into the agarose surface and become stationary. The volume of dissection buffer may need to be reduced slightly if the embryos drift during image acquisition (*see* **Note 17**).

**3.2 Image
Acquisition**

Imaging in the Midplane Using MAS37

EBP-2 is a *C. elegans* EB1-like tip-tracking protein; it is only visible on the growing tips of microtubules. This allows the direct measurement of plus-end growth of an individual microtubule in vivo, provided the minus end of the microtubule is fixed. Microtubules in the one-cell *C. elegans* embryos do not seem to exhibit microtubule flux or treadmilling (the removal of tubulin subunits from the minus ends with concomitant addition of subunits to the plus ends) [15–18]. Therefore, we assume that the minus ends of the astral microtubules are relatively stable and anchored to the centrosomes. By tracking the EB1-GFP particles that travel in a single optical plane and in a straight line away from the centrosome, one can determine the growth rate of an individual microtubule.

1. After placing the slide on the spinning disk confocal, identify a suitable early embryo (e.g., in early pronuclear migration) at low (10×) magnification (*see* **Note 18**).

2. Switch to high magnification (e.g., 60×, NA 1.42) and begin a DIC time-lapse acquisition (10 s interval). Continue the DIC imaging until nuclear envelope breakdown, recognizable when the distinct boundary between nucleus and cytoplasm becomes obscure (Fig. 2).

3. At the start of NEB, set the timer to 120 s.
   If imaging at the cortex with MAS94, skip to **step 7**.

4. After 120 s, begin stream acquisition of GFP (300 ms/frame, 300 frames, no binning; *see* **Note 19**).

5. At the completion of this stream movie, switch back to DIC time lapse. Continue the time lapse until the end of cytokinesis (*see* **Note 20**).

6. After image acquisition, lower the objective and immediately measure the temperature of the immersion oil. Insert the temperature probe into the oil that remains on the bottom surface of the coverslip, being very careful to avoid touching the probe to the objective lens. For most experiments, we reject movies that were acquired outside of the temperature range of 20 ± 0.5 °C (*see* **Note 21**).

Imaging at the Cortex Using MAS94

Imaging of microtubules at the inner cell cortex is often warranted, for instance, to quantify microtubule contact/residency time [16, 19] or to detect dynein-dependent microtubule gliding [8]. In the latter case, the microtubules must be released from the centrosome, to allow dynein motors to pull them along the cortex. For this, use the MAS94 strain, which contains a temperature-sensitive mutation in katanin that causes ectopic microtubule severing at the centrosomes in mitosis. Dynein-dependent microtubule gliding movement occurs in the microtubule-plus direction and thus adds

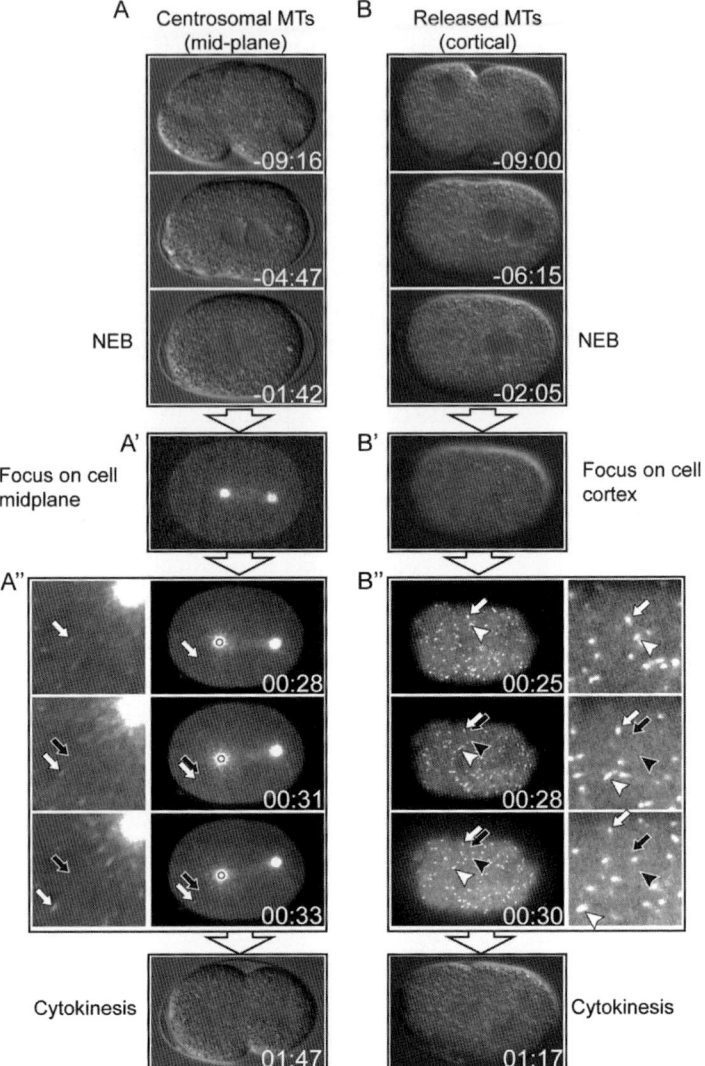

Fig. 2 Imaging of EB1-GFP at the mid-plane and cortex to measure microtubule growth and gliding. (**a**) Single frames from a DIC time-lapse prior to stream acquisition (*upper panels*); and frames from an EB1-GFP stream (*lower panels*) in the *midplane*. The start of stream acquisition is $t = 0$. In order to achieve proper focus without excessive photobleaching prior to stream acquisition, low laser power and 2×2 camera binning are used (**a′**). Movements of single EB1 dots are shown in the enlargements (**a″**); the moving dot (*white arrow*) is shown relative to its starting position (*black arrow*) and its centrosome of origin (black ring). (**b**) Similar setup as in (**a**) but the focus is at the cell cortex (**b′**). Tracking of two EB1 dots is shown (**b″**). White markers indicate the positions of the dots over time relative to their starting positions (*black markers*). The two dots exhibit expected EB1 movements, consistent with microtubule growth (*arrows*) and growth + dynein-dependent gliding (*arrowheads*)

to the speed of the growing microtubule end (baseline growth rate can be estimated from the centrosome as described above). We have found that centrosomal microtubule EB1-GFP velocities at 20 °C rarely (<2 %) exceed 1.0 μm/s. Therefore, speeds above this value likely indicate gliding motion in the plus direction, and this has been attributed to cortical dynein activity [8]. If such EB1-GFP particle velocity comparisons are used, it is important to record centrosome-based and cortical EB1 velocities at the same cell cycle stage and temperature, as both variables influence microtubule growth rates (and thus overall EB1-GFP velocity).

7. While acquiring DIC time-lapse images, change the focus to the inner cell cortex nearest the coverslip (Fig. 2 and *see* **Note 22**).

8. After focusing on the cortex, stop the DIC time lapse (120 s after NEB), and begin the GFP stream acquisition (300 ms/ frame, 300 frames, no binning).

9. At the completion of this stream movie, switch back to DIC time lapse and while images are being acquired, return the focus to the midplane. Continue the time lapse until the end of cytokinesis.

10. Take a temperature reading as in **step 6**.

3.3 Post Image Analysis and EB1-GFP Particle Tracking

There are a number of commercial and free software packages that provide a method to automatically track objects within image stacks. As with any computer-based tracking method, one must be prepared to manually verify the output and confirm that the software accurately tracked the objects of interest. An example of a common tracking error is when a switch is made from one moving particle to another, especially if the trajectories intersect. Although some programs are better at incorporating vector history in the tracking algorithm, any mistracked events should be rejected after manually inspecting the output. Some details on tracking and track quantifications relevant to this protocol are described in [8, 20–24]. Most programs offer graphical output with analysis of trajectories and calculation of average velocities. Alternatively, the raw data can be exported into other platforms such as MATLAB or MS Excel for these purposes. Here we present some details on the use of the freeware ImageJ [25]:

1. To detect and track EB1-GFP dots, install the ImageJ program and the particle detector and tracker plugin [26].

2. Import image sequences into ImageJ (*see* **Note 23**); then launch particle detector and tracker from the plugin menu. Select constraints for optimal particle detection (Fig. 3a; *see* **Note 24**).

3. To see track output and quality select "Visualize All Trajectories" in the ImageJ results menu (Fig. 3b, c). Impose further constraints to eliminate dot trajectories lasting less than ten consecutive frames (*see* **Note 25**).

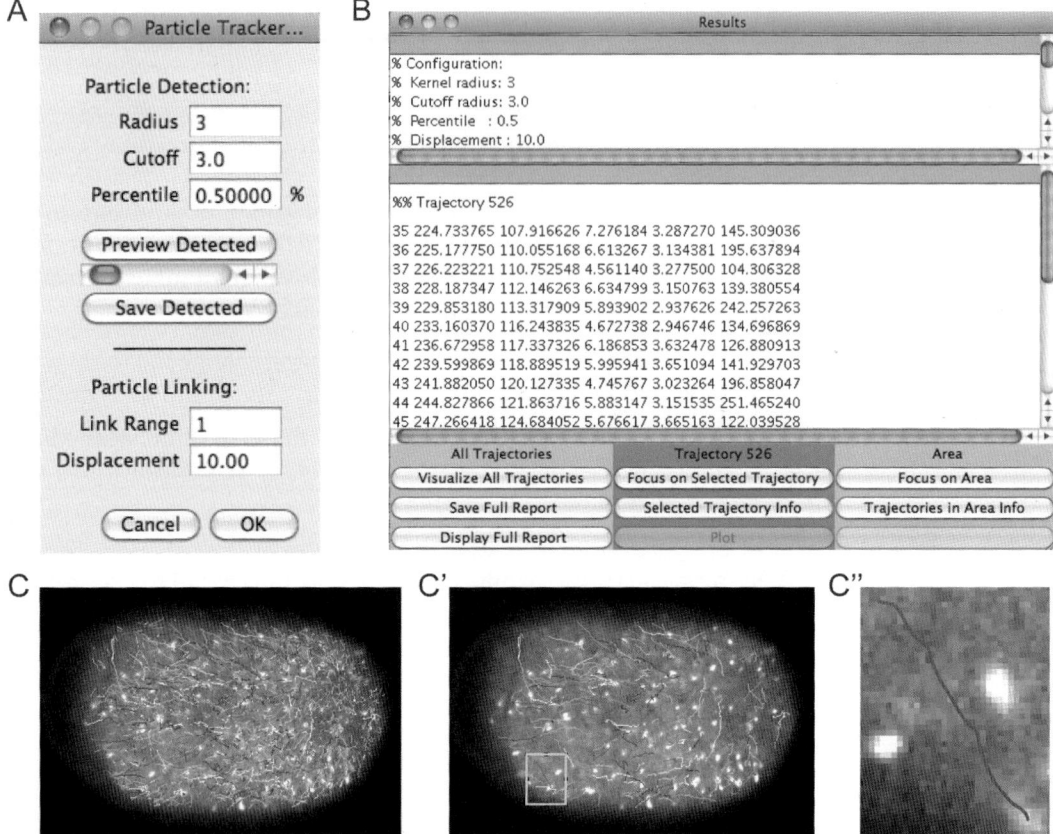

Fig. 3 Tracking of EB1-GFP at the cell cortex using ImageJ. Screenshots from the ImageJ particle tracking program showing: (**a**) Particle tracker input parameters and (**b**) output results are displayed for trajectory number 526. Only lines 35–45 are shown. (**c**) All output trajectories are superimposed on a single frame of the GFP stream movie. (**c′**) A map indicates the trajectories remaining after filtration of tracks shorter than ten consecutive frames. (**c″**) A magnification of the *yellow box* in **c′** shows a specific track (526) used for the calculation of dot velocity (Color figure online)

4. Save the full report in the result window and, if desired, export X and Y coordinates for selected tracks into MS Excel and use the trajectory points to calculate the average velocity of individual dots (dynein-dependent microtubule velocity measurements at the cortex; *see* **Note 26**; for microtubule growth measurements at midplane, *see* **Note 27**). Track 526 is shown as an example (Fig. 3c′, yellow rectangle and magnified in Fig. 3c″).

5. In the example of track 526, one could calculate the velocity of the microtubule tip. The dot was trackable throughout frames 35–47 (a subset is shown in Fig. 3b), thus, the total time was 3.6 s ($300 \text{ ms} \times 12$). The total distance travelled by the dot was 4.39 μm ($37.92 \text{ pixels} \times 0.1157$) where 0.1157 is the conversion factor from pixels to microns (*see* **Note 28**). The velocity of dot 526 was therefore 1.22 μm/s.

4 Notes

1. Test a few different needle gauges and lengths for the dissection when performing this step for the first time. Some researchers prefer scalpels for this procedure; however, we find that scalpels displace too much liquid to allow accurate cutting of the worms. Attaching the needle to a 1 mL syringe body may improve comfort and control during the dissection.

2. Plastic pasteur pipettes can be reused to transfer melted agarose to the glass slide; periodically compress pipette with finger pressure to squish out solidified agarose.

3. Use HPLC grade or ultrapure Milli-Q water and only analytical grade reagents. In experiments involving embryos with a permeabilized eggshell, we have observed alarmingly specific cellular defects (e.g., failure in cytokinesis) attributable to using Milli-Q water with a resistivity well below 18 MΩ.

4. We typically set the temperature of the temp block to 90–93 °C for about 10 min to melt 2 % agarose. Once melted, maintain the agarose at 80 °C. Agarose breaks down after a couple of hours at high temperature, so add a fresh tube to the heat block periodically (approx. every hour of imaging) to provide a continuous supply.

5. Always carefully open the hot microfuge tube as molten agarose may boil out (delayed boiling), especially the first time opening.

6. Make an eyelash tool by pulling out one's own (or a willing participant's) eyelash. Then glue the eyelash to the end of a suitable holder. As an alternative to glue, place the narrow end of a yellow pipette tip near a flame just until it begins to melt (do not let it catch on fire) and, working quickly, insert the root of the eyelash into the molten plastic. Wait for plastic to cool completely before using.

7. A simple (but fragile) worm pick can be made by placing 5 mm of a 15 mm long platinum wire (30–32 gauge, 0.010 in. in diameter) into the end of a glass pasteur pipette. Place the tip near a flame until it starts to melt, and then rotate the pipette to limit bending of the tip. Once the glass at the tip has completely melted with the wire embedded, let cool.

8. This strain is used to measure microtubule growth rates. The *ebp-2* gene (*e*nd-*b*inding *p*rotein) encodes an EB1 plus-tip tracker that locates specifically to growing plus ends of microtubules. The *ebp-2::gfp* fusion DNA was integrated into the *C. elegans* genome via the microparticle bombardment method [27]. Transcription is controlled by the *pie-1* promoter to facilitate germline and early embryonic expression. MAS37 exhibits stable EB1-GFP expression and is useful for crossing to mutant or transgenic lines.

9. This strain is used to measure microtubule growth rates and dynein-dependent microtubule gliding along the cortex. This strain contains the EB1-GFP marker as well as a temperature-sensitive mutation *mei-1*(*ct46*). *mei-1* encodes the worm orthologue of the catalytic subunit of the microtubule-severing enzyme, katanin (reviewed in [28]). In wild-type worms, katanin is required for female meiotic spindle assembly, but the protein is degraded prior to the first mitotic division. The *ct46* mutation results in MEI-1 protein persistence into mitosis, resulting in ectopic severing of microtubules during mitosis. The severing is particularly evident near the centrosomes, as visualized by EM tomography [29], which is consistent with immunofluorescence experiments that showed the ectopic MEI-1 is concentrated at centrosomes and chromosomes [30]. In this assay, the ectopic katanin activity is used to generate free microtubules in the one-cell mitotic *C. elegans* embryo. Dynein-dependent microtubule movements at the cortex can then be monitored using the EB1-GFP probe.

10. To maintain temperature during imaging, we crafted a simple temperature control collar by bending a ¼″ copper tubing twice around a cylinder (cylinder diameter should be slightly smaller than the diameter of the imaging objective). Once a smooth coil has been made, remove the cylinder. Attach one length of rubber tubing to one end of the coil and another tubing to the other end. Connect the tubing to the input and output ports of the circulating water bath (e.g., Thermo Haake DC30-K10) and pump the temperature-controlled liquid through the copper collar. It may be necessary to insulate the tubing between the pump and the objective collar with foam wrap. Double-check all fittings for leaks before attaching the apparatus to the microscope objective. Small-diameter clamps can be used to further fix the tubing to the copper coil. Carefully mount the coil onto the objective (the coil should act as a spring to grip onto the objective nose) and make sure that no moving parts are obstructed. To prevent damage to the tubing or microscope, we avoid using the objective turret motor to switch objectives once the temperature control coil is installed. Some initial tests will be necessary to find the Haake temperature range at which the desired temperature of the objective oil is achieved. In this way, the objective acts as an effective temperature source/sink to heat/cool the immersion oil, which is in contact with the coverslip in the vicinity of the sample. We find that this method allows a more rapid transfer of heat/cooling than other methods that utilize temperature-controlled slide holders, where heat conduction occurs over much larger distances. This is especially important when imaging temperature-sensitive processes within the rapidly developing *C. elegans* embryos.

11. The adult 1 mm long *C. elegans* worms (Fig. 1) are normally grown on 60 mm petri dishes containing Nutrient Growth Media agar seeded with OP50 *E. coli*. The life cycle is 3.5 days at 20 °C, with a short embryonic development (14 h) followed by four larval stages (L1–L4) separated by molts. L4 larvae are recognizable by a distinctive white patch on the ventral side of the body, halfway along the length. L4 develop into adults in ~8 h at 20 °C. Embryos, larvae, and adults are entirely transparent, which allows the observation of all cells with DIC optics. *C. elegans* exist as self-fertilizing hermaphrodites and males, making *C. elegans* easy to maintain and conduct genetic crosses. The reference wild-type strain is called N2 (or Bristol, from the English town where it was isolated).

12. *C. elegans* hermaphrodites continue to fertilize eggs until they deplete their sperm stock; by starting with relatively young adults in the morning, you can easily recover freshly fertilized embryos for imaging throughout most of the day. If longer incubations are necessary, or if the adults are older, one can add wild-type *C. elegans* males to the plate, as they will fertilize hermaphrodites and allow an extended period of embryo production, without interfering with maternal phenotypes caused by the RNAi.

13. If the number of worms is not limiting for your experiment, it is often advantageous to dissect two or three worms at a time to maximize the chance of isolating an early embryo for imaging. Exceeding this number usually only increases the time spent browsing the field of view for the optimum embryo and offers limited benefit in terms of imaging efficiency.

14. The optimal volume of egg buffer will depend on the rate of evaporation and efficiency of dissection.

15. After each dissection, we place the needles on the dissection microscope stage or a safe place beside the scope. We do not re-sheath the needles between dissections—most accidental piercings occur this way. Usually one pair of needles will last for the entire experiment (or longer). Old needles should be disposed of according to local biohazard and/or sharp waste regulations.

16. Active swimming by the worms in liquid makes precise cutting a challenge. Within the adult *C. elegans* worm, the one-cell embryos are usually farthest away from the vulva and are often still in the body cavity after a single cut through the body of the worm. If the embryos are not released into the medium, use the eyelash tool to gently press on the body cavity until all embryos are free.

17. For imaging experiments that can be completed within 20 min, there is usually no need to seal the coverslip. In cases where it is necessary to maintain a constant focal plane over a longer period of time, it is advisable to put a thin ring of vacuum grease around the agarose pad to prevent desiccation (and a corresponding change in focus).

18. Fortunately, the one-cell *C. elegans* embryo has many visible hallmarks of this early development [13]. Although simple brightfield may suffice to identify the transparent embryos, DIC works best for accurate staging. There are many useful online resources to aid setting up DIC optics (e.g., http://www.olympusmicro.com).

19. The efficiency of light transmission through different optical systems varies, and spinning disk heads are notorious for loss of light. These settings are based on a 50 mW diode-pumped solid state laser, Yokogawa spinning disk CSU10, and Olympus IX81 microscope with 60× 1.42NA objective; however, user settings for optimal signal to noise with this GFP probe should be empirically determined. In order to achieve good signal to noise for the individual dots, the settings might produce saturated pixels at the centrosome, due to the high number of growing microtubules there. This is not an issue unless quantification of the centrosomal fluorescence intensity is desired.

20. At 20 °C, this approach should yield an EB1-GFP stream movie that occurs at the end of metaphase/beginning of anaphase. Use the DIC images of NEB and cytokinesis as reference points to judge whether the GFP stream was acquired at the desired cell cycle stage, and adjust timing if necessary.

21. When the microscope is in a well-ventilated room at a constant temperature, it may not be necessary to further control the temperature of the imaging environment. However, lasers and computers generate considerable heat, and we note that our microscope, in a well-ventilated temperature-controlled room, shows a 4 °C rise in temperature over the course of a day's imaging, making it necessary to monitor temperature and adjust the water bath accordingly throughout the experiment. Microtubule growth is extremely sensitive to changes in temperature [22]; therefore, if temperature control is not possible, we recommend to at least monitor the sample temperature after image acquisition and keep only those datasets that fall within 1 °C of each other for microtubule dynamics measurements. If possible, choose a digital thermometer and a wire probe that has a flat tip for easier access to the coverslip oil (e.g., Fluke 51-II digital thermometer with K-type beaded wire thermocouple probe).

22. The correct focal plane via DIC is when a few cytoplasmic yolk granules remain in focus but the 2D area of the embryo is reduced by about 5 μm around the entire outer edge compared to the mid-embryo focal plane. A live DIC display is also useful to find the correct focal plane; however, one must remember to acquire (and save) at least one DIC image to verify the position of the focus prior to the GFP stream.

23. Convert the image stack into 8-bit grayscale and save as a TIFF. This could be done using the microscope imaging software or ImageJ.

24. For EB1-GFP imaging with no camera binning, set the radius of the dot to be detected to 3 pixels and a search radius of 10 pixels. Adjust the threshold values for individual streams to maximize the number of EB1-GFP dots detected in the first frame. Use the "Preview Detected" tab to make sure that a maximum number of dots are detected while minimizing errors in grouping dots or detecting cytoplasmic background fluorescence.

25. To minimize errors in estimating dot velocities, filter the resulting tracks to keep only those that span at least ten consecutive frames (Fig. 3c′). It is important to manually inspect each track to ensure a single EB1-GFP dot was tracked. Track output is depicted in a rainbow of colors to show different trajectories.

26. For EB1 tracking at the cortex in embryos that have ectopic mitotic katanin, we assume that the majority of EB1 dots traveling along the surface belong to free MTs [8]. The distance travelled between two frames is simply

$$\sqrt{\left[\left(X_{pos.in\ frame1} - X_{pos.in\ frame2}\right)^2 + \left(Y_{pos.in\ frame1} - Y_{pos.in\ frame2}\right)^2\right]}.$$

27. For EB1 dot tracking in the midplane of wild-type embryos, centrosome movement must also be taken into account since this will affect the dot position and resulting velocity. The centrosome center can be estimated from the EB1 fluorescence. The distance between an EB1 dot and its centrosome of origin can be determined by $\sqrt{\left[\left(X_{cen} - X_{dot}\right)^2 + \left(Y_{cen} - Y_{dot}\right)^2\right]}$, where X_{cen} and Y_{cen} represent the position of the center of the centrosome, as determined from the EB1-GFP fluorescence. Because the minus end of the microtubule is presumed to be stable and anchored to the centrosome [17, 22, 31], the change in distance over time provides the microtubule growth rate.

28. The pixel-to-μm conversion should be done for each objective used. This can be determined manually by acquiring an image of a stage micrometer at the same magnification as the experimental setup and measuring the number of pixels in a straight line between two grid marks of known distance. Divide the distance (μm) by the number of pixels along that line in the acquired image to get the conversion factor.

Acknowledgements

We thank the Srayko Lab for discussions and comments on this article. This work was supported by the Canadian Institutes of Health Research (CIHR) and Alberta Innovates Technology Futures. M.S. was supported by scholar awards from the Alberta Heritage Foundation for Medical Research and CIHR.

References

1. Sulston JE, Horvitz HR (1977) Post-embryonic cell lineages of the nematode, *Caenorhabditis elegans*. Dev Biol 56:110

2. Gennerich A, Vale RD (2009) Walking the walk: how kinesin and dynein coordinate their steps. Curr Opin Cell Biol 21:59

3. Vale RD et al (1996) Direct observation of single kinesin molecules moving along microtubules. Nature 380:451

4. Howard J, Hunt AJ, Baek S (1993) Assay of microtubule movement driven by single kinesin molecules. Methods Cell Biol 39:137

5. Billger MA, Bhattacharjee G, Williams RC Jr (1996) Dynamic instability c microtubules assembled from microtubule-associated protein-free tubulin: neither variability of growth and shortening rates nor "rescue" requires microtubule-associated proteins. Biochemistry 35:13656

6. Cassimeris LU, Walker RA, Pryer NK, Salmon ED (1987) Dynamic instability of microtubules. Bioessays 7:149

7. Mitchison T, Kirschner M (1984) Dynamic instability of microtubule growth. Nature 312:237

8. Gusnowski EM, Srayko M (2011) Visualization of dynein-dependent microtubule gliding at the cell cortex: implications for spindle positioning. J Cell Biol 194:377

9. Maddox AS, Maddox PS (2012) High-resolution imaging of cellular processes in *Caenorhabditis elegans*. Methods Cell Biol 107:1

10. Bossinger O, Cowan CR (2012) Methods in cell biology: analysis of cell polarity in *C. elegans* embryos. Methods Cell Biol 107:207

11. Boyd L, Hajjar C, O'Connell K (2011) Time-lapse microscopy of early embryogenesis in *Caenorhabditis elegans*. J Vis Exp Aug 25;(54) pii 2852. doi: 10.3791/2852. PMID: 21897352

12. Fabritius AS, Ellefson ML, McNally FJ (2011) Nuclear and spindle positioning during oocyte meiosis. Curr Opin Cell Biol 23:78

13. Oegema K, Hyman AA (2006) Cell division. WormBook 1

14. Edgar LG, McGhee JD (1986) Embryonic expression of a gut-specific esterase in *Caenorhabditis elegans*. Dev Biol 114:109

15. Kwok BH, Kapoor TM (2007) Microtubule flux: drivers wanted. Curr Opin Cell Biol 19:36

16. Labbe JC, Maddox PS, Salmon ED, Goldstein B (2003) PAR proteins regulate microtubule dynamics at the cell cortex in *C. elegans*. Curr Biol 13:707

17. Labbe JC, McCarthy EK, Goldstein B (2004) The forces that position a mitotic spindle asymmetrically are tethered until after the time of spindle assembly. J Cell Biol 167:245

18. Waterman-Storer CM, Salmon ED (1997) Microtubule dynamics: treadmilling comes around again. Curr Biol 7:R369

19. Kozlowski C, Srayko M, Nedelec F (2007) Cortical microtubule contacts position the spindle in *C. elegans* embryos. Cell 129:499

20. Piehl M, Cassimeris L (2003) Organization and dynamics of growing microtubule plus ends during early mitosis. Mol Biol Cell 14:916

21. Ozlu N et al (2005) An essential function of the *C. elegans* ortholog of TPX2 is to localize activated aurora A kinase to mitotic spindles. Dev Cell 9:237

22. Srayko M, Kaya A, Stamford J, Hyman AA (2005) Identification and characterization of factors required for microtubule growth and nucleation in the early *C. elegans* embryo. Dev Cell 9:223

23. Sironi L et al (2011) Automatic quantification of microtubule dynamics enables RNAi-screening of new mitotic spindle regulators. Cytoskeleton (Hoboken) 68:266

24. Matov A et al (2010) Analysis of microtubule dynamic instability using a plus-end growth marker. Nat Methods 7:761

25. Schneider CA, Rasband WS, Eliceiri KW (2012) NIH image to ImageJ: 25 years of image analysis. Nat Methods 9:671

26. Sbalzarini IF, Koumoutsakos P (2005) Feature point tracking and trajectory analysis for video imaging in cell biology. J Struct Biol 151:182

27. Praitis V, Casey E, Collar D, Austin J (2001) Creation of low-copy integrated transgenic lines in *Caenorhabditis elegans*. Genetics 157:1217

28. Sharp DJ, Ross JL (2012) Microtubule-severing enzymes at the cutting edge. J Cell Sci 125:2561

29. Srayko M, O'Toole ET, Hyman AA, Muller-Reichert T (2006) Katanin disrupts the microtubule lattice and increases polymer number in C. elegans meiosis. Curr Biol 16:1944

30. Clark-Maguire S, Mains PE (1994) Localization of the mei-1 gene product of *Caenorhabditis elegans*, a meiotic-specific spindle component. J Cell Biol 126:199

31. O'Toole E, Greenan G, Lange KI, Srayko M, Muller-Reichert T (2012) The role of gamma-tubulin in centrosomal microtubule organization. PLoS One 7:e29795

Chapter 8

Xenopus Egg Extracts as a Simplified Model System for Structure–Function Studies of Dynein Regulators

Eliza Żyłkiewicz and P. Todd Stukenberg

Abstract

Many proteins act in multiple pathways which complicates phenotypic analysis. *Xenopus* egg extracts reconstitute complex reactions in vitro, and this can be used to develop assays that isolate a single function of a multifunctional protein. We have applied this system to study regulators of cytoplasmic dynein (dynein), which has numerous roles in the cell including trafficking, nuclear migration, and mitotic spindle formation. Here we describe a functional assay to specifically study the regulation of spindle pole self-organization by dynein and summarize an experimental approach that was used to perform a structure–function analysis of its regulator Ndel1. The approaches presented here can be generalized to isolate a single function of other multifunctional proteins.

Key words Dynein, Structure, Function, Protein, *Xenopus*

Abbreviations

Σ   Extinction coefficient
MW   Molecular weight

1   Introduction

Studying multifunctional proteins presents a major challenge in knockdown animals and tissue culture cells, since it can be difficult to distinguish if a protein directly affects the process under investigation or if an observed phenotype is an indirect consequence of an earlier function. *Xenopus* egg extracts perform complex cellular reactions in vitro, such as mitotic spindle formation or DNA replication [1]. They support these reactions without transcription, translation, or cell cycle changes which greatly diminishes the concerns of indirect effects. Moreover, for a reaction as complex as mitotic spindle formation, it can be greatly beneficial to break the

David J. Sharp (ed.), *Mitosis: Methods and Protocols*, Methods in Molecular Biology, vol. 1136,
DOI 10.1007/978-1-4939-0329-0_8, © Springer Science+Business Media New York 2014

process down into simpler sub-reactions that can be easily modeled and experimentally manipulated.

One example of a multifunctional protein complex is cytoplasmic dynein (dynein). Dynein is involved in organelle trafficking, cilia function, nuclear migration, and mitosis [2]. During mitosis dynein participates in cell cycle control where it turns off spindle checkpoint signaling, organizes bundles of microtubules at both kinetochores and spindle poles, is required for initial attachments between kinetochores and spindle microtubules, and also positions spindles in response to polarity clues. Dynein is the major minus-end-directed microtubule motor and must work in opposition to at least 45 plus-end-directed kinesins [3]. How dynein is regulated to perform so many cellular tasks is a critical unanswered question.

Dynein associates with a number of regulators. Dynactin is a 1.2 mDa complex that is required for most dynein functions [4]. In addition, Nde1, Ndel1, and LIS1 have been characterized as dynein regulators and implicated in high load-bearing dynein functions such as nuclear or chromosome movement. Ndel1 and Nde1 are highly similar (55 % identity) and probably share a role of a scaffold that links dynein to the WD40 protein LIS1, which regulates dynein processivity. It is unclear if Ndel1 and Nde1 move with dynein or simply load LIS1 onto the motor [2].

Historically, *Xenopus* egg extracts have been a powerful system to study the mitotic spindle [1, 5]. These extracts are stockpiled with mitotic proteins and allow for reconstruction of mitotic reactions in a test tube, but at the same time retain the complexity of an in vivo system. More importantly, there are established assays to study a number of sub-reactions of spindle formation including kinetochore assembly, spindle checkpoint signaling, and spindle formation around chromatinized beads, where bipolar spindle structures form in the absence of either centrioles or kinetochores [6]. In this chapter, we focus on formation of self-organized asters in extracts after the addition of a constitutively active Ran GTPase which reconstitutes reactions that gather microtubules at the poles of mitotic spindles [7–10]. Active Ran stimulates microtubule nucleation, and these individual microtubules are self-organized by microtubule-associated proteins (MAPs) and motors into an astral structure with focused minus ends and splayed plus ends. Dynein is required for the focusing activity, where it is believed to carry cross-linkers such as NuMA to microtubule minus ends and accumulate in the center of an aster [5]. This assay allows separation of a single dynein function in cross-linking microtubules independent of its role at the kinetochores and centrosomes.

Based on this assay, we designed a three-step strategy to study the role of Ndel1 as a dynein regulator in cross-linking microtubules in *Xenopus* egg extracts [11]. Our workflow is summarized in Fig. 1. First, we used a functional microtubule aster assembly assay to determine phenotypes associated with Ndel1 depletion and verified

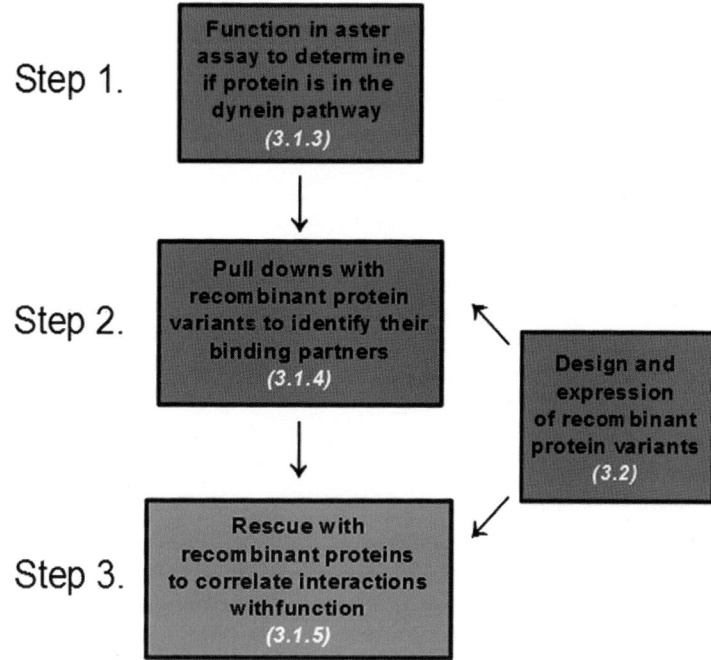

Fig. 1 Summarized workflow for structure–function studies in *Xenopus* egg extracts. *Numbers in brackets* refer the reader to sections in this chapter where respective experiments are described in detail

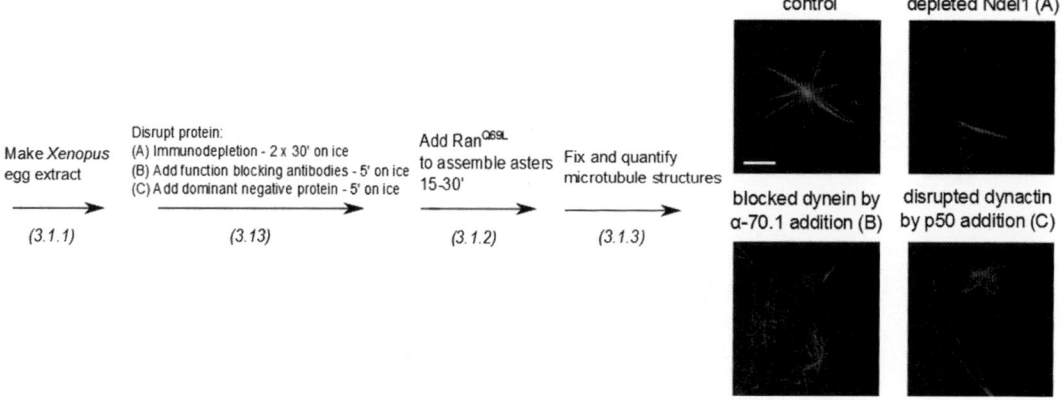

Fig. 2 Summarized workflow for step 1—testing whether depletion phenocopies dynein or other know proteins in the dynein regulatory pathway. *Numbers* in *brackets* indicate sections where experimental methods are described in detail. Representative phenotypes of disrupting dynein by addition of function-blocking antibodies, dynactin by addition of dominant negative protein p50, and immunodepleting Ndel1 are shown, bar 10 μm. Similar phenotypes were shown previously in [11]

that they resemble the phenotype of blocking dynein and its other regulator dynactin in extracts [9, 12] (step 1, Fig. 2, described in detail in Subheading 3.1.3). Second, we biochemically characterized binding partners of different domains within Ndel1 and, using structural information, generated mutants of these protein variants

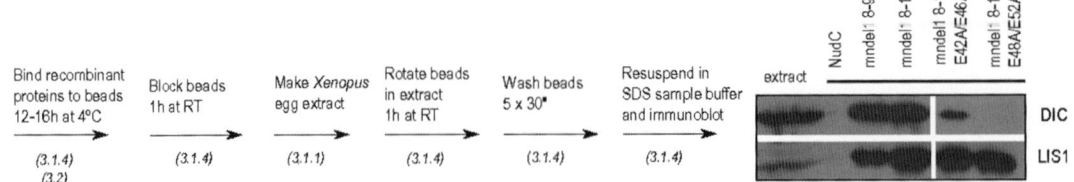

Fig. 3 Summarized workflow for step 2—pull down to identify binding partners and map interaction sites. Binding of LIS1 and dynein intermediate chain (DIC) to Ndel1 proteins is shown on the *right*. mNdel1 8-192$^{E42A/E46A}$ partially post ability to interact with dynein (probably due to the fact that it is not properly folded, *see* Fig. 5b), whereas mNdel1 8-192$^{E48A/E52A}$ completely lost this ability [11]. mNdel1 protein 8-99 does not directly interact with LIS1 and levels of LIS1 binding are low. NudC was used as a negative control and binds neither DIC nor LIS

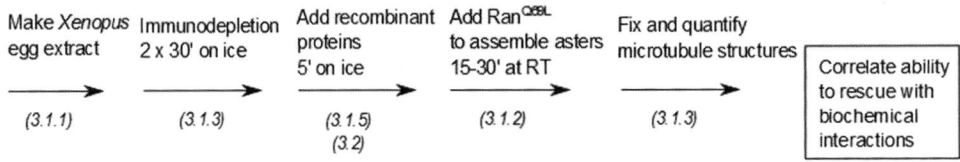

Fig. 4 Summerized workflow for step 3—rescue experiments with recombinant proteins. Results from these experiments (rescue or no rescue of aster formation, other intermediate phenotypes are possible, should be correlated with biochemical results). For example, mNdel18-192$^{E48A/E52A}$ from Fig. 3 does not bind dynein and does not rescue aster formation, suggesting that dynein binding to that domain is critical for Ndel1 function as a dynein regulator

that lost these interactions (step 2, Fig. 3, described in detail in Subheading 3.1.4). Third, we characterized the functional consequence on dynein function by rescuing extracts depleted of Ndel1 protein with recombinant variants (step 3, Fig. 4, Subheading 3.1.5). In addition, in Subheading 3.2, we summarize methods that we used to express and purify proteins and verify that mutations do not destabilize protein folding.

A number of issues should be considered before embarking upon a structure–function study in *Xenopus* egg extracts. First and foremost, one must be able to produce functional recombinant protein that will be used to rescue depletion of endogenous proteins. There are two options: expression and purification from recombinant systems such as *E. coli* or insect cells or rescue with proteins in vitro translated in extract from mRNA. The first method is preferred as one has control over the amount, quality, and folding status of the protein, and the protein can be subsequently used in biochemical assays. However, where expression is difficult, a number of protocols for rescue with in vitro translated proteins can be used [13]. Second, an antibody must be available to deplete the endogenous protein from extracts and verify depletion levels by Western blotting. Since large quantities of antibodies are required for these experiments, we generate our own antibodies. We have previously published protocols that yield high-quality antibodies, half of which are able to efficiently deplete its antigen from extracts [14]. Of note, most proteins form complexes with other proteins and it must be experimentally tested whether they are co-depleted by antibodies. If so, one might need to add back multiple proteins to rescue in the functional assay.

Our approach combines biochemical analysis of protein interactions and functional rescue experiments and presents a powerful tool to perform structure–function experiments with recombinant protein variants produced in *E. coli* or other expression systems. We used this workflow to characterize mutants that could separate functions of Ndel1 in LIS1 and dynein binding and identified the minimal functional unit of the protein within the N-terminal coiled-coil domain. A similar approach can be used to study other complex and multifunctional proteins, for which simplified assays have been developed in extracts, e.g., kinesins or microtubule depolymerases [15].

2 Materials

All reagents and materials were purchased from Sigma unless noted otherwise.

2.1 Xenopus Egg Extracts

Preparation of *Xenopus* egg extracts has previously been described in detail, so we will not provide a summary of this procedure. Instead we refer the reader to two resourceful publications [1, 6] for a complete list of reagents required for generating high-quality mitotically arrested (CSF-arrested) *Xenopus* egg extracts. Extracts supplemented with 250 mM sucrose can be frozen in liquid nitrogen and stored at −80 °C; however, we recommend using only freshly prepared extracts for depletion and rescue experiments.

2.2 Aster Assembly Experiments in Xenopus Egg Extracts

1. Constitutively active Ran mutant (we have successfully used Q69L variant) subcloned into a protein expression vector and purified from *E. coli* (any other expression system can be used).

2. Normal mouse and rabbit IgG used as control.

3. Antibodies against dynein intermediate chain (70.1, as ascites; Sigma).

4. Dynactin subunit p50 or CC1 (p150) subcloned into a protein expression vector and purified from *E. coli* (any other expression system can be used).

5. Specific antibodies against protein of interest (for a detailed protocol used in our laboratory to generate high-quality antibodies against mitotic proteins, *see* [14]).

6. Phosphate-buffered saline (PBS): 10 mM phosphate, 3.7 mM KCl, 137 mM NaCl, pH 7.4.

7. Slide-A-Lyzer MINI dialysis devices, 10 K and 20 K MWCO (Pierce).

8. Amicon Ultra-0.5 protein concentrators, 10 kDa and 50 kDa (Millipore).

9. Rhodamine-labeled porcine tubulin (Cytoskeleton), stored at 4 °C in a dark and dry place.

10. GPEM buffer: 80 mM PIPES, 1 mM MgCl$_2$, 1 mM EGTA, 1 mM GTP, pH 6.8, stored at –80 °C in small aliquots.

11. Extract fix: 50 % glycerol, 4 % *para*formaldehyde (made fresh from 16 % stock; Electron Microscopy Sciences) in 1× MMR (5 mM Na-HEPES, 0.1 mM EDTA, 100 mM NaCl, 2 mM KCl, 1 mM MgCl$_2$, 2 mM CaCl$_2$, pH 7.8), stored at –20 °C.

12. 18 mm round cover glasses.

13. Dynabeads® Protein A (Invitrogen) (*see* **Note 1**).

14. DynaMag™ for magnetic bead collection (Invitrogen).

15. 2× SDS sample buffer: 100 mM Tris, 4 % SDS, 4 % DTT, 20 % glycerol, 0.02 % Brilliant Blue, pH 6.8.

16. SDS-PAGE gel running apparatus, semidry transfer apparatus, and standard reagents for immunoblotting.

17. Primary antibodies for Western blotting and secondary antibodies coupled to HRP.

18. Clear nail polish.

19. Fluorescent microscope with 63× or 40× objective.

2.3 Pull Down Experiments with Wild-Type and Mutant Proteins

1. Recombinant proteins purified from *E. coli* or other expression system.

2. CNBr-activated Sepharose 4 Fast Flow.

3. Sepharose binding buffer: 100 mM Na$_2$HCO$_3$, 500 mM NaCl, pH 8.3.

4. Sepharose blocking buffer: 100 mM Tris–HCl, pH 8.0.

5. Sepharose wash buffer: CSF-XB (10 mM HEPES, 100 mM KCl, 0.1 mM CaCl$_2$, 2 mM MgCl$_2$, 5 mM EGTA, 50 mM sucrose, pH 7.8) with additional 125 mM NaCl, 0.1 % Triton X 100.

6. 2× SDS sample buffer: 100 mM Tris, 4 % SDS, 4 % DTT, 20 % glycerol, 0.02 % Brilliant Blue pH 6.8.

7. SDS-PAGE gel running apparatus, semidry transfer apparatus, and standard reagents for immunoblotting.

8. Appropriate primary antibodies for Western blotting, including anti-dynein intermediate chain antibodies and anti-LIS1 antibodies, and secondary antibodies coupled to HRP.

2.4 CD Spectroscopy of Recombinant Proteins

1. Different variants of protein of interest purified from *E. coli* or other expression systems.

2. 10 mm quartz cuvette, 3,500 μl.

3. Circular dichroism spectropolarimeter.

4. 1 mm quartz cuvette, 350 μl.

5. CD buffer (typically low-salt phosphate buffer free of imidazole and DTT).

3 Methods

3.1 Assays to Study Function of Dynein Regulators in Xenopus Egg Extracts

3.1.1 Preparation and Handling of Xenopus Egg Extracts

Xenopus egg extracts prepared as described before [1] should be first tested to ensure that they are mitotically (CSF) arrested as stated. We recommend performing all extract manipulations at 18–20 °C. Supplemented and tested extracts should be stored on ice. Extracts are sensitive to shearing forces and it is important to always use cut pipette tips to increase the bore size (remove ~ 3 mm) to transfer extracts between tubes.

3.1.2 Ran Aster Assembly Reactions

1. Purify constitutively active Ran^{Q69L}, dialyze into a buffer that mimics salt content of extract (*see* **Note 2**; e.g., 10 mM K-HEPES, 100 mM KCl pH 7.8), and concentrate to 750 μM using Amicon concentrators with 10,000 MW pore cutoff or smaller (use parameters MW = 28,545 g/mol, $\Sigma = 24,452$ M^{-1} cm^{-1} to determine protein concentration using UV spectrometry).

2. Resuspend 20 μg of Rhodamine-labeled tubulin in 10 μl of GPEM to achieve final concentration of 2 mg/ml (*see* **Note 3**). Final concentration of labeled tubulin in extract should be 15 μg/ml (133× stock dilution). To supplement extract with fluorescent tubulin, first remove 1/10 volume of previously aliquoted extracts that will be required for the entire experiment. Add 1/13 volume of fluorescent tubulin, let sit on ice for 5 min, then add back to the remaining 9/10 volumes of extract, and gently mix. Leave on ice for 5 min. All extracts containing fluorescent tubulin should be protected from light.

3. Warm extract to 18 °C for 5 min by letting them sit on a bench or in a water bath.

4. Add Ran^{Q69L} to the final concentration of 25 μM.

5. At 15 min, 30 min, and 45 min, fix 1 μl of each reaction in 3 μl of the extract fix spotted on a microscope slide. After fixing all reactions for one time point, immediately squash with 18 mm cover glasses and close with nail polish.

6. Examine and quantify asters under confocal microscope. In 1 μl of extract, depending on its quality, we usually observe a

few hundred asters with central bright foci and microtubules extending in all directions (Fig. 2b). Over time different microtubule structures can be observed (Fig. 2b; *see* **Note 4**).

7. Optionally, asters can me fixed and spun onto *poly*lysine-coated cover glasses and proteins associated with microtubules can be immunolabeled and be examined by fluorescence microscopy (for detailed protocols for spinning extract reactions onto cover glasses, *see* [14]).

3.1.3 Disrupting Function of Dynein, Dynactin, and Protein of Interest

Cell-free systems such as *Xenopus* egg extracts are easy to manipulate. Reagents to disrupt the dynein pathway in extracts (e.g., function-blocking antibodies and dominant negative subunits) have been characterized previously [9, 12]. There are three methods commonly used to disrupt protein function in extracts: addition of function-blocking antibodies, addition of dominant negative proteins, and immunodepletion. Immunodepletion is preferred, as one can directly verify the efficiency of protein removal but by Western blotting, but here we will describe how all three methods are applied to interfere with dynein function in *Xenopus* egg extracts. This section is schematically summarized in Fig. 2:

1. To interfere with dynein function by disrupting dynactin, express and purify p50, a dominant negative subunit of the dynactin complex (*see* **Note 5**); dialyze into 10 mM K-HEPES, 100 mM KCl, pH 7.2; and concentrate to 15 mg/ml (*see* **Note 2**). Determine protein concentration by UV spectroscopy and the following parameters (for human protein): MW = 44,819 g/mol, $\Sigma = 17,545$ M^{-1} cm^{-1}.

2. To directly block dynein with anti-dynein intermediate chain antibodies (70.1) (*see* **Note 6**), dialyze anti-70.1 antibodies into PBS to remove azide and determine concentration after dialysis. Concentrate to 15 mg/ml.

3. To deplete Ndel1 or other protein of interest, use 50 μl of Dynabead slurry per 100 μl of extract per one round of depletion. Typically, at least two rounds of depletion are required to effectively remove protein from extract.

4. Place beads on the magnet, let sit for 30 min, remove the buffer, and add fresh PBST. Wash three times with PBST to remove azide.

5. Add 60 μg of polyclonal rabbit antibody against Ndel1 (or other protein of interest) to 500 μl of PBST, add to beads, and mix.

6. Make an additional set of beads with normal rabbit IgG at the same concentration.

7. Allow antibodies to bind to beads for 12–16 h on a rotator at 4 °C.

8. Place beads on the magnet and remove the unbound antibody.

9. Measure unbound IgG concentration to determine binding efficiency.

10. Wash beads three times with PBST and then with CSF-XB.

11. Take sample of pre-depletion extract and resuspend in 9 volumes of 2× SDS-loading buffer. Add extract and gently mix by tapping on the sides of the tube. Incubate for 30 min on ice, gently mixing the contents of the tube every few minutes and not allowing beads to settle at the bottom of tube.

12. Place both tubes on the magnet and allow beads to bind to the side of the tube for at least 2 min. Gently remove depleted extracts and transfer to clean labeled tubes. Make sure that no beads remain in extract. If desired, repeat depletion with fresh beads. After each round of depletion, take a gel sample and immediately resuspend in 9 volumes of SDS-loading buffer.

13. Depleted extracts can be used in further experiments. However, when antibodies against a protein of interest are first tested, it is necessary to determine depletion efficiency. Some proteins yield phenotypes after only 70 % depletion, but in most cases, more efficient depletion is required to unambiguously interpret resulting phenotypes.

14. To test the depletion efficiency, run an SDS-PAGE gel with a control and depleted reactions and immunoblot with antibodies against Ndel1 or other protein of interest. Use α-tubulin or another abundant protein that is not co-depleted with protein of interest as a loading control. Perform serial dilutions of pre-depletion extract from 100 % to 10 % and determine depletion efficiency (*see* **Note 7**). In addition, run a serial dilution with known quantities of the recombinant full-length protein of interest and compare to the amount of protein in 1 μl of extract. Estimate concentration of your protein in extract. Mitotic proteins are often present in extracts at 10–100 nM concentrations [16].

15. If the protein is not significantly depleted, either do another round of depletion or increase the amount of antibody coupled to Dynabeads. In some cases, even partial depletion yields a phenotype.

16. Determine whether known binding partners of protein of interest are co-depleted.

17. Supplement extracts with Rhodamine tubulin as described in Subheading 3.1.2 and aliquot 20 μl of extract into individual labeled tubes using cut pipette tips.

18. To individual tubes, add anti-70.1 antibodies to the final concentration of 1 mg/ml and p50 to the final concentration of 1 mg/ml and, as a control, use a mixture of rabbit IgG at 1 mg/ml and BSA at 1 mg/ml. Make additional tubes for control-depleted extracts and extracts depleted of protein of interest.

19. Warm up reactions to room temperature and add Ran^{Q69L} as described in Subheading 3.1.2. Fix reactions at different time points and examine microtubule structures under confocal microscope. Blocking dynein and dynactin typically yields single and cross-linked short microtubules, but rarely asters (Fig. 2). Depleting Ndel1 mostly yields bundles of microtubules, but not asters (Fig. 2). This observed phenotype is not identical, but very similar to blocking dynein, confirming that Ran aster assay could be used to perform a structure–function study of Ndel1 as a dynein regulator.

3.1.4 Pull Down Experiments to Study Protein Interactions in Xenopus Egg Extracts

Low-affinity interactions are critical in biology but often difficult to detect. When used with the correct controls, pull down assays in extracts can be a powerful method to detect low-affinity and indirect interactions, for example, between dynein and its regulators. What is more, residues responsible for these interactions can be identified by mutagenesis and their function verified in extract-based assays such as Ran aster assembly. Many different commercially available resins can be used for pull down experiments. In our experience, cyanogen bromide-activated Sepharose is an inexpensive agarose matrix which requires mild coupling conditions and yields highly repeatable results. CNBr-activated Sepharose contains cyanate esters and imidocarbonates, which react with primary amines on proteins. Schematic summary of pull down experiments is provided in Fig. 3.

1. Dialyze protein(s) into Sepharose binding buffer (*see* **Note 8**). Perform multiple rounds of dialysis at 4 °C. Best coupling efficiency is achieved at pH 8.3–8.5. Other buffers can be used, except primary amines such as Tris which block binding sites and should be avoided. Include a control protein, for example, GST, IgG, or any other protein that does not interact with dynein. BSA can be used, but we find that in many cases, it yields high levels of nonspecific binding.

2. Weight out appropriate amount of Sepharose. Allow to swell for 15 min in cold 1 mM HCl to yield ~1 ml of gel matrix from 0.3 mg of powder. Place a fritted funnel on Erlenmeyer flask with vacuum connector. Connect to vacuum. Wash with 50 volumes of cold 1 mM HCl for 15 min, followed with a quick wash with 50 volumes of coupling buffer.

3. Mix dialyzed proteins and washed Sepharose. We typically add proteins to the final concentration of 30 μM on beads. This value has to be determined experimentally for every protein, but we recommend first testing concentrations in 10–50 μM range.

4. Allow proteins to bind to beads for 12–16 h on rotator at 4 °C. Spin in a centrifuge for 1 min at $800 \times g$ at 4 °C.

Remove supernatants and check protein concentration to confirm that protein is completely bound to beads (*see* **Note 9**). Wash once with coupling buffer and once with PBS. Spin after every wash.

5. Add fresh or thawed *Xenopus* egg extract. Use 50 μl of Sepharose per 100 μl of extract. Gently mix to ensure that Sepharose does not accumulate at the bottom of the tube. Rotate end over end or nutate for 1 h at room temperature.

6. Wash five times with Sepharose wash buffer by spinning in for 30 s at $800 \times g$. Aspirate supernatant without drying out the resin and resuspend in 500 μl of fresh buffer.

7. Resuspend beads in equal volume of 2× SDS sample buffer. Run an SDS-PAGE gel.

8. Proteins bound to beads can be identified by immunoblotting or mass spectrometry (Fig. 3).

9. Different wash conditions (e.g., high ionic strength) can be used to disrupt protein interactions and provide information about the nature of these interactions. For example, if electrostatic interactions are disrupted by high ionic strength, residues potentially responsible for these interactions can be identified and mutated (example in Fig. 3). Mutant variants of the protein can be further tested in pull down and rescue experiments.

3.1.5 Rescue Experiments in Xenopus Egg Extracts

In order to perform a function analysis in *Xenopus* egg extracts, the depleted endogenous protein can be rescued with recombinant variants and compared to the full-length wild-type protein. The full-length wild-type protein should fully rescue the depletion phenotype to confirm that depletion of protein of interest is specifically associated with the phenotype. Rescue phenotypes can be correlated with interactions of different protein variants, for example, in case of Ndel1, we have determined that different portions of the protein bind LIS1 and dynein and together constitute the minimal functional domain of the protein which can rescue aster formation. Point mutations in both domains that disrupted LIS1 and dynein binding in pull down experiments could not rescue aster formation (Fig. 3), confirming that these interactions are critical for Ndel1 function. Schematic summary of rescue experiments is provided in Fig. 4:

1. Deplete endogenous protein as described in Subheading 3.1.3 and supplement with a recombinant variant of the protein.

2. Test different concentrations, including estimated endogenous concentration. Induce Ran aster formation as described in Subheading 3.1.3, fix, and process for fluorescent microscopy.

3.2 Preparation of Recombinant Proteins for Structure–Function Studies

3.2.1 Design and Expression of Recombinant Proteins for Xenopus Egg Extract Experiments

In structure–function analysis, different truncated portions (variants) of the protein are tested to identify their functions and ability to interact with other proteins. In order to express these protein variants, one must first design appropriate constructs to maximize chances that protein will be expressed in high quantities, soluble, and properly folded in solution. In general, entire domains should be expressed, at least during initial analysis, to stabilize proteins. In order to do this, at least basic knowledge of the protein architecture is critical. Structural information about the protein is most accurate when based on atomic resolution structure available in the Protein Data Bank (http://pdb.org). In case of Ndel1, the crystal structure of the protein was used to isolate the coiled-coil domain and design further truncations within this domain [17], and we found this to be a very efficient approach. However, if a structure is not available, based on the primary sequence of the protein, many structural features (e.g., presence of structural motifs such as coiled-coils, similarity to annotated domains) can be predicted using bioinformatic analysis with software such as SMART (Simple Modular Architecture Research Tool; http://smart.embl.de/) [18, 19]. In addition, secondary structure prediction programs, such as PredictProtein (http://expasy.org), can be used to further narrow down the structured regions in the protein. Importantly, constructs of unstructured fragments (e.g., loops) should be designed with caution; such proteins tend to be unstable and very susceptible to protease activity. If loops need to be preserved between multiple domains, limited proteolytic digestion can be used to generate stable fragments.

Multiple purification systems are available from different manufacturers, and protocols were previously published for obtaining high yields of recombinant proteins using a variety of tags and purification resins. Therefore, these protocols will not be summarized in this chapter. However, we want to point out that it is important to adjust pH of all protein purification buffers depending on the theoretical isoelectric point (pI) of a specific protein. Ideally, purification should be performed at least two pH units away from the theoretical pI to prevent precipitation. In many cases, even one pH unit difference helps increase stability of the protein. A number of biophysical parameters of proteins, including pI, can be determined based on the amino acid sequence using ProtParam tool at the ExPaSy portal (http://www.expasy.org).

For any mutated protein domain, it is particularly important to verify proper overall fold of the mutant protein. Misfolded or aggregated proteins are not useful in functional experiments. Methods such as gel filtration, multi-angle light scattering (MALS), nuclear magnetic resonance (NMR), and circular dichroism (CD) spectroscopy can be used to compare the folding and oligomeric state of the wild-type and mutant proteins. Summary of all these

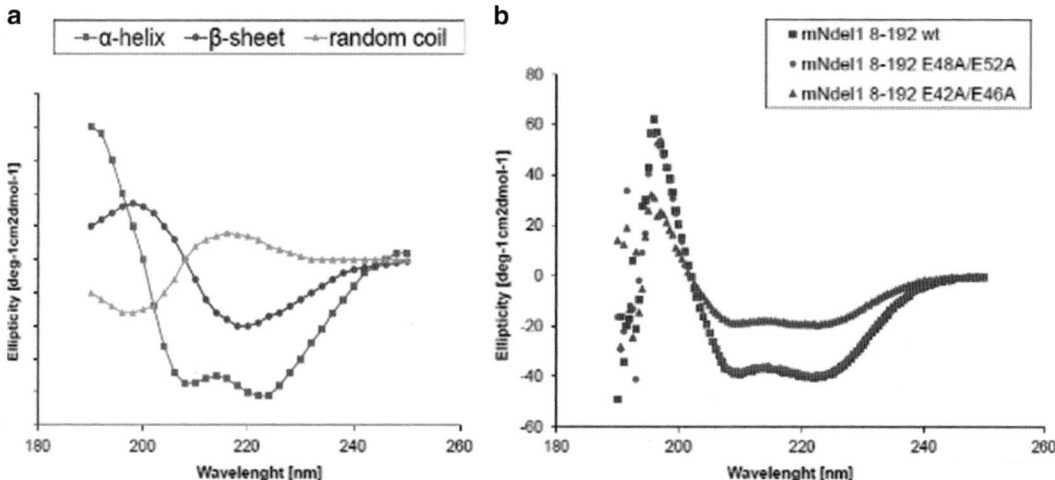

Fig. 5 Comparison of folding status of recombinant wild-type proteins and recombinant mutant proteins by CD spectroscopy. (**a**) Typical spectra of helical (*blue*), β-sheet (*red*), and random coil (*green*) structures. (**b**) Spectra for a wild-type Ndel1 coiled-coil domain (*blue*) and two proteins with mutations in this domain (*red* and *green*) (Figure adapted from [11]). Lower helical content of mNdel1 8-192[E42A/E46A] indicates that this protein is not properly folded in solution (Color figure online)

methods is beyond the scope of this chapter; however, below we provide a brief experimental outline for using CD spectroscopy as it's one of the fastest and simplest methods to use.

3.2.2 Circular Dichroism Spectroscopy of Wild-Type and Mutant Proteins

Circular dichroism spectropolarimetry (CD) is a fast and inexpensive method to assess proper protein folding. It is particularly a powerful tool to verify whether introduced mutations in the protein affect its proper folding in solution. CD in the far ultraviolet region (190–260 nm) of light spectrum results from the peptide bond located in a folded protein (for more information on using CD, *see* [20, 21]). In Fig. 5, typical spectra of α-helix, β-sheet, and random coil structures are shown. Destabilization of the α-helical protein fold results in decreased positive ellipticity at 192 nm and decreased negative ellipticity at 208 and 222 nm, respectively (example in Fig. 5b). Destabilized β-sheets show decreased positive ellipticity at 198 nm and decreased negative ellipticity at 218 nm. Below, we summarize how structural content of the wild-type and mutated protein variants can be compared using this method:

1. While deciding on a buffer composition for measurements at 190–260 nm, consider the following: (1) low absorbant buffers are recommended, e.g., 10 mM potassium fluoride or 50 mM potassium phosphate, and (2) NaCl should be avoided or, if necessary, used at low concentration. Additives such as EGTA, EDTA, or DTT should be avoided as well.

2. Dialyze all proteins to be analyzed together at 4 °C. Place all dialysis cassettes into the same beaker with buffer, as CD can

be sensitive even to very small changes in buffer composition. If removing absorbant additives from protein solution, perform multiple rounds of dialysis. Take a sample of the buffer from the last round of dialysis to use as blank in CD measurements.

3. Measure protein concentration at 280 nm in a 10 mm cuvette. If ellipticity values will be directly compared between different protein variants, it is critically important to precisely adjust all proteins to the same concentration (*see* **Note 10**).

4. Follow manufacturer's instructions on proper operation of the Xenon lamp. Warm up water bath to desired temperature (we typically perform measurements at 20 °C, but other conditions can be used as long as protein is stable).

5. If using AVIV software, we typically set the following parameters in the Experiment Configuration tab: Type: Wavelength scan, Bandwidth: 1 nm, Auto Slit Closure: on, Setpoint: 20 °C. In Wavelength Configuration in the Experiment Configuration menu, set the following: Wavelength Start, 250 nm; Wavelength End, 190 nm; Sample Every, 1 nm; Averaging Time, 1 s; Number of Scans, 5; Wait Time Between Scans, 0 s.

6. Thoroughly wash 1 mm cuvette with ethanol and then with ddH_2O.

7. Put 350 µl of the dialysis buffer as a blank sample into a 1 mm cuvette and collect data.

8. Wash cuvette with water and then with ethanol and again with water.

9. Load protein sample and collect data. Using a 350 µl cuvette allows reducing the sample volume but requires relatively high protein concentration. We recommend starting with protein concentration of 0.2 mg/ml and adjusting it as needed (*see* **Note 11**). CD signal is dependent on the secondary structure content, and therefore the optimal amount of protein has to be determined experimentally.

10. Repeat data collection for every protein sample with multiple cuvette washes between measurements.

11. Plot data (saved as *.txt files, *see* **Note 12**) with wavelength (λ) values on the X axis [nm] and ellipticity (θ) values on the Υ axis [$deg^{-1}cm^2dmol^{-1}$]. Plotted data can be directly compared for wild-type and mutated variants of the same protein/domain to determine changes in the secondary structure content. Figure 5b shows experimentally determined ellipticity values for the coiled-coil fragment of Ndel1 and different mutants. Mutating residues E48 and E52 to alanines (mNdel1 8-192$^{E48A/E52A}$) does not affect protein folding; however mutations in residues E42 and E46 (mNdel1 8-192$^{E42A/E46A}$) result

in a lower α-helical content. mNdel1 8-192$^{E42A/E46A}$ is considered poorly folded and should not be used in functional experiments, since potential lack of activity of this protein would be difficult to interpret.

4 Notes

1. Protein A Dynabeads® effectively bind rabbit antibodies. For mouse antibodies, protein G beads should be used instead. See Invitrogen website for binding capacities of different antibody-binding proteins (http://www.invitrogen.com/dynal).

2. All proteins that will be added to *Xenopus* egg extracts should be dialyzed into a buffer that mimics salt content of extracts (100 mM), preferably HEPES or phosphate buffer. Tris is not recommended.

3. Other fluorophores can be used instead of Rhodamine (568 nm). Extracts are fairly autofluorescent in the Rhodamine channel, but we never found this to be a major problem when fixed reactions were examined under confocal microscope. Other fluorophores excited at longer wavelengths, such as X-Rhodamine (588 nm), can be used if high background is a problem. We do not recommend using FITC (488 nm, very high background since extracts have high autofluorescence at this wavelength).

4. We typically quantify microtubule structures in a set number of microscope fields (100–200 structures) and classify them as focused or unfocused (Fig. 2). It is very important to choose the right time point for quantifying asters (usually 15 or 30 min) when most of them are separated from each other. Over time, asters start forming bipolar and multipolar spindles, to eventually collapse into a big structure where separate bright foci are impossible to distinguish and quantify.

5. Other dominant negative subunits of the dynactin complex have been characterized, such as the coiled-coil 1 (CC1) portion of p150 [22], and can be used instead of p50.

6. In 2012 first specific small molecule inhibitors (ciliobrevins) that block dynein ATPase activity were published [23]. At the time this chapter was published, ciliobrevin A was commercially available from Tocris Bioscience.

7. *Xenopus* egg extracts have very high protein content (50 mg/ml) and running more than 1 μl of extract per well will result in overloading a gel.

8. Proteins expressed and purified under unfolding ("insoluble") conditions can be coupled to Sepharose and used as bait in pull

down experiments (for a detailed protocol for insoluble protein preps, *see* [14]). Insoluble protein is dialyzed into coupling buffer supplemented with 6 M urea. After coupling is completed, beads are washed with coupling buffer supplemented with urea and all subsequent steps are performed in urea-free buffers. Not all proteins will properly fold on beads after urea is removed, so experiments performed with insoluble proteins should be interpreted with caution.

9. Some proteins have better binding efficiency when coupled to beads for 1–4 h at room temperature.

10. We find that concentration measurements by NanoVue are not accurate enough and highly recommend using a spectrometer with a longer light path instead.

11. If sample at this concentration is not available, lower concentration can also be used, but light path length and sample volume need to be increased accordingly, e.g., for measurements in 10 mm cuvettes, 3,500 µl of sample is required.

12. In AVIV software, averaged trace can be saved by clicking "Save Average Trace" in Axis Definitions Menu on the main screen.

Acknowledgements

We want to acknowledge Dr. Zygmunt S. Derewenda (University of Virginia) for insightful discussions on the structure of Ndel1 and generously sharing reagents. We also thank Drs. Won-Chan Choi and Ankoor Roy (University of Virginia) for generating Ndel1 constructs and help with CD spectroscopy experiments. This work was made possible by NIH grants R01-NS036267 and R01-GM063045.

References

1. Murray AW (1991) Cell cycle extracts. Methods Cell Biol 36:581–605

2. Vallee RB, McKenney RJ, Ori-McKenney KM (2012) Multiple modes of cytoplasmic dynein regulation. Nat Cell Biol 14(3):224–230

3. Hirokawa N et al (2009) Kinesin superfamily motor proteins and intracellular transport. Nat Rev Mol Cell Biol 10(10):682–696

4. Schroer TA (2004) Dynactin. Annu Rev Cell Dev Biol 20:759–779

5. Walczak CE, Heald R (2008) Mechanisms of mitotic spindle assembly and function. Int Rev Cytol 265:111–158

6. Hannak E, Heald R (2006) Investigating mitotic spindle assembly and function in vitro using Xenopus laevis egg extracts. Nat Protoc 1(5):2305–2314

7. Carazo-Salas RE et al (1999) Generation of GTP-bound Ran by RCC1 is required for chromatin-induced mitotic spindle formation. Nature 400(6740):178–181

8. Kalab P, Pu RT, Dasso M (1999) The ran GTPase regulates mitotic spindle assembly. Curr Biol 9(9):481–484

9. Ohba T et al (1999) Self-organization of microtubule asters induced in Xenopus egg extracts by GTP-bound Ran. Science 284(5418):1356–1358

10. Wilde A, Zheng Y (1999) Stimulation of microtubule aster formation and spindle assembly by

the small GTPase Ran. Science 284(5418):1359–1362

11. Zylkiewicz E et al (2011) The N-terminal coiled-coil of Ndel1 is a regulated scaffold that recruits LIS1 to dynein. J Cell Biol 192(3):433–445

12. Wittmann T, Hyman T (1999) Recombinant p50/dynamitin as a tool to examine the role of dynactin in intracellular processes. Methods Cell Biol 61:137–143

13. Tymms MJ (1995) In vitro transcription and translation protocols. Methods in molecular biology 1995. Humana Press, Totowa NJ, xii, 432 p

14. Emanuele MJ, Stukenberg PT (2009) Probing kinetochore structure and function using Xenopus laevis frog egg extracts. Methods Mol Biol 545:221–232

15. Zhang X, Ems-McClung SC, Walczak CE (2008) Aurora A phosphorylates MCAK to control ran-dependent spindle bipolarity. Mol Biol Cell 19(7):2752–2765

16. Emanuele MJ et al (2005) Measuring the stoichiometry and physical interactions between components elucidates the architecture of the vertebrate kinetochore. Mol Biol Cell 16(10):4882–4892

17. Derewenda U et al (2007) The structure of the coiled-coil domain of Ndel1 and the basis of its interaction with Lis1, the causal protein of Miller-Dieker lissencephaly. Structure 15(11):1467–1481

18. Schultz J et al (1998) SMART, a simple modular architecture research tool: identification of signaling domains. Proc Natl Acad Sci U S A 95(11):5857–5864

19. Letunic I, Doerks T, Bork P (2012) SMART 7: recent updates to the protein domain annotation resource. Nucleic Acids Res 40(Database issue):D302–D305

20. Greenfield NJ (2006) Using circular dichroism spectra to estimate protein secondary structure. Nat Protoc 1(6):2876–2890

21. Greenfield NJ (2006) Determination of the folding of proteins as a function of denaturants, osmolytes or ligands using circular dichroism. Nat Protoc 1(6):2733–2741

22. Quintyne NJ et al (1999) Dynactin is required for microtubule anchoring at centrosomes. J Cell Biol 147(2):321–334

23. Firestone AJ et al (2012) Small-molecule inhibitors of the AAA+ATPase motor cytoplasmic dynein. Nature 484(7392):125–129

Part III

Specialized Biophysical and Microscopy Approaches for Analyzing Microtubules and Microtubule Associated Proteins

Chapter 9

Covalent Immobilization of Microtubules on Glass Surfaces for Molecular Motor Force Measurements and Other Single-Molecule Assays

Matthew P. Nicholas, Lu Rao, and Arne Gennerich

Abstract

Rigid attachment of microtubules (MTs) to glass cover slip surfaces is a prerequisite for a variety of microscopy experiments in which MTs are used as substrates for MT-associated proteins, such as the molecular motors kinesin and cytoplasmic dynein. We present an MT-surface coupling protocol in which aminosilanized glass is formylated using the cross-linker glutaraldehyde, fluorescence-labeled MTs are covalently attached, and the surface is passivated with highly pure beta-casein. The technique presented here yields rigid MT immobilization while simultaneously blocking the remaining glass surface against nonspecific binding by polystyrene optical trapping microspheres. This surface chemistry is straightforward and relatively cheap and uses a minimum of specialized equipment or hazardous reagents. These methods provide a foundation for a variety of optical tweezers experiments with MT-associated molecular motors and may also be useful in other assays requiring surface-immobilized proteins.

Key words Microtubules, Protein immobilization, Plasma cleaning, Silanization, Aminosilane functionalization, Chemical cross-linking, Glutaraldehyde, Fluorescence labeling, Single-molecule assays, Microtubule motor proteins

1 Introduction

In vitro, single-molecule microscopy studies of molecular motor proteins can provide valuable insights into how they generate forces essential to mitosis and other critical cellular functions. Single-molecule studies of microtubule (MT) motors and other MT-associated proteins (MAPs) frequently employ surface-immobilized MTs. While axonemes (bundles of MTs and other associated proteins) have been used frequently for this purpose [1–5] and have the advantage of adsorbing strongly to untreated glass, they contain other proteins in addition to tubulin, and they are not the natural substrates for cytosolic molecular motors. Moreover, the large diameter of axonemes (~160 nm) may be

David J. Sharp (ed.), *Mitosis: Methods and Protocols*, Methods in Molecular Biology, vol. 1136,
DOI 10.1007/978-1-4939-0329-0_9, © Springer Science+Business Media New York 2014

undesirable in certain assays. Therefore, MTs are generally preferable, and so a reliable method for fixing them to a glass surface is of utility.

Rigid MT attachment is important in order to achieve high-quality measurements: unwanted motion makes it impossible to precisely track the positions of MT-bound MAPs and introduces compliance to the motor-MT system that complicates interpretation of forces measured by optical tweezers and other methods. Unfortunately, MTs adsorb very poorly to untreated glass. Any MTs that do bind tend to bow and agitate on the surface and often detach when a force is applied. Therefore, specialized methods of MT attachment to glass are required. For optical trapping, it is also required that the remaining surface be well passivated (usually using a blocking agent such as bovine serum albumin (BSA), casein, or polyethylene glycol (PEG) derivatives), so that the trapped microspheres ("beads") do not bind to the glass via non-specific adsorption. This is important because attractive bead-surface interactions could, for example, impair molecular motor movement and/or give false indications of MAP-MT binding.

We sought a method to eliminate essentially all nonspecific microsphere-surface interactions. A variety of techniques have been reported in the literature for immobilizing MTs on glass. We experimented with several approaches, such as coating glass with nonspecifically adsorbed poly-L-lysine [6–9], antibodies [10], rigor kinesins (both G234A [11, 12] and T93N [13, 14] mutants), or BSA-biotin (with streptavidin and biotinylated MTs) [15, 16]. We also experimented with silane-PEG-N-hydroxysuccinimide (silane-PEG-NHS) reagents. In our hands, several of these techniques were effective at *either* immobilizing MTs *or* preventing trapping beads from binding to the surface but did not accomplish both simultaneously (*see* Note 1). For example, although rigor kinesin is very effective at binding MTs (we use this technique in other single-molecule assays in our laboratory), microspheres frequently bind irreversibly to the kinesin-coated surface, even in the presence of blocking agents.

Aminosilanization (coating of surfaces, typically containing free hydroxyl groups, with aminated alkoxysilane molecules) has been used for years to promote binding of a variety of biological materials to glass (*see*, e.g., [17] and references therein), including MTs [18–20]. Reported aminosilanization conditions and protocols vary widely, and optimization of this surface chemistry is still an active area of research [21–26]. However, we found that, when used in combination with chemical cross-linking [27–31] and appropriate surface blocking [20], aminosilanized cover slips perform exceptionally well, both in terms of binding MTs and limiting microsphere adhesion. Certain parameters, such as incubation times and blocking reagents, are critical to achieving good results with MTs. Therefore, we thought it would be of general benefit to provide the protocols we established, based on current literature and optimization in the laboratory. The procedures given here are cheap and easy and do not require hazardous cleaning reagents. MTs attach rigidly to the glass and support robust movement by a variety of MT-associated molecular motors. Trapping beads, whether coated in protein or not, show no binding (even transiently) to the blocked cover slip surface when pushed into or dragged against the glass (even after days in the chamber at room temperature).

The method works as follows. First, the glass is prepared through a series of cleaning procedures that remove contaminants and hydroxylate the surface. Next, the surface is aminosilanized, thereby coating the glass with positively charged amine groups that attract MTs (due to the strongly negative MT surface charge). The treated glass cover slip is then used to assemble a microscopy flow chamber. This flow chamber is treated with glutaraldehyde and extensively washed, leaving a mixture of surface amines and reactive formyl groups on the glass. Fluorescent tubulin is then prepared and polymerized into MTs. The MTs are bound to the surface, attracted initially by the surface charge, and then covalently coupled to the glass via reaction between exposed lysyl residues on the MT and the surface-bound formyl groups. Finally, the remaining surface is blocked with β-casein, which physically adsorbs to the glass, thereby passivating it. The casein also reacts with any remaining formyl groups to inactivate them. Figure 1 summarizes how the individual steps contribute to the overall procedure. The reader is encouraged to consult each of the Notes, which contain important practical information, as well as explanations of theoretical principles.

The procedures presented here yield a flow chamber suitable for a number of optical tweezers studies involving MAP function,

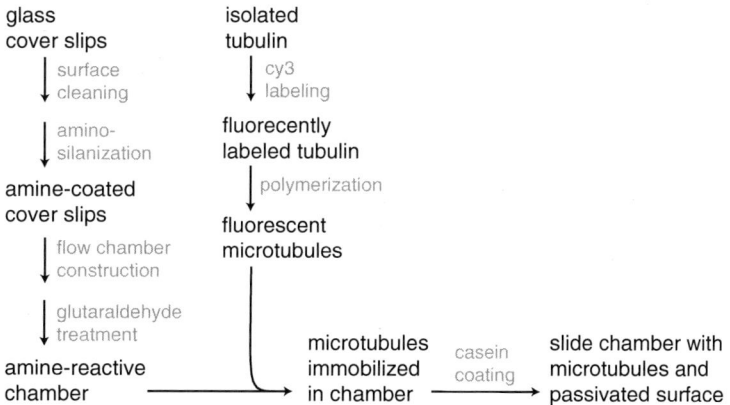

Fig. 1 Protocol scheme. Each pathway summarizes the major steps in preparing a slide chamber with MTs attached and the surface passivated against nonspecific microsphere binding

such as our accompanying protocol for measurement of kinesin force generation (*see* Chapter 10). These methods are also likely to be useful in other single-molecule studies of MAP-MT interactions and other assays requiring immobilization of proteins on glass surfaces.

2   Materials

2.1   Cover Slip Aminosilanization

1. Cover slips (18 mm × 18 mm × 0.170 mm) (No. 1.5, Zeiss Cat. No. 474030-9000-000; *see* **Note 2**).

2. Porcelain cover glass holders (Thomas, Part No. 8542E40-TS; *see* **Note 3**).

3. Clean blunt-nosed forceps (clean in ethanol and flame prior to use).

4. Double-deionized water (ddH$_2$O) purification system (*see* **Note 4**).

5. Mucasol™ detergent solution, 2 % (v/v): 98 % double-deionized water, 2 % Mucasol fast alkaline cleaning agent (Merz Hygiene GmbH), by volume.

6. APTES (3-aminopropyl triethoxysilane, also called APS; Sigma Cat. No. 440140 or A3648) or AEAPTES (*N*-(2-aminoethyl)-3-aminopropyl triethoxysilane, Gelest Cat. No. SIA0590.5) (*see* **Note 5**).

7. Acetone (Sigma, cat. no. 270725; *see* **Note 6**).

8. Spectrophotometric grade ethanol, 200 proof (Sigma, Cat. No. 459828; *see* **Note 6**).

9. Five plastic jars, 250 mL each, made of polymethylpentene (PMP) (Nalgene). Label each of the jars and lids with the following: *ddH$_2$O, Mucasol, Ethanol, Acetone, Silane* (*see* **Note 7**).

10. Bath sonicator with temperature control (Fisher Scientific, FS30D) and holder for jars (*see* **Note 8**).

11. Plasma cleaner (PDC-001, Harrick Plasma).

12. Oven or hot plate (*see* **Note 9**).

13. Vacuum desiccator.

14. Personal protective equipment: nitrile gloves, eye protection, lab coat, etc.

2.2   Microscope Slide Chamber Construction

1. Microscope slides (3″ × 1″ × 1.0 mm).

2. Parafilm® "M" (Bemis, *see* **Note 10**).

3. Forceps (tweezers): one sharp-nosed and one blunt-nosed.

4. Heat gun (e.g., Fit Gun-3, Alpha Wire) or heating block.

2.3 Slide Chamber Glutaraldehyde Treatment

1. 25 % glutaraldehyde (Sigma): divided in ~10 μL aliquots and stored at –20 °C (*see* **Note 11**).

2. Kimwipes® delicate task wipes (Kimtech).

3. Filtered, compressed air or nitrogen (*see* **Note 12**).

2.4 Labeling of Tubulin with Cy3 Dye

See **Note 13**:

1. Bovine brain tubulin (1 mg/vial, Cytoskeleton, Inc.; *see* **Note 14**).

2. Cy3™ monofunctional *N*-hydroxysuccinimide (NHS) ester reactive dye (10 mg/mL): dissolve one vial of Cy3 monofunctional NHS ester dye (0.2 mg, GE Healthcare) in 20 μL anhydrous DMSO. Store at –20 °C in the dark (*see* **Note 15**).

3. PM buffer: 100 mM PIPES and 1 mM $MgCl_2$, pH ~7.

4. Beckman Coulter Optima™ TLX Ultracentrifuge (120,000 maximum rpm).

5. Beckman TLA-100 fixed angle rotor (7×20 mm, 0.2 mL tube).

2.5 Cy3-Labeled MT Polymerization

1. BRB80 buffer: 80 mM PIPES (Sigma P6757), 2 mM $MgCl_2$, and 1 mM EGTA, pH ~7.

2. Unlabeled bovine brain tubulin (10 μL aliquots of 10 mg/mL): dissolve 1 vial of 1 mg lyophilized powder (Cytoskeleton, Inc.) in 100 μL BRB80 buffer (final concentration 10 mg/mL). Aliquot and flash freeze. Store aliquots at –80 °C (*see* **Note 16**).

3. Cy3-labeled tubulin (1.6 μL aliquots of 30 μg tubulin, labeled/unlabeled ~ 1:2; each preparation will be slightly different - adjust accordingly).

4. Guanosine-5'-triphosphate (GTP; 25 mM, with equimolar $MgSO_4$).

5. Anhydrous DMSO.

6. Paclitaxel (Taxol) (2 mM): dissolve 1 vial of lyophilized powder (0.2 μmol/vial; Cytoskeleton, Inc.) in 100 μL anhydrous DMSO. Store at –20 °C. The working concentration is 10–20 μM.

7. Dithiothreitol (DTT) (1 M): dissolve DTT in ddH_2O and store at –20 °C. The working concentration is 1 mM.

8. Glycerol cushion (BRB80 with 60 % glycerol): 80 mM PIPES, 2 mM $MgCl_2$, 1 mM EGTA, and 60 % glycerol (v/v).

9. Beckman Coulter Optima™ TLX Ultracentrifuge (120,000 maximum rpm).

10. Beckman TLA-100 fixed angle rotor (7×20 mm, 0.2 mL tube).

2.6 β-Casein Preparation

1. Bovine β-casein, 1 g (Sigma, Cat. No. C6905), store at –20 °C upon arrival.

2. 50 mL of 20 mM Tris-HCl buffer, pH ~7, chilled.

3. 0.5 M NaOH for pH adjustment.

4. pH test strips (Fisher Scientific).

5. Ultracentrifuge with Beckman Ti 70.1 rotor and 10.4 mL centrifuge tubes (Beckman, Cat No. 355603).

6. 50 mL syringe with 0.22 μm filter.

7. Glass Pasteur pipette with long tip.

2.7 Sample Preparation: MT Attachment and Surface Passivation

1. BRB80 buffer (*see* Subheading 2.5).

2. 25 mg/mL bovine β-casein (*see* **Note 17** and Subheading 3.6).

3. 10 mM paclitaxel (Sigma) in DMSO.

4. 1 M DTT (*see* Subheading 2.5).

5. Glutaraldehyde-functionalized slide chamber.

6. Cy3-labeled MTs (*see* Subheading 3.5).

7. Pieces of filter paper cut into strips ~2 in. long and ~0.5 in. wide.

8. Vacuum grease (*see* **Note 18**) and cotton-tipped applicator.

3   Methods

3.1 Cover Slip Aminosilanization

This protocol describes how to amino-functionalize glass cover slips for subsequent treatment with glutaraldehyde. These cover slips are used as substrates for MT immobilization. The glass is first cleaned and activated (Fig. 2) and then treated with an

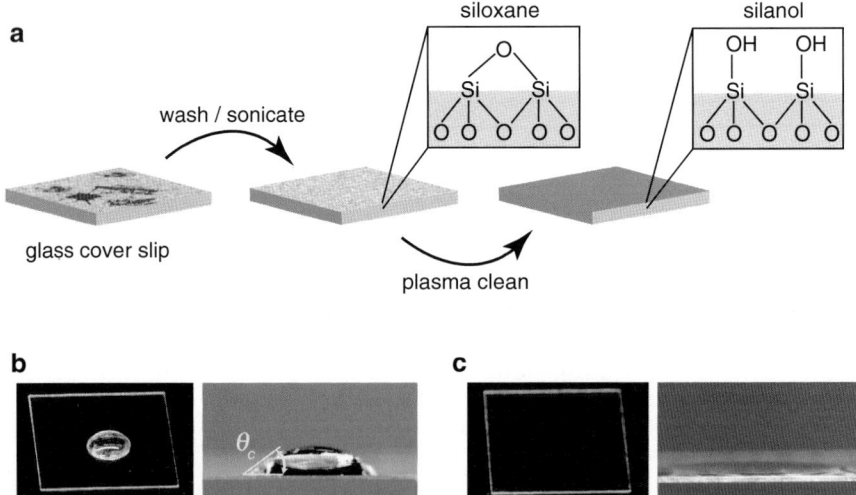

Fig. 2 Glass preparation for aminosilanization. (**a**) Borosilicate glass cover slips are sonicated in an alkaline (pH ~11.5) phosphate detergent to remove gross surface contaminants (and possibly a very thin surface layer of the silicon dioxide network that forms the glass structure). Next, plasma cleaning removes any residual organic contaminants and converts surface siloxanes to silanol groups that are more reactive with aminopropyltriethoxysilanes. (**b**) Prior to plasma cleaning, 20 μL of ddH$_2$O deposited on the cover slip surface forms a bead (*left*), which, when viewed from the side (*right*), forms a dome shape with a non-negligible contact angle, θ_c (even if one attempts to spread the drop over the surface). (**c**) Following plasma cleaning, 20 μL of ddH$_2$O flows evenly over the highly hydrophilic glass surface and does not form a bead (*left*). The contact angle is greatly reduced (and difficult to observe; *right*)

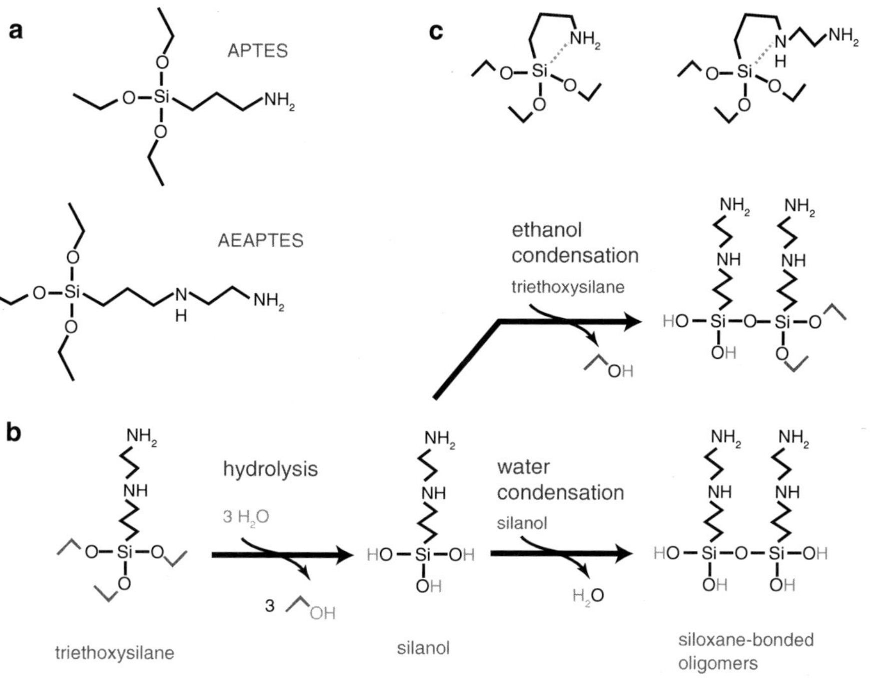

Fig. 3 Aminopropyltriethoxysilane structures and reaction scheme. (**a**) Chemical structures of 3-aminopropyltriethoxysilane (APTES) and *N*-(2-aminoethyl)-3-aminopropyltriethoxysilane (AEAPTES). (**b**) General reaction scheme for aminopropyltriethoxysilanes, shown for the specific case of AEAPTES. Exposure to water leads to hydrolysis of the ethoxy groups, yielding two alcohols, a silanol and an ethanol leaving group. Silanols can undergo condensation reactions either with each other (water condensation) or with unhydrolyzed silanes (ethanol condensation) to form siloxane bonds and yield oligomers. The water released by the condensation reaction can participate in hydrolysis of additional ethoxy groups. Thus, even small amounts of water can catalyze silane polymerization. Uncontrolled polymerization yields complex, disordered networks that form a viscous sol–gel [36]. Note that the reaction scheme shown here is a simplified, conceptual summary and that hydrolysis and condensation can potentially occur on different parts of each molecule simultaneously. Silane polymerization is catalyzed by the addition of ammonia (not shown). (**c**) Aminopropyltriethoxysilanes for five-membered ring structures (*left*: APTES, *right*: AEAPTES), allowing the amino group to intramolecularly catalyze silane polymerization, even in the absence of water

aminopropyltriethoxysilane. The relevant chemistry [21, 26, 29, 32–36] is presented in Fig. 3 and the aminosilane surface coating strategy is outlined in Fig. 4 (*see* **Note 19**):

1. Fill bath sonicator with water and set temperature to 45 °C.

2. Set oven to 110 °C.

3. Rinse out the *Mucasol* jar with ddH₂O and fill it nearly to the top with 2 % Mucasol.

4. Rinse off the cover slip holders and put in the cover slips. Wear gloves to avoid getting finger oils on the glass (handle only the edges) and use forceps to place the cover slips in the holders.

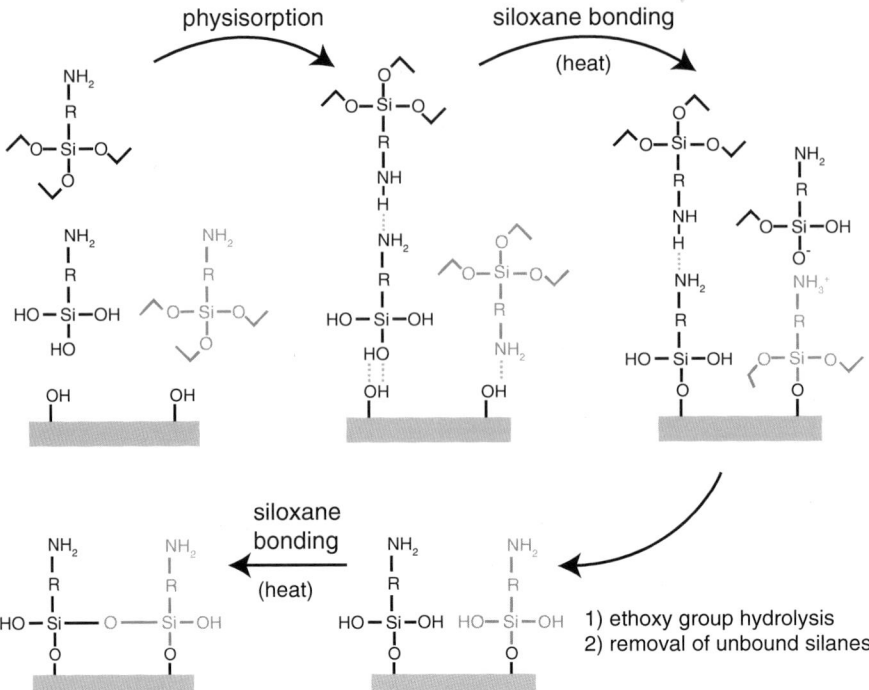

Fig. 4 Aminopropyltriethoxysilane functionalization of glass using APTES (R=−(CH₂)₃−) or AEAPTES (R=−(CH₂)₃−NH(CH₂)₂−). An appropriately cleaned glass surface containing a high density of silanol groups is exposed to the aminosilane in nearly anhydrous acetone, yielding a diverse mixture of free silanes and silanols in solution, which then physisorb onto the surface via hydrogen bonding and/or ionic interactions (several configurations in addition to the ones shown are possible). The addition of heat drives the formation of siloxane bonds between the physisorbed silanes/silanols and the glass surface by supplying energy and removing condensation products (water and ethanol) by evaporation. This reaction is catalyzed by the terminal amine group. Rinsing in ethanol and water removes any remaining physisorbed aminosilane deposits and leads to hydrolysis of remaining ethoxy groups on the bound silanes, converting them to silanols. A final heating/drying helps these silanol groups form intramolecular siloxane linkages (siloxane bonds may also form via reaction of adjacent ethoxy and silanol groups, as during the initial bonding to the surface). This conceptual scheme does not illustrate several concurrent pathways that also lead to stable binding of aminosilanes to the glass surface (e.g., aminosilane oligomerization in solution, followed by physisorption and binding to the surface). The final product of this treatment is a glass surface densely covered in covalently attached amines

From this point forward, avoid any direct contact with the cover slips, and handle only the holders. We typically prepare two racks (**24** cover slips) at a time.

5. Put cover slip holders in the *Mucasol* jar (one holder on top of the other), fill to the top with 2 % Mucasol, and cap the jar. *Optional*: For stability, insert a plastic "wedge" (a piece of a 15-mL conical tube works well) in the jar to prevent the cover slip holders from moving.

6. Submerge the body of the *Mucasol* jar in the bath sonicator (just up to the cap; *see* **Note 8**).

7. Run sonicator on degas setting for 5 min. Then sonicate for 25 min.

8. Rinse the *ddH₂O* jar well with ddH₂O and then fill it. Using forceps, transfer cover slip racks directly from the *Mucasol* jar to the *ddH₂O* jar. Rinse under running ddH₂O for about 3–5 min, gently swirling and periodically dumping the water from the jar. Minimize exposure to the air (*see* **Note 20**).

9. After rinsing, fill the *ddH₂O* jar and sonicate for 5 min. Rinse again briefly. *This is an acceptable place to stop and resume at a later time (leave the cover slips in water until proceeding).*

10. Set a paper towel on a flat, stable surface next to a clean piece of aluminum foil that is big enough to wrap around the cover slip holder. Remove one cover slip holder at a time from the jar and tap it on the paper towel a few times to remove water (firmly, but without knocking the cover slips out of their slots). There should be very little water left. Immediately place the holder on the aluminum foil and wrap the foil around it. The holder should be protected from dust, while still allowing some air to circulate. Keep the holder in the foil until **step 17**.

11. Put all racks in the oven until all cover slips are completely dry (~30 min). They should look completely clean by eye.

12. Remove racks and let them cool for a few minutes. Then place them in the plasma cleaner (still in the aluminum foil covers). Evacuate the chamber to ~800 μtorr and turn on the RF coil on "high" for 5 min (*see* **Note 21**). Then turn off the cleaner, return the chamber to atmospheric pressure, and remove the cover slips.

13. *Optional*: test one cover slip by placing a drop of approximately 20 μL of ddH₂O on the center of the glass surface. It should spread evenly over the glass without beading at all, indicating the highly hydrophilic nature of the plasma-treated surface (Fig. 2c).

14. Leave the racks (still covered) on the bench for 15 min (*see* **Note 22**).

Perform steps 15–22 *in the fume hood*:

15. Fill the *Acetone* jar with 125 mL acetone and the *Silane* jar with 100 mL acetone. Measure an additional 125 mL in a graduated cylinder for use in **step 18**.

16. Add 2 mL aminosilane (APTES or AEAPTES) to make a 2 % v/v solution (~80 mM). Gently swirl the liquid in the jar to mix well (*see* **Note 23** *regarding safety!*).

17. Using a metal forceps, dip each cover slip rack in the *Acetone* jar several times. Immediately dip each rack in the *Silane* jar for 10 s, gently agitating the jar by hand (*see* **Note 24**). After treating the first rack, immediately proceed to **step 18** before repeating **steps 17** and **18** for the second rack.

18. Remove the cover slip rack and again dip repeatedly in the Acetone jar to remove excess, weakly physisorbed aminosilane.

19. Dump the contents of the *Silane* jar into an appropriate waste container. Transfer the contents of the *Acetone* jar to the *Silane* jar, and fill the *Acetone* jar with the remaining 125 mL of acetone in the graduated cylinder. Dip the cover slip racks repeatedly in the *Acetone* jar.

20. Tap each cover slip rack dry on a paper towel, cover in aluminum foil, and bake for 1 h at 110 °C.

21. Dispose of the contents of the *Silane* jar. Swirl the acetone in the *Acetone* jar to rinse the walls thoroughly, and then transfer it to the *Silane* jar. Rinse the walls well. Rinse forceps in the acetone briefly to remove aminosilane, and dispose of the acetone (*see* **Note 25**).

22. Remove cover slip racks from the oven and let them cool for a few minutes. Then place them in the *Ethanol* jar and fill it with spectrophotometric ethanol (ethanol can be reused over several preparations). Sonicate for 10 min (*see* **Note 26**).

23. Transfer the cover slip racks to the ddH$_2$O jar and rinse well in ddH$_2$O. Sonicate 5 min in ddH$_2$O and rinse again briefly.

24. Repeat **steps 10** and **11**.

25. Store the aminosilanized cover slips in the cover slip holders in a vacuum desiccator containing Drierite (anhydrous CaSO$_4$) (*see* **Note 27**).

3.2 Microscope Slide Chamber Construction

This protocol describes constructing a microscope slide chamber for use in single-molecule microscopy. The chamber consists of an aminosilanized cover slip fixed to a standard microscope slide by two strips of Parafilm. The chamber is subsequently treated with glutaraldehyde to facilitate the covalent binding of MTs to the cover slip. After flowing reagents into the chamber, it is sealed using vacuum grease:

1. *Optional*: clean microscope slides in 2 % Mucasol, as done for cover slips prior to silanization.

2. Cut a sheet of Parafilm into strips of ~5 mm in width (*see* **Note 28**).

3. Cut two pieces from the Parafilm strips of ~1 inch (2.5 cm) in length.

4. Wear gloves. Place a slide on top of a photocopy of the template provided in Fig. 5i. Hold it firmly in position with one hand. With the other hand, put each piece of Parafilm (film side down, so it sticks) carefully along the edge of the gray rectangle that demarcates the chamber (Fig. 5a). Once the pieces are in the proper position, push down lightly in a few spots with a finger to help them stick.

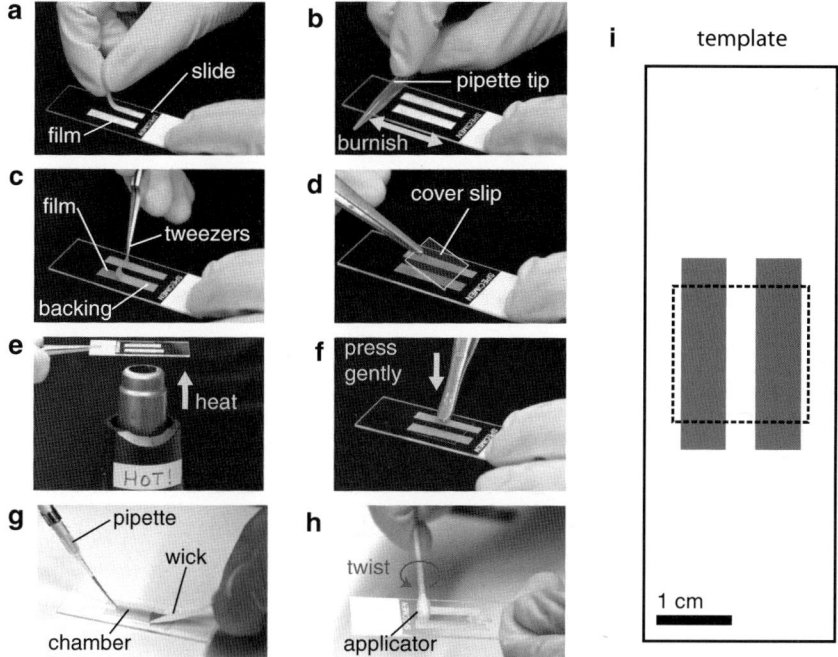

Fig. 5 Slide chamber preparation. (**a**) Parafilm is applied to a glass slide and (**b**) burnished with a clean pipette tip so it adheres to the glass. (**c**) The paper backing is removed with a pair of sharp-nosed tweezers (paper backing is pseudo-colored pink in this image to enhance visibility). (**d**) An aminosilanized glass cover slip is carefully placed on top of the exposed film strips (the cover slip is pseudo-colored pink and outlined in *yellow* in this image to enhance visibility) and lightly pressed down (not shown), and (**e**) the slide chamber is heated briefly (cover-slip-side-up) until the Parafilm becomes transparent. (**f**) The cover slip is then gently pressed with a tweezers in order to form a tight bond with the parafilm. After treating the chamber with glutaraldehyde (not shown), (**g**) microtubule solution and then trapping assay solution are introduced from one end of the chamber while simultaneously using a filter paper "wick" on the opposite end to help draw the solution through (in the photograph, a blue dye is used instead of trapping solution in order to enhance visibility). (**h**) The chamber is then sealed using a cotton-tipped applicator saturated in vacuum grease. By twisting the applicator at the entrance of the chamber, grease is swept into the mouth of the chamber, forming a perfect seal. (**i**) Using a template during steps (**a**–**d**) helps ensure consistent chamber volumes and cover slip placement. The template provided has a chamber width of 4 mm, yielding a volume of ~10 µL (Color figure online)

5. Use a small cylinder (e.g., a 1 mL plastic pipette tip) to burnish the film onto the glass (Fig. 5b). Rub with even pressure across both strips several times to make good adhesion (*see* **Note 29**).

6. Use the pincers on a pair of sharp-nosed forceps to "pick" at the paper backing on one of the strips of Parafilm until it begins to separate. Once the backing is separated from the Parafilm,

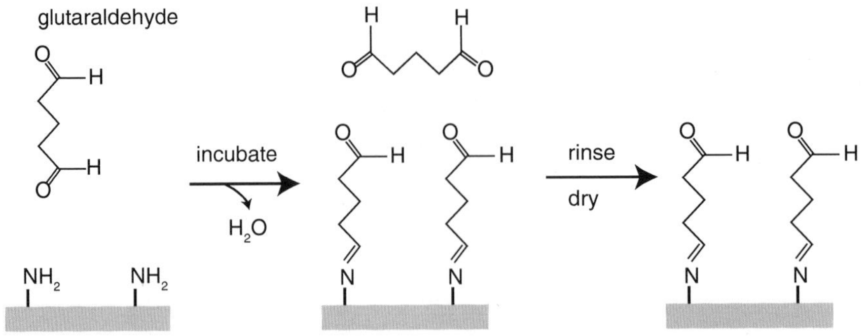

Fig. 6 Glutaraldehyde treatment of amino-functionalized glass. The bifunctional glutaraldehyde binds surface amine groups via a hydrolysis reaction, thus functionalizing the surface with formyl groups that will bind surface-exposed lysines on proteins. Extensive rinsing removes any free glutaraldehyde that could yield unwanted reactions in the final assay. Finally, the glass is thoroughly dried and stored under vacuum

use the tweezers to pull it off completely (Fig. 5c; *see* **Note 30**). Remove the backing from both pieces of Parafilm.

7. Position the slide over the template again. Use a forceps to pick up an aminosilanized cover slip by the corner (so as not to contaminate the "useful" area that will form the surface of the chamber). Carefully place it in position on the template (Fig. 5d, *see* **Note 31**).

8. Lightly tap the cover slip with the back of the tweezers over the part touching the Parafilm (this is just to make it stick weakly so it does not fall off).

9. Pick up the whole slide chamber with the tweezers, cover-slip-side-up. With the heat gun on the table pointing upward, turn it on "high." Hold the center of the slide over the heat gun, about ½ in. away (Fig. 5e). As the Parafilm heats, it becomes transparent. Wait ~1 s after this happens and remove the slide quickly.

10. Hold the slide chamber firmly with forceps (or set on a clean surface). Then quickly use the back of the tweezers to *lightly* press or tap on the cover slip regions above the Parafilm (Fig. 5f; this forms a tight seal between the two pieces of glass). Set the slide chamber aside to cool (Parafilm will turn opaque).

11. Store the chamber in the vacuum desiccator for later use, or proceed immediately to glutaraldehyde treatment (*see* **Note 32**).

3.3 Slide Chamber Glutaraldehyde Treatment

Here we provide a protocol for activating the aminosilanized surface of the slide chamber with glutaraldehyde (Fig. 6), to enable subsequent covalent attachment of MTs (*see* **Note 33**). This is a widely used method for protein immobilization [30], and similar methods have been used by others for optical trapping assays [31]. We generally prepare several slides at once and store them dry for later use (similar to other reports [37, 38]), but this procedure may also be done immediately prior to an experiment:

1. Prepare a "humidity chamber" by placing a damp Kimwipe in a tight-sealing box large enough to hold several slide chambers. An empty pipette tip box works well (*see* **Note 34**).

2. Remove the 25 % glutaraldehyde from the freezer (one 10 μL aliquot per 3 slide chambers to be treated). Centrifuge briefly and then dilute 1:2 in chilled, ultrapure ddH$_2$O (e.g., 10 μL glutaraldehyde + 20 μL ddH$_2$O) to yield ~8 % glutaraldehyde (*see* **Note 35**).

3. Set each slide chamber to be treated cover-slip-side-up in the humidity chamber.

4. Fill each slide chamber with glutaraldehyde solution using a pipette. Pipette the solution into the mouth of the chamber, and it will be drawn in via capillary action (*see* Fig. 5g; no filter paper is needed for this step). Use approximately 10 μL per chamber, adding solution just to the point that ~1 μL starts to pool at the outlet of the chamber.

5. Cover the humidity chamber and incubate at room temperature for 30 min.

6. After the incubation, use a folded Kimwipe to wick the solution out of each chamber (it helps to hold the slide upright, drawing the solution out of the bottom end). Remove all of the solution (*see* **Note 36**).

7. To remove residual unreacted glutaraldehyde, rinse each chamber three times with 200 μL of ultrapure ddH$_2$O, wicking it through with Kimwipes (*see* **Note 37**).

8. Dry each chamber with filtered, compressed air. Use a pipette tip on the end of the air line to direct air through the chamber and hold a Kimwipe on the opposite end to collect the water. Blow air only in the direction of solution flow during **steps 6** and **7**. When completely dry, store the chambers in the vacuum desiccator.

3.4 Labeling of Tubulin with Cy3 Dye

This protocol explains how to label purified tubulin with the cyanine dye Cy3 (Fig. 7). This enables visualization of fluorescent MTs in the microscope using 532 nm laser excitation. The protocol is modified from the one given by Peloquin et al. [39]:

1. Add 20 μL cold PM buffer to 1 mg bovine brain tubulin (1 vial of Cytoskeleton tubulin). Dissolve and mix well; keep the tube on ice.

2. Add 1 μL of 25 mM Mg-GTP and 2.5 μL DMSO to the tubulin; incubate at 37 °C for 15 min.

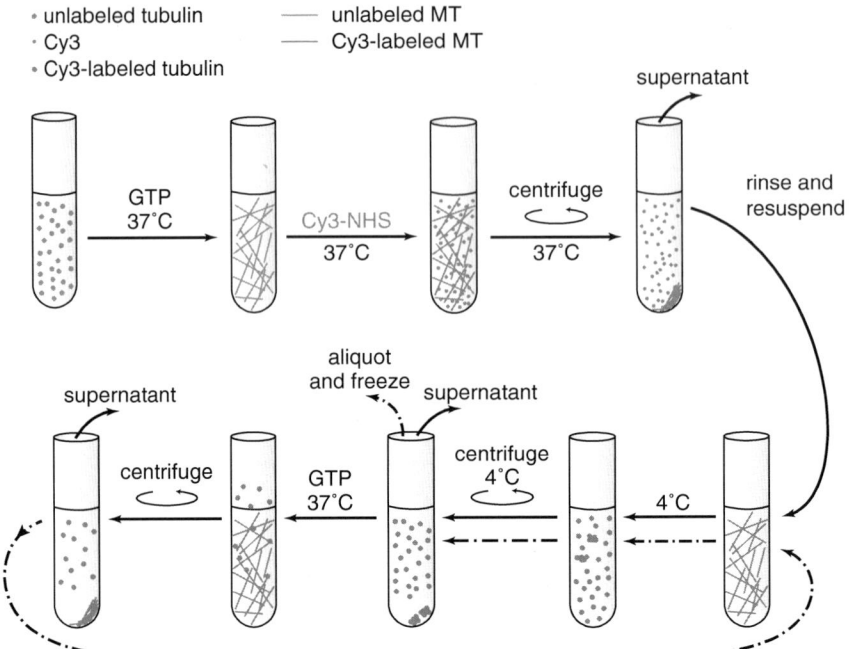

Fig. 7 Labeling of tubulin with Cy3 dye. Tubulin is first polymerized into MTs. Cy3 conjugated to the reactive *N*-hydroxysuccinimide (NHS) is then added, which facilitates attachment of the dye to amine groups on the surface of the MT. After pelleting the MTs and removing excess dye, the MTs are resuspended and depolymerized in the cold. The insoluble fraction is then pelleted and the supernatant transferred to a new tube in which the tubulin is repolymerized, pelleted, and depolymerized again, followed by aliquotting and snap freezing. This strategy ensures that the tubulin is labeled in regions other than the key interfaces required for polymerization and that the final product contains only tubulin capable of cyclic polymerization and depolymerization (and thus unperturbed by the attached dye)

3. Mix 2 μL Cy3 monofunctional NHS ester with the polymerized tubulin and incubate at 37 °C for 40 min in the dark (*see* **Note 38**).

4. Transfer the mixture to a clean 0.2 mL TLA-100 tube. Remove the excess dye by centrifugation for 5 min at 37 °C at 37,000 rpm (53,000 rcf, average; *k*-factor 48.1).

5. Carefully remove the supernatant, then add 40 μL cold PM buffer to resuspend the pellet. Incubate on ice for 15 min to depolymerize the MTs. Cool rotor to 4 °C.

6. Spin the solution for 5 min at 4 °C at 37,000 rpm (53,000 rcf, average; *k*-factor 48.1) to remove insoluble fraction.

7. Move supernatant to a clean new tube, add 2 μL of 25 mM Mg-GTP and 5 μL DMSO, and then incubate at 37 °C for 15 min in the dark. Warm rotor to 37 °C.

8. Spin the solution for 5 min at 37 °C at 37,000 rpm (53,000 rcf, average; *k*-factor 48.1).

a

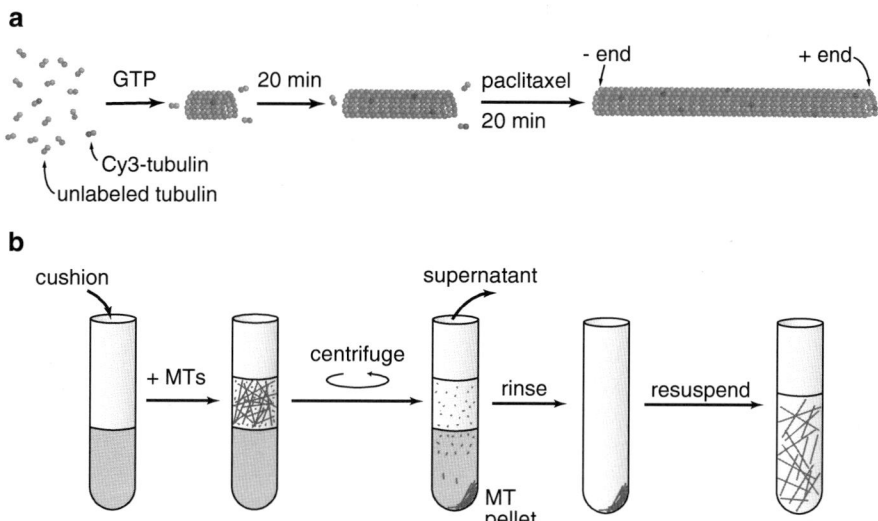

b

Fig. 8 MT preparation. (a) Addition of GTP to free tubulin (αβ-tubulin dimers) induces MT polymerization (promoted by addition of glycerol [40, 41] and 37 °C temperature). When preparing fluorescent MTs, the small amount (<5 %) of tubulin labeled with the organic fluorophore Cy3 incorporates randomly into the MT lattice and is distributed sparsely enough that the dye molecules do not affect motor interaction with the MTs in the optical trapping assay. Initially, MTs exhibit dynamic instability [42, 43]. Paclitaxel greatly enhances polymerization and stabilizes the MTs [44–49] (the DMSO in which paclitaxel is dissolved also enhances MT polymerization [50]). (b) Removal of residual-free tubulin and very short MT fragments is accomplished by sedimentation through a 60 %-glycerol "cushion." First, MTs are layered carefully on top of the cushion. Following centrifugation, the free tubulin and very short MTs remaining in the supernatant are removed. The MT pellet is then resuspended in buffer. These MTs are stable for days to weeks at room temperature

9. Discard supernatant, add 25 μL cold PM buffer, then incubate on ice for 15 min. Cool rotor to 4 °C.

10. Spin the solution at 37,000 rpm (53,000 rcf, average; k-factor 48.1) for 5 min at 4 °C.

11. Move supernatant to a clean 1.5 mL microcentrifuge tube. Use Nanodrop or other microliter UV spectrometer to determine the concentration of both tubulin and Cy3 dye, and calculate the labeling ratio (equivalent to labeled/unlabeled tubulin). Aliquot to appropriate amount, flash freeze, and store at −80 °C.

3.5 Cy3-Labeled MT Polymerization

This procedure yields fluorescently labeled MTs suitable for optical trapping and other single-molecule assays (Fig. 8).

Skip steps 1–4 if 20 μL aliquots of 10 mg/mL unlabeled tubulin is already prepared.

1. Set ultracentrifuge to 4 °C with TLA-100 rotor inside.

2. Spin a vial of lyophilized, unlabeled tubulin (1 mg) briefly in tabletop centrifuge at maximum speed. Put the vial on ice, and

add 100 μL BRB80 (to make ~10 mg/mL tubulin solution). Pipette gently to dissolve, avoiding formation of air bubbles.

3. Transfer the tubulin solution to a Beckman centrifuge tube (343775 or 342303), and centrifuge at 80,000 rpm (250,000 rcf, average; k-factor 10.3) for 10 min. Transfer the supernatant to a 0.5 mL microcentrifuge tube on ice, leaving any pellet (precipitated and/or polymerized tubulin) behind.

4. Make 20 μL aliquots, flash freeze in liquid nitrogen, and store at –80 °C.

Skip step 5 *if 5 μL aliquots of 10 mg/mL (labeled/unlabeled ≈ 1:22) tubulin is available.*

5. Remove 1 aliquot each of Cy3-labeled tubulin and unlabeled tubulin from the freezer. Thaw quickly in the hand and immediately place on ice. Combine the labeled and unlabeled tubulin (this gives approximately 10 mg/mL total tubulin, labeled/unlabled ≈ 1:22; *see* **Note 39**). Make 5 μL aliquots, flash freeze, and store at –80 °C.

6. Thaw a 5 μL tubulin aliquot from **step 5**.

7. Dilute 1 μL of 1 M DTT in 49 μL ddH$_2$O, to yield 20 mM DTT.

8. Prepare polymerization buffer: 2 μL glycerol cushion (*see* **Note 40**), 1 μL of 25 mM Mg-GTP, 6.5 μL BRB80, 0.5 μL of 20 mM DTT (from **step 7**). Add 5 μL polymerization buffer to the 5 μL tubulin aliquot. Mix gently with the pipette (avoid bubbles).

9. Incubate the tubulin solution for 20 min in a 37 °C water bath to start polymerizing the tubulin into MTs. Protect from light to avoid photobleaching of the fluorescent dye.

10. Dilute 1 μL of 2 mM paclitaxel in 9 μL DMSO (200 μM paclitaxel). Add 1.1 μL of this to the MTs (20 μM paclitaxel final). Mix gently.

11. Incubate MTs at 37 °C for an additional 20 min (*see* **Note 41**). During the incubation, set the ultracentrifuge to 25 °C (*see* **Note 42**) with the TLA-100 rotor inside.

12. In a Beckman centrifuge tube (343775 or 342303), add 60 μL of glycerol cushion and 0.7 μL of 2 mM paclitaxel and mix (for a balance tube, use 60 μL glycerol cushion and 10 μL H$_2$O).

13. Carefully layer the 10 μL of MTs onto the glycerol cushion (do not disturb the cushion with the pipette tip). Mark the outside edge of the tube with a permanent marker to help find the pellet position after the centrifugation.

14. Centrifuge at 80,000 rpm (250,000 rcf, average; k-factor 10.3) for 10 min. During the centrifugation, prepare the wash buffer: 150 μL BRB80 plus 1.52 μL of 2 mM paclitaxel.

15. Following the centrifugation, the pellet, containing the MTs, will be toward the outside wall of the tube (the side marked in **step 13**) and should have a slightly pink color. Remove ~50 μL of supernatant/cushion, keeping the pipette at the top of the liquid and moving down as it is withdrawn (*see* **Note 43**).

16. Rinse the walls of the tube above the remaining cushion with 50 μL of wash buffer, and carefully remove it. Then remove the remaining cushion, pipetting from the side of the tube opposite the pellet.

17. Gently rinse the pellet with an additional 50 μL of wash buffer, and carefully remove it without disturbing the pellet.

18. Resuspend the pellet in 10 μL of wash buffer (~5 mg/mL tubulin; *see* **Note 44**). The final mixture will be fairly viscous. Transfer it to a small microcentrifuge tube shielded from the light (use a black tube or cover it in aluminum foil). These MTs are stored at room temperature and can be used in optical trapping assays for a week or more.

3.6 β-Casein Preparation

This procedure yields 25 mg/mL β-casein for use in passivating the aminosilanized cover slip surface:

1. Remove the β-casein bottle from the freezer and place it on ice.

2. Add 25 mL of Tris-HCl buffer to bottle. Stir gently with a serologic pipette to dissolve as much as possible. Avoid making bubbles.

3. Use ~15 μL of solution to check the pH on a pH strip. Add NaOH dropwise (~18 drops) until the pH reaches ~8 (*see* **Note 45**).

4. Add another 5 mL of Tris buffer. As solution is added, rinse off the pipette from **step 1** to prevent loss of protein. Swirl to dissolve and readjust pH to ~8 with NaOH (~12 drops).

5. Place the bottle on a rocker in a 4 °C cold room for 1 h. Readjust pH to 8 with NaOH (~4 drops).

6. Return bottle to rocker in cold room for 2 h more. Readjust pH to 8.

7. Gently swirl bottle and rotate to get any residue off the walls and into solution. Let the bottle sit upright for 5 min and wash the walls with 5 mL Tris buffer.

8. Using a serologic pipette, transfer the entire protein solution from the bottle to a 50 mL conical tube. Wash the walls of the bottle with 3 mL of Tris buffer and add it to the conical tube. Then add Tris buffer to bring total volume in tube to 40 mL.

9. Centrifuge the solution in the ultracentrifuge at 65,000 rpm (388,000 rcf, average; *k*-factor 42.3) for 20 min at 4 °C.

10. Carefully remove the tubes from the centrifuge. There will be a faint cloudy layer at the top and a very small gray pellet at the

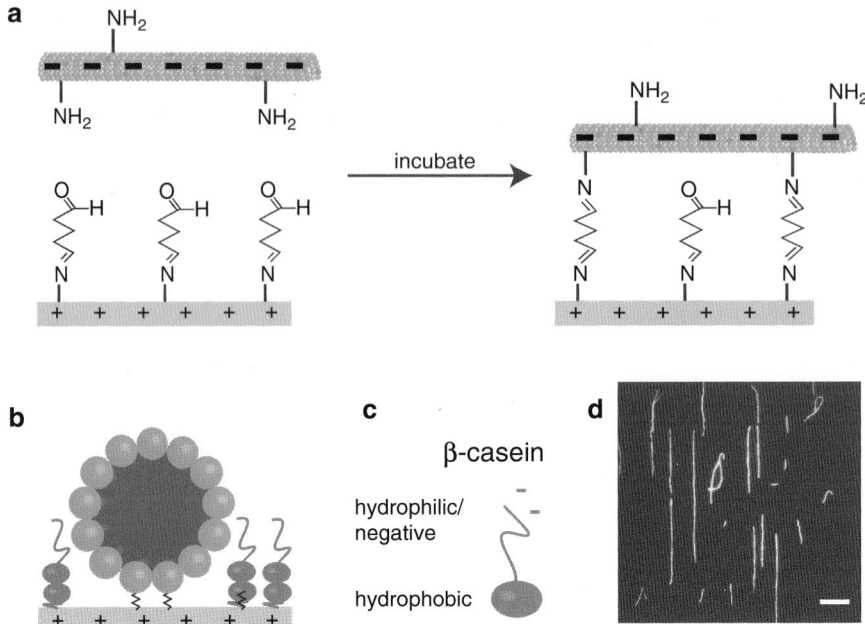

Fig. 9 MT immobilization in aminosilane-/glutaraldehyde-treated slide chambers. (**a**) The negatively charged MT is attracted to the positively charged glass surface. Following incubation, reactive formyl groups attached to the glass bind to amines on surface-exposed lysines on the MT, thus covalently linking the MT to the cover slip. (**b**) End-on view of MT attached to glass, with remaining surface passivated with β-casein (introduced to the chamber at 2 mg/mL). The β-casein, an amphiphile capable of adsorption on a variety of surfaces, forms a bilayer on the glass surface that prevents unwanted interactions with trapping microspheres, and its lysine residues react with any remaining formyl groups. (**c**) The β-casein N-terminal hydrophilic region is negatively charged, allowing favorable interactions with the positively charged glass surface. The top layer extends this hydrophilic region into the solution, forming a "brush" layer on the surface that prevents trapping beads from sticking. (**d**) Fluorescence image of Cy3-labeled MTs covalently bound to the cover slip in a microscope flow chamber. They are very well aligned with the long axis of the chamber due to combination of the laminar flow induced when filling the chamber and the highly favorable initial adhesion to the positively charged glass. Scale bar: 10 μm

bottom. Use the Pasteur pipette to withdraw the middle, clear layer to a clean 50 mL conical tube on ice. This solution should look completely clear.

11. Sterile filter the solution to a clean conical tube on ice. Aliquot, snap freeze, and store at –80 °C (large aliquots can be stored for subsequent thawing and realiquotting).

3.7 Sample Preparation: MT Attachment and Surface Passivation

The procedure below describes how to attach MTs to the glass cover slip and block the remaining surface to prevent unwanted adsorption of microspheres during subsequent optical trapping experiments (Fig. 9). At the near-neutral pH of most assay buffers, MTs have negative surface charge (especially at the C-terminal, glutamate-rich tubulin "E-hook" with which kinesin interacts) [51–54], whereas aminosilanized surfaces ($pK_a = 7$–10 [55, 56]) are protonated and thus positively charged. The MTs are thus rapidly adsorbed to the glass, whereupon they covalently bind the formyl groups left from previous glutaraldehyde treatment. During the MT incubation, reagents for the desired assay should be prepared (e.g., molecular motors bound to optical trapping beads in the presence of ATP; *see* Chapter 10). Following blocking with β-casein, the reagents are introduced to the flow chamber, the chamber ends are sealed, and the sample is taken to the microscope for the experiment.

Prepare the following freshly at the beginning of a set of experiments:

1. "BRB/Tx": 350 µL of BRB80, 0.5 µL of 10 mM paclitaxel. Keep at room temperature.

2. Blocker: 156 µL of BRB/Tx, 14 µL of 25 mg/mL β-casein. Keep at room temperature.

3. "MT30": 0.5 µL MT stock (10 mg/mL), 14.5 µL BRB/Tx (i.e., 30× dilution). Keep at room temperature and protect from the light (can be used for several days).

Do the following for each experiment:

4. Mix the MT30 suspension gently to evenly distribute the MTs. Add 0.5–0.8 µL MT30 (depending on the desired MT density on the cover slip) to 10 µL BRB/Tx. Flow this into a glutaraldehyde-treated flow chamber, using a piece of filter paper (*see* Fig. 5g) waiting on the opposite end in order to get very good flow through the chamber (this aligns the MTs with the chamber's long axis) (*see* **Note 36** regarding the direction of flow).

5. Incubate the MTs in the chamber for 20–30 s and flush the chamber with 40 µL of BRB/Tx. Incubate 20 min or longer (*see* **Notes 46** and **47**).

6. During the MT incubation, prepare the reagents for the main assay (*see*, e.g., Chapter 10 for a protocol to measure kinesin force generation), timed so that all incubations end at approximately the same time.

7. About 1 min before the end of the incubations, flush the flow chamber with 40 µL of Blocker and leave it to incubate.

8. Flow the assay solution into the flow chamber. Dry the ends with a Kimwipe, wiping away from the center of the chamber and taking care not to suck solution out of the chamber itself. Seal the chamber with vacuum grease as shown in Fig. 5h (*see* **Note 48**). Avoid getting any grease on the surface of the cover slip that will contact the objective. When finished, wipe away any excess grease (*see* **Note 49**).

4   Notes

1. An additional technique [57], reported to be very effective at attaching MTs while preventing nonspecific absorption, is coating the glass surface with a highly hydrophobic silane. Anti-tubulin antibodies are then adsorbed to the glass, and the remaining surface blocked with Pluronic® F-127 (Sigma), a poloxamer (triblock copolymer) of polypropylene glycol (which is hydrophobic and sticks to the hydrophobic surface) flanked on either side by polyethylene glycol (which is hydrophilic and forms a protective "brush" layer in the solvent above the glass) [57]. F-127 has been used to minimize bead-surface interactions in optical trapping experiments in the absence of protein [58, 59], but we have not tested this method.

2. Zeiss cover slips are manufactured by Schott/Marienfeld Superior (Lauda-Königshofen, Germany). While other cover slips will probably work acceptably, we use these because of the high quality of the glass and more importantly because of their highly precise thickness of 170 ± 5 μm. Individual No. 1.5 cover slips within typical lots from most manufacturers can vary in thickness from 160 to 190 μm, potentially leading to a significant deviation from the 170 μm cover slip thickness for which most oil-immersion microscope objectives are designed. This can induce unwanted spherical aberration that broadens the width of the focused laser beams (e.g., optical trapping beams) in all dimensions, as well as degrading imaging quality in fluorescence microscopy. This problem can be avoided by manually measuring each cover slip with a micrometer before use (and discarding those that are too thin or thick) or by purchasing cover slips like the ones recommended, for which there is very little variability in thickness.

3. We find porcelain cover slip holders the most convenient because they fit nicely in our jars and tend to sit very stably on the bench. However, they are expensive. Cheaper racks made of other materials may be substituted as long as they are resistant to 100 % acetone and ethanol and can withstand temperatures up to 110 °C. Polytetrafluoroethylene (Teflon®) and polypropylene (e.g., Wash-N-Dry™ polypropylene racks) are acceptable alternatives.

4. Many water purifiers have a low flow rate, making the rinsing steps in the cover slip preparation protocol more time consuming and less effective. If the flow rate is less than ~10 mL/s, we advise filling a *clean* polypropylene carboy with ddH$_2$O and using the spigot on the carboy to dispense the water.

5. We have had excellent results with both APTES and AEAPTES, but we prefer AEAPTES due to its increased stability against

self-catalyzed hydrolysis by its terminal amine [26]. It is critical to avoid contact with humidity during storage, to prevent polymerization of the aminosilane (even small amounts of water can catalyze extensive polymerization over time). Preferably, the reagents should be stored under nitrogen or argon in a bottle sealed with a rubber septum (e.g., Sigma Cat. No. 440140), so that no direct contact between the stock solution and room air occurs. Alternatively, we aliquot the aminosilanes into 2 mL volumes, flash freeze in liquid nitrogen, and store in an airtight plastic bag with Drierite desiccant (anhydrous $CaSO_4$) at −80 °C until use. If this method is used, the aliquot should be brought to room temperature in a desiccator, to prevent water condensation. Some authors recommend distilling APTES regularly to remove polymers [60]. However, we have obtained satisfactory results with aminosilanes properly stored for several months. *See also* **Note 24**.

6. We use very high (>99 %) purity reagents for cover slip cleaning and surface preparation, in order to avoid contaminating the glass surface with undesired chemicals (especially organic molecules, which may interfere with surface chemistry by binding the substrate and/or lead to background fluorescence during microscopy).

7. Any non-glass jars that are resistant to the chemicals used can be substituted (e.g., other jars composed of PMP). We recommend using a separate jar for each reagent (water, Mucasol solution, ethanol, acetone, and silane solution) to minimize cross-contamination and any potential unwanted reactions that might result. Keep jars tightly closed when not in use.

8. To suspend the jars in the sonicator bath, we use a simple homemade holder fashioned from a polystyrene box lid. Use a scalpel to cut a recess in the lid and a hole through which the jar body (but not the lid) can pass. Set this holder on top of the sonicator, and insert the jar, such that as much of the jar as possible is submerged in the bath without making physical contact with any of the walls of the sonicator.

9. The cover slips are baked at approximately 110 °C a few different times during preparation. If an oven is unavailable, we have found that a hot plate or other heater can be used as an "oven" by simply placing an aluminum foil "tent" overtop of it. Simply adjust the heating element such that the air temperature inside the tent is approximately 110 °C. For whatever baking method is used, it is important that the inside of the oven be clean. Baking cover slips in an oven with dust, or other contaminants that get vaporized during heating, can foul the glass surface and thereby ruin the effects of the prior cleaning steps.

10. We prefer to use Parafilm in place of the double-sided tape often employed in constructing microscope slide chambers

because it attracts fewer contaminants (e.g., dust), it is easier to place on the slide, and it allows slides to be prepared in large numbers for later use (versus having to apply the tape immediately before placing the cover slip). In addition, we have observed glue dissolving along the edges of the chamber over time in flow chambers made with tape (visible as whitish streaks along the chamber edges). This is in agreement with work by Schäffer et al. [58], who reported direct evidence of solution contamination in flow chambers constructed with tape.

11. Glutaraldehyde can polymerize extensively in aqueous solution (*see* [27, 28] for a comprehensive review). We store it at −20 °C to limit this behavior.

12. The compressed air typically available at the lab bench is generally not suitable for drying chambers, as it may contain abundant contaminants (e.g., particulate matter, water vapor, and oil droplets from air pumps). Instead, highly pure, compressed nitrogen gas is generally used, with an oil-free filter system at the output of the air line. However, in practice, we have found that the bench air in our lab can be used by employing the following system: A ~1 m hose filled with Drierite is attached to the air line, with two 0.2 μm syringe filters in series at the output. This provides air of sufficient dryness and absence of contaminants for preparing chambers for routine optical trapping studies.

13. Labeling tubulin with Cy3 is not absolutely necessary, and Cy3-labeled tubulin may be substituted with another fluorescently labeled tubulin, e.g., carboxytetramethylrhodamine (TAMRA)-labeled tubulin (Cytoskeleton, Inc.). However, Cy3 is brighter and more photostable than rhodamine derivatives, allowing a lower labeled/unlabeled tubulin ratio to be used while still easily visualizing MTs. Finally, whereas rhodamine MTs suffer light-induced structural damage resulting from the formation of reactive oxygen species (ROS) [61, 62], we have found Cy3 MTs to be less affected (we also use an oxygen scavenging system to eliminate ROS in either case).

14. Tubulin may be purified in the laboratory relatively easily, which is generally far more economical. Traditionally, the approach is to purify tubulin by cycles of polymerization, sedimentation, and depolymerization [63, 64] (followed by removal of MAPs with a phosphocellulose column), and abundant protocols are available both via laboratory websites and the literature [65–67]. Recently, a one-step, high-efficiency purification method was developed [68] that may also prove useful (and may permit the isolation of tubulin from a wider variety of sources).

15. DMSO is preferred over ddH$_2$O as solvent: NHS ester has a fast hydrolysis rate in water, especially at high pH (after hydrolysis, it can no longer react with NH$_2$ groups of amino acids); moreover, the dye is more soluble in DMSO.

16. The vial should always be kept on ice, since the tubulin will polymerize at higher temperature.

17. Caseins (chemical properties reviewed in refs. [69, 70]), the major proteins in milk [71], are highly amphipathic, giving them surfactant-like properties [72, 73]. As reported previously [20], casein is very effective in passivating an aminosilanized surface to prevent protein binding. In our hands, 1 mg/mL β-casein was superior to BSA (even up to 10 mg/mL) and Pluronic F127 in blocking the aminosilanated cover slip surface and preventing trapping microspheres from binding. Free microspheres do not bind to the cover slip surface, and there are no measureable binding forces when a trapped bead is dragged along the surface. We use β-casein rather than whole casein because it is available in higher purity and we found it dissolves much better. Moreover, it is resistant to denaturation/ precipitation induced by DTT (β-casein contains no cysteine residues [71]) or elevated temperatures [74], and forms stable micelles even at low pH [75]. Although it is a natively unstructured protein [76, 77], β-casein nevertheless has a domain structure, with a highly negatively charged N-terminal region and a hydrophobic C-terminus. This allows it to interact favorably with both hydrophilic and hydrophobic surfaces [72, 76, 78, 79], forming a "brush" layer on the surface with the hydrophilic N-termini pointing up into the solvent (Fig. 9a, b). These properties make it an excellent choice for blocking APTES surfaces, since different parts of the protein can interact favorably with both the positively charged amine and the aliphatic chain of APTES.

18. Slide chamber ends should be sealed to prevent solution evaporation (which can markedly change concentrations), limit exposure to oxygen (which can produce ROS, especially in the presence of microsphere and fluorescent dye interactions with laser light), and provide physical closure of the chamber. Vacuum grease is highly preferred to nail polish for this purpose, since it can be applied more precisely (thus limiting the possibility for damage to valuable microscope optics by accidental contact) and because it is immiscible in aqueous solution (as opposed to the acetone solvent used in nail polish). When using nail polish as sealant, Schäffer et al. [58] reported contamination of the chamber solution, and we observe regions of casein precipitation near the chamber edges that grow toward the center over minutes (presumably with similar effects on other proteins). This is accompanied by nonspecific binding of trapping microspheres to the cover slip and MTs. These problems are not encountered when using vacuum grease.

19. The optimal conditions (e.g., temperature, solvent vs. vapor deposition, solvent type, silane concentration, etc.) for silane

deposition are still a topic of debate and ongoing research [25, 26]. Based on our review of the literature, and our desire to keep our protocol as simple as possible, we use a solution-phase deposition in acetone solvent. Although aqueous deposition protocols are available, we chose to avoid this method because silanes can polymerize extensively when exposed to water. In addition, given the pK_a of surface silanol groups (pK_a ~4 or ~9, depending on the silanol configuration) on glass [80–82] and amino groups on aminosilanes (pK_a~10) [55, 83], in aqueous solution near neutral pH, the surface silanols carry a negative charge, while the aminosilane carries a positive charge [24, 32, 60, 83]. This causes the silane to physisorb with its amino group toward the glass, which prevents the associated silanol groups on the glass from participating in condensation reactions with APTES silanol groups (though possibly limiting unwanted silane polymerization on the glass surface [24]).

20. The purpose of the cleaning steps is to remove gross surface contaminants (e.g., dust, oil, and glass particles remaining from the manufacturing process), leaving the glass surface as clean as possible prior to the final plasma cleaning. During the rinsing step, the main goal is to remove all traces of Mucasol, while preventing exposure to any dust that may be in the air.

21. Plasma cleaning (*see* refs. [84, 85] for very informative reviews and ref. [86] for discussion of the effects on glass) removes surface contaminants (particularly hydrocarbons) from the glass and induces silanol formation [86] on the surface, thereby rendering it very hydrophilic [87] and amenable to silanization. Plasma cleaners work by exciting the molecules in a low-pressure gas with a radio-frequency electromagnetic wave. The energized gas emits ultraviolet radiation and eventually forms a plasma consisting of ions, free electrons, and reactive radical species that bombard contaminated surfaces. As contaminants are broken down, they rapidly vaporize, leaving the surface extremely clean. Various mixtures of gas can be used, including room air (i.e., ~78 % nitrogen, ~21 % oxygen, ~1 % argon), which is what we use. Historically, similar cleaning processes have been accomplished by chemical means (reviewed by Kern [88]), in particular "piranha" solution (a 3:1 mixture of sulfuric acid and hydrogen peroxide). This solution is corrosive, violently reactive with organic matter, and potentially explosive [26, 56, 57]. Although piranha is still widely used and can replace plasma cleaning, we highly recommend the latter based on its efficacy, ease, safety, and environmental friendliness.

22. Following plasma cleaning, the cover slips are left in room air briefly in order to allow a layer of water to adsorb on the surface. During the subsequent treatment with aminosilane, this promotes hydrolysis of aminosilane ethoxy groups to silanols specifically near the surface, rather than throughout the solution,

thereby favoring surface binding rather than polymerization of unbound aminosilane.

23. Aminosilanes are corrosive to the skin, eyes, and mucous membranes. Be familiar with the material safety data sheet (MSDS). Handle only in the fume hood and wear appropriate personal protective equipment (nitrile gloves, eye/face protection, and lab coat). If storing under nitrogen or argon (e.g., using a container with a rubber septum seal), use extreme care to prevent spraying bottle contents or overpressurizing the bottle with compressed gas. Any spills or droplets of aminosilane solutions should be cleaned promptly with a Kimwipe with acetone or ethanol. If allowed to dry, spills form white films that are difficult to remove. We perform all aminosilane work on absorbent pads with plastic backing.

24. Deposition times for aminosilanes vary considerably in the literature and in the technical documentation provided by manufacturers, from minutes to many hours. Prolonged reaction times may lead to formation of complex, branched silane networks and agglomerates on the treated surface [22, 24, 60, 89] that expand upon exposure to water [60]. These networks constitute a dynamic gel that can restructure continuously [36]. In our hands, this short 10 s incubation yields excellent results. For longer incubations (e.g., 30 min, or even 1 min, if using improperly stored aminosilanes, which may have partially polymerized), we have observed behavior consistent with the formation of an aminosilane gel on the surface that hinders interactions between subsequently attached MTs and motors during both optical trapping and single-molecule fluorescence experiments. Motors are unable to bind to and move along particular MTs when cover slips are treated with aminosilane for these longer incubation times. Interestingly, in the absence of initial glutaraldehyde surface treatment, it is precisely the MTs that remain stably bound after introducing β-casein that are unable to support movement, possibly due to the MTs embedding within the silane gel network (*see also* **Note 46**). We therefore recommend as short an incubation time as needed to support reliable MT attachment and molecular motor motility on all MTs tested (prior to glutaraldehyde treatment, β-casein should release virtually all MTs from the glass, consistent with the absence of any significant gel network on the surface).

25. This protocol creates a significant amount of acetone waste. This could be decreased using vapor-phase silanization (rather than the solution-phase silanization employed here), which has also been reported to be less sensitive to ambient conditions (e.g., humidity) [26]. However, this requires additional equipment (e.g., a vacuum chamber compatible with organic vapors). Since the waste acetone should be free of significant

contaminants except for aminosilane, and since aminosilanes boil at a much higher temperature than acetone (>200 °C), the acetone could be recycled, e.g., via purification using a combination of Drierite (anhydrous $CaSO_4$, to remove water) and fractional distillation [90, 91].

26. Washing the silanized cover slips in ethanol and water has three purposes: (1) removal of loosely bound aminosilane (sometimes visible as small white deposits after the initial baking), (2) hydrolysis of any remaining ethoxy groups in order to form silanols and promote intramolecular siloxane bond formation, and (3) final cleaning to remove any surface contaminants.

27. Aminosilanized surfaces are easily contaminated and thereby inactivated by organic impurities adsorbed from the air [92]. In addition, carbon dioxide has a well-studied capacity to adsorb onto and react with these surfaces (e.g., refs. [93, 94]), forming carbamates and bicarbonates that will not facilitate glutaraldehyde attachment. In a protocol similar to ours, Jeney et al. [19] recommend storing aminosilanized cover slips for no more than 2 days in order to ensure strong MT attachment.

    Although storing the treated glass in water or other solution may limit adsorption of airborne contaminants and CO_2, the aminosilane amine groups can catalyze siloxane bond hydrolysis and release from the surface (i.e., the reverse of the condensation reaction they catalyze during binding) [23, 26, 95], which is more facile in solution (the use of AEAPTES rather than APTES somewhat limits this problem [26]).

    Storage in vacuum significantly prolongs the activity of aminosilanized surfaces [92, 96]. We store aminosilanized cover slips in a vacuum desiccator with Drierite to remove any residual humidity and leave the vacuum line on in order to ensure the lowest pressure possible over extended periods. Avoid opening the desiccator unnecessarily. Cover slips stored in this manner retain activity for days to weeks. In fact, we have used cover slips stored for over one month. If cover slips are stored for more than 1–2 weeks, we usually clean them by repeating **steps 22–25** (sonication in ethanol and water) before using them.

28. A paper cutter works well to make long, straight, clean cuts. Whatever cutting tool is used, it should be very sharp. Dull cuts will crush the film into the paper backing along the edge and make it difficult to separate them. Furthermore, ragged edges can disrupt smooth fluid flow through the chamber, thereby interfering with good MT alignment on the glass. When cutting, keep the film side up, with the paper backing on the table, so the Parafilm stays clean. These strips can be stored in a clean plastic bag until ready for use.

29. When burnishing the Parafilm, do not press too hard, as this can either deform the film or make it difficult to remove the

backing later. Many slides with two pieces of Parafilm attached can be prepared at once and stored indefinitely in a dust-free box or bag for later use.

30. Pull parallel to the glass surface more than upward, to avoid pulling the film itself off the glass. If the film does begin to separate from the slide, this indicates it was not burnished sufficiently. Simply press it back down and burnish it again.

31. Place one edge down over the template and then carefully remove the tweezers, letting the cover slip drop into place. Make an effort to position it perfectly on the first attempt, to avoid sliding the cover slip around on the Parafilm and contaminating the aminosilanized surface. To avoid contamination, it is also advisable to use separate forceps for removing the Parafilm backing and for handling the cover slips.

32. We prefer to immediately treat the chambers with glutaraldehyde, since the density of amine groups on the surface with which the glutaraldehyde can react will be greatest for newer slide chambers. In addition, treatment with glutaraldehyde may prolong the activity of the chamber by preventing loss of active amines, as happens slowly for an untreated aminosilanized surface. If the slide chambers will be stored for more than a few days, we prefer to seal the ends with small pieces of plastic cling wrap (carefully burnished with a cotton-tipped applicator to form a tight seal). This limits exposure to any residual gases or contaminants in the desiccator and is easily removed prior to use.

33. The use of cross-linkers such as glutaraldehyde is often a matter of concern. We wish to emphasize that the procedure employed here immobilizes glutaraldehyde on the glass surface, with all free glutaraldehyde subsequently removed. During MT immobilization, only the bottom surface of the MT (which the motor would not make contact with anyhow) is chemically altered, leaving the upper surface in its native form. In addition, it is worth noting that previous reports have demonstrated kinesin motility on glutaraldehyde-fixed MTs at rates similar to those untreated for MTs [97]. Finally, all remaining glutaraldehyde on the glass surface is blocked with β-casein (which contains several lysine residues [71] capable of reacting with the glutaraldehyde) before any motors are introduced.

34. The purpose of the humidity chamber is to limit evaporation of the slide chamber contents during incubation.

35. We measure the pH of the 8 % glutaraldehyde solution to be approximately 4.5. This pH is favorable for preventing glutaraldehyde polymerization [27, 28, 30], but is more acidic than that commonly employed for glutaraldehyde reactions, and somewhat unfavorable for reaction with aminosilane surface amine groups ($pK_a \approx 7$–10 [55, 56]), which must be deprotonated in order to react. On the other hand, it is beneficial if a

large portion of the surface amines do not react, so that the aminosilanized surface retains its positive charge. It also is likely unnecessary to create a very high density of surface aldehydes for MT binding, and possibly even undesirable, given the effects of glutaraldehyde on MT morphology in tissue fixation [98].

36. The most effective method for using a Kimwipe to withdraw the solution is to fold the wipe several times in a rectangular shape and then press the corner of this rectangle firmly into the corner of one of the mouths of the slide chamber. Do not remove the wipe until all solution has been withdrawn. *From this point forward, always flow solutions in the same direction through each chamber.* This avoids drawing any contaminants from the wicking paper into the chamber.

37. Use a 200 μL pipette so that flow remains constant during each rinse. Try to flow the water through the chamber as rapidly as possible. If persistent bubbles form (thereby preventing adequate rinsing of parts of the cover slip surface), use the filtered air or a nitrogen stream to blow all the solution out of the chamber, and rinse it again. Make sure the entire surface of the cover slip is washed.

38. Increasing the concentration of functionalized dye in solution does not necessarily lead to better labeling yields. Tubulin with multiple dyes attached may not polymerize well or may irreversibly aggregate [39].

39. If using rhodamine-labeled tubulin rather than Cy3-labeled tubulin, a ratio of 1:10 labeled/unlabeled is recommended.

40. Glycerol enhances MT polymerization and stability [40, 41, 99].

41. Paclitaxel, which is a potent stimulator of MT polymerization, is added after initially polymerizing with only glycerol and GTP, because adding paclitaxel initially leads to the formation of many short (<1 μm) MTs [44], rather than fewer longer filaments. The latter is more desirable for the optical trapping assay. MT length can be controlled to some extent by adjusting the incubation period prior to adding paclitaxel.

42. MTs polymerize best at 37 °C but are stable in the presence of paclitaxel at room temperature.

43. The purpose of centrifuging the MT solution through the glycerol cushion is to remove any unpolymerized tubulin (which will remain in the supernatant) and very short MT fragments (which will remain in the cushion). These "contaminants" can interfere with MT binding to the cover slip and clutter the surface of the cover slip with unwanted fluorescence. Following the centrifugation, the objective is to remove all protein except that in the pellet before resuspending the MTs. Take care not to (a) disturb the pellet or (b) mix the upper and lower levels of the supernatant.

44. Resuspension of the MT pellet can take several minutes. Pipette solution up and down above the pellet, and avoid sticking the pipette tip directly into the pellet, especially in the beginning, as this can cause shearing of the MTs. Avoid making bubbles.

45. Casein is acidic and will lower the pH as it dissolves. It becomes less soluble as the pH approaches the casein isoelectric point of ~4.6.

46. Although the MTs immediately and rigidly attach to the charged glass surface, the covalent linkage via glutaraldehyde is slower (minutes to hours [28, 100]). This reaction must be allowed to complete before adding the Blocker solution (2 mg/mL β-casein). Otherwise, β-casein tends to compete with the MTs for attachment to the cover slip (and likely also shields the positive charge of the aminosilane). This results in floppy attachment of the MTs to the glass surface, which makes finding a well-aligned and stationary MT very difficult. On the other hand, if the β-casein is added after the MTs have attained a covalent linkage, they remain immobilized upon and rigidly attached to the glass. With planning, the MT incubation on the cover slip can be done during other tasks (e.g., microscope alignment) so that it does not introduce a delay. When performing consecutive experiments, we usually take a 2-min break from the microscope to start the MT incubation for the next slide. Alternatively, several slide chambers can be prepared with MTs at once and kept in a humidity chamber until ready for use. We have stored slide chambers with bound MTs for several hours, or even overnight, with no loss of capacity to support motility.

47. If surface-bound microspheres are needed for calibration or as fiducial markers, a very low concentration of protein-coated microspheres (e.g., the same ones used for optical trapping) can be added directly after the MTs. These will also form covalent linkages with the glass and remain very stably attached.

48. The best seal is made when a small amount of vacuum grease enters the mouth of the chamber. This seal can last for days or even weeks without significantly affecting the volume of liquid in the chamber. Perhaps nonintuitively, forcing more grease into the chamber mouth actually yields a worse seal: the accompanying increase in pressure in the chamber eventually forces buffer through the weakest part of the seal, thus breaking it and allowing buffer evaporation and oxygen entry.

49. If any grease is accidentally left of the cover slip surface, it can be cleaned using a very small amount of spectrophotometric grade ethanol with a Kimwipe or cotton-tipped applicator. Wipe in a single motion away from the center of the chamber.

Acknowledgements

We thank Johan Andreasson, Robert Coleman, and Wei Chen for helpful discussions regarding aminosilanization and Laura E.K. Nicholas for assistance with photography and figure illustration. The authors are supported by the National Institutes of Health grant R01GM098469. M.P.N. received support from the NIH-funded Medical Scientist Training and Molecular Biophysics Training programs (NIH grants T32GM007288 and T32GM008572, respectively) at the Albert Einstein College of Medicine.

References

1. Yildiz A, Tomishige M, Vale RD et al (2004) Kinesin walks hand-over-hand. Science 303: 676–678

2. Reck-Peterson SL, Yildiz A, Carter AP et al (2006) Single-molecule analysis of dynein processivity and stepping behavior. Cell 126: 335–348

3. Gennerich A, Carter AP, Reck-Peterson SL et al (2007) Force-induced bidirectional stepping of cytoplasmic dynein. Cell 131:952–965

4. Cho C, Reck-Peterson S, Vale R (2008) Cytoplasmic dynein's regulatory ATPase sites affect processivity and force generation. J Biol Chem 283:25839–25845

5. Kardon JR, Reck-Peterson SL, Vale RD (2009) Regulation of the processivity and intracellular localization of Saccharomyces cerevisiae dynein by dynactin. Proc Natl Acad Sci USA 106:5669–5674

6. Guydosh NR, Block SM (2009) Direct observation of the binding state of the kinesin head to the microtubule. Nature 461:125–128

7. Visscher K, Schnitzer MJ, Block SM (1999) Single kinesin molecules studied with a molecular force clamp. Nature 400:184–189

8. Barak P, Rai A, Rai P et al (2013) Quantitative optical trapping on single organelles in cell extract. Nat Methods 10:68–70

9. Vershinin M, Carter BC, Razafsky DS et al (2006) Multiple-motor based transport and its regulation by Tau. Proc Natl Acad Sci USA 104:87–92

10. Helenius J, Brouhard G, Kalaidzidis Y et al (2006) The depolymerizing kinesin MCAK uses lattice diffusion to rapidly target microtubule ends. Nature 441:115–119

11. Rice S, Lin AW, Safer D et al (1999) A structural change in the kinesin motor protein that drives motility. Nature 402:778–784

12. Gestaut DR, Graczyk B, Cooper J et al (2008) Phosphoregulation, lattice diffusion, and depolymerization-driven movement of the Dam1 complex do not require ring formation. Nat Cell Biol 10:407–414

13. Nakata T, Hirokawa N (1995) Point mutation of adenosine triphosphate-binding motif generated rigor kinesin that selectively blocks anterograde lysosome membrane transport. J Cell Biol 131:1039–1053

14. Crevel IM-TC, Nyitrai M, Alonso MC et al (2003) What kinesin does at roadblocks: the coordination mechanism for molecular walking. EMBO J 23:23–32

15. Adio S, Jaud J, Ebbing B et al (2009) Dissection of kinesin's processivity. PLoS One 4:e4612

16. Schroeder HW III, Mitchell C, Shuman H et al (2010) Motor number controls cargo switching at actin-microtubule intersections in vitro. Curr Biol 20:687–696

17. Taylor AP, Webb RI, Barry JC et al (2000) Adhesion of microbes using 3-aminopropyl triethoxy silane and specimen stabilisation techniques for analytical transmission electron microscopy. J Microsc 199:56–67

18. Turner DC, Chang C, Fang K et al (1995) Selective adhesion of functional microtubules to patterned silane surfaces. Biophys J 69: 2782–2789

19. Jeney S, Florin EL, Hörber JK (2001) Use of photonic force microscopy to study single-motor-molecule mechanics. Methods Mol Biol 164:91–108

20. Brown TB, Hancock WO (2002) A polarized microtubule array for kinesin-powered nanoscale assembly and force generation. Nano Lett 2:1131–1135

21. Kanan SM, Tze WTY, Tripp CP (2002) Method to double the surface concentration and control the orientation of adsorbed (3-aminopropyl)dimethylethoxysilane on silica powders and glass slides. Langmuir 18: 6623–6627

22. Howarter JA, Youngblood JP (2006) Optimization of silica silanization by 3-aminopropyltriethoxysilane. Langmuir 22: 11142–11147

23. Asenath SE, Chen W (2008) How to prevent the loss of surface functionality derived from aminosilanes. Langmuir 24:12405–12409

24. Kim J, Seidler P, Wan LS et al (2009) Formation, structure, and reactivity of amino-terminated organic films on silicon substrates. J Colloid Interface Sci 329:114–119

25. Aissaoui N, Bergaoui L, Landoulsi J et al (2012) Silane layers on silicon surfaces: mechanism of interaction, stability, and influence on protein adsorption. Langmuir 28:656–665

26. Zhu M, Lerum MZ, Chen W (2012) How to prepare reproducible, homogeneous, and hydrolytically stable aminosilane-derived layers on silica. Langmuir 28:416–423

27. Walt DR, Agayn VI (1994) The chemistry of enzyme and protein immobilization with glutaraldehyde. Trends Anal Chem 13:425–430

28. Migneault I, Dartiguenave C, Bertrand MJ et al (2004) Glutaraldehyde: behavior in aqueous solution, reaction with proteins, and application to enzyme crosslinking. Biotechniques 37:790–802

29. Hermanson GT (2008) Bioconjugate techniques, 2nd edn. Academic, New York

30. Betancor L, López-Gallego F, Alonso-Morales N et al (2006) Glutaraldehyde in protein immobilization. Methods Biotechnol 22:57–64

31. Clancy BE, Behnke-Parks WM, Andreasson JOL et al (2011) A universal pathway for kinesin stepping. Nat Struct Mol Biol 18: 1020–1027

32. Iler RK (1979) The chemistry of silica: solubility, polymerization, colloid and surface properties and biochemistry of silica. Wiley-Interscience, Hoboken, NJ

33. Rother D, Sen T, East D et al (2011) Silicon, silica and its surface patterning/activation with alkoxy- and amino-silanes for nanomedical applications. Nanomedicine 6: 281–300

34. Der Voort PV, Vansant EF (1996) Silylation of the silica surface: a review. J Liq Chrom Relat Tech 19:2723–2752

35. Witucki GL (1993) A silane primer: chemistry and applications of alkoxysilanes. J Coating Tech Res 65:57–60

36. Brinker C (1988) Hydrolysis and condensation of silicates: effects on structure. J Non-Cryst Solids 100:31–50

37. Tan W, Desai TA (2004) Layer-by-layer microfluidics for biomimetic three-dimensional structures. Biomaterials 25:1355–1364

38. Diao J, Ren D, Engstrom JR et al (2005) A surface modification strategy on silicon nitride for developing biosensors. Anal Biochem 343:322–328

39. Peloquin J, Komarova Y, Borisy G (2005) Conjugation of fluorophores to tubulin. Nat Methods 2:299–303

40. Keates RAB (1980) Effects of glycerol on microtubule polymerization kinetics. Biochem Biophys Res Commun 97:1163–1169

41. O'Brien ET, Erickson HP (1989) Assembly of pure tubulin in the absence of free GTP: effect of magnesium, glycerol, ATP, and the nonhydrolyzable GTP analogues. Biochemistry 28: 1413–1422

42. Mitchison T, Kirschner M (1984) Dynamic instability of microtubule growth. Nature 312:237–242

43. Gardner MK, Zanic M, Howard J (2013) Microtubule catastrophe and rescue. Curr Opin Cell Biol 25:14–22

44. Carlier MF, Pantaloni D (1983) Taxol effect on tubulin polymerization and associated guanosine 5′-triphosphate hydrolysis. Biochemistry 22:4814–4822

45. Orr GA, Verdier-Pinard P, McDaid H et al (2003) Mechanisms of taxol resistance related to microtubules. Oncogene 22: 7280–7295

46. Schiff PB, Fant J, Horwitz SB (1979) Promotion of microtubule assembly in vitro by taxol. Nature 277:665–667

47. Schiff PB, Horwitz SB (1980) Taxol stabilizes microtubules in mouse fibroblast cells. Proc Natl Acad Sci USA 77:1561–1565

48. Thompson WC, Wilson L, Purich DL (1981) Taxol induces microtubule assembly at low temperature. Cell Motil 1:445–454

49. Kumar N (1981) Taxol-induced polymerization of purified tubulin. Mechanism of action. J Biol Chem 256:10435–10441

50. Robinson J, Engelborghs Y (1982) Tubulin polymerization in dimethyl sulfoxide. J Biol Chem 257:5367–5371

51. Baker NA, Sept D, Joseph S et al (2001) Electrostatics of nanosystems: application to microtubules and the ribosome. Proc Natl Acad Sci USA 98:10037–10041

52. Minoura I, Muto E (2006) Dielectric measurement of individual microtubules using the electroorientation method. Biophys J 90:3739–3748

53. Sackett DL, Bhattacharyya B, Wolff J (1986) Promotion of tubulin assembly by carboxy-terminal charge reduction. Ann NY Acad Sci 466:460–466

54. Redeker V, Melki R, Promé D et al (1992) Structure of tubulin C-terminal domain

obtained by subtilisin treatment. The major α and β tubulin isotypes from pig brain are glutamylated. FEBS Lett 313:185–192

55. Bezanilla M, Manne S, Laney DE et al (1995) Adsorption of DNA to mica, silylated mica, and minerals: characterization by atomic force microscopy. Langmuir 11:655–659

56. Van der Maaden K, Sliedregt K, Kros A et al (2012) Fluorescent nanoparticle adhesion assay: a novel method for surface pKa determination of self-assembled monolayers on silicon surfaces. Langmuir 28:3403–3411

57. Gell C, Bormuth V, Brouhard GJ et al (2010) Microtubule dynamics reconstituted in vitro and imaged by single-molecule fluorescence microscopy. Methods Cell Biol 95:221–245

58. Schäffer E, Nørrelykke SF, Howard J (2007) Surface forces and drag coefficients of microspheres near a plane surface measured with optical tweezers. Langmuir 23:3654–3665

59. Tolić-Nørrelykke SF, Schäffer E, Howard J et al (2006) Calibration of optical tweezers with positional detection in the back focal plane. Rev Sci Instrum 77:103101–103111

60. Vandenberg ET, Bertilsson L, Liedberg B et al (1991) Structure of 3-aminopropyl triethoxy silane on silicon oxide. J Colloid Interface Sci 147:103–118

61. Vigers GP, Coue M, McIntosh JR (1988) Fluorescent microtubules break up under illumination. J Cell Biol 107:1011–1024

62. Guo H, Xu C, Liu C et al (2006) Mechanism and dynamics of breakage of fluorescent microtubules. Biophys J 90:2093–2098

63. Weisenberg RC (1972) Microtubule formation in vitro in solutions containing low calcium concentrations. Science 177:1104–1105

64. Borisy GG, Olmsted JB (1972) Nucleated assembly of microtubules in porcine brain extracts. Science 177:1196–1197

65. Miller HP, Wilson L (2010) Preparation of microtubule protein and purified tubulin from bovine brain by cycles of assembly and disassembly and phosphocellulose chromatography. Methods Cell Biol 95:3–15

66. Castoldi M, Popov AV (2003) Purification of brain tubulin through two cycles of polymerization–depolymerization in a high-molarity buffer. Protein Expr Purif 32:83–88

67. Gell C, Friel CT, Borgonovo B et al (2011) Purification of tubulin from porcine brain. Methods Mol Biol 777:15–28

68. Widlund PO, Podolski M, Reber S et al (2012) One-step purification of assembly-competent tubulin from diverse eukaryotic sources. Mol Biol Cell 23:4393–4401

69. Fox PF, McSweeney PLH (2003) Advanced dairy chemistry: protein. Springer, New York

70. Fox PF, McSweeney PLH (1998) Dairy chemistry and biochemistry. Springer, New York

71. Eigel WN, Butler JE, Ernstrom CA et al (1984) Nomenclature of proteins of cow's milk: fifth revision. J Dairy Sci 67:1599–1631

72. Nylander T, Wahlgren NM (1997) Forces between adsorbed layers of β-casein. Langmuir 13:6219–6225

73. Mikheeva LM, Grinberg NV, Grinberg VY et al (2003) Thermodynamics of micellization of bovine β-casein studied by high-sensitivity differential scanning calorimetry. Langmuir 19:2913–2921

74. Thorn DC, Meehan S, Sunde M et al (2005) Amyloid fibril formation by bovine milk κ-casein and its inhibition by the molecular chaperones αS- and β-casein. Biochemistry 44:17027–17036

75. Portnaya I, Ben-Shoshan E, Cogan U et al (2008) Self-assembly of bovine β-casein below the isoelectric pH. J Agric Food Chem 56:2192–2198

76. Evers CHJ, Andersson T, Lund M et al (2012) Adsorption of unstructured protein β-casein to hydrophobic and charged surfaces. Langmuir 28:11843–11849

77. Tompa P (2002) Intrinsically unstructured proteins. Trends Biochem Sci 27:527–533

78. Kull T, Nylander T, Tiberg F et al (1997) Effect of surface properties and added electrolyte on the structure of β-casein layers adsorbed at the solid/aqueous interface. Langmuir 13:5141–5147

79. Verma V, Hancock WO, Catchmark JM (2008) The role of casein in supporting the operation of surface bound kinesin. J Biol Eng 2:14

80. Ong S, Zhao X, Eisenthal KB (1992) Polarization of water molecules at a charged interface: second harmonic studies of the silica/water interface. Chem Phys Lett 191:327–335

81. Fan H-F, Li F, Zare RN et al (2007) Characterization of two types of silanol groups on fused-silica surfaces using evanescent-wave cavity ring-down spectroscopy. Anal Chem 79:3654–3661

82. Hair ML, Hertl W (1970) Acidity of surface hydroxyl groups. J Phys Chem 74:91–94

83. Balladur V, Theretz A, Mandrand B (1997) Determination of the main forces driving DNA oligonucleotide adsorption onto aminated

silica wafers. J Colloid Interface Sci 194: 408–418

84. Belkind A, Gershman S (2008) Plasma cleaning of surfaces. Vacuum Coating and Technology November, 46–57

85. Isabell TC, Fischione PE, O'Keefe C et al (1999) Plasma cleaning and its applications for electron microscopy. Microsc Microanal 5:126–135

86. DeRosa RL, Schader PA, Shelby JE (2003) Hydrophilic nature of silicate glass surfaces as a function of exposure condition. J Non-Cryst Solids 331:32–40

87. Bhattacharya S, Datta A, Berg JM et al (2005) Studies on surface wettability of poly(dimethyl) siloxane (PDMS) and glass under oxygen-plasma treatment and correlation with bond strength. J Microelectromech Syst 14:590–597

88. Kern W (1990) The evolution of silicon wafer cleaning technology. J Electrochem Soc 137: 1887–1892

89. Williams EH, Davydov AV, Motayed A et al (2012) Immobilization of streptavidin on 4H-SiC for biosensor development. Appl Surf Sci 258:6056–6063

90. Armarego WLF, Chai C (2012) Purification of laboratory chemicals. Butterworth-Heinemann, Oxford

91. Weires NA, Johnston A, Warner DL et al (2011) Recycling of waste acetone by fractional distillation. J Chem Educ 88: 1724–1726

92. Siqueira Petri DF, Wenz G, Schunk P et al (1999) An improved method for the assembly of amino-terminated monolayers on SiO_2 and the vapor deposition of gold layers. Langmuir 15:4520–4523

93. Leal O, Bolívar C, Ovalles C et al (1995) Reversible adsorption of carbon dioxide on amine surface-bonded silica gel. Inorg Chim Acta 240:183–189

94. Serna-Guerrero R, Da'na E, Sayari A (2008) New insights into the interactions of CO_2 with amine-functionalized silica. Ind Eng Chem Res 47:9406–9412

95. Etienne M, Walcarius A (2003) Analytical investigation of the chemical reactivity and stability of aminopropyl-grafted silica in aqueous medium. Talanta 59:1173–1188

96. Möller R, Csáki A, Köhler JM et al (2000) DNA probes on chip surfaces studied by scanning force microscopy using specific binding of colloidal gold. Nucleic Acids Res 28:e91

97. Turner D, Chang C, Fang K et al (1996) Kinesin movement on glutaraldehyde-fixed microtubules. Anal Biochem 242:20–25

98. Cross AR, Williams RC Jr (1991) Kinky microtubules: bending and breaking induced by fixation in vitro with glutaraldehyde and formaldehyde. Cell Motil Cytoskeleton 20:272–278

99. Shelanski ML, Gaskin F, Cantor CR (1973) Microtubule assembly in the absence of added nucleotides. Proc Natl Acad Sci USA 70:765–768

100. Habeeb AFSA, Hiramoto R (1968) Reaction of proteins with glutaraldehyde. Arch Biochem Biophys 126:16–26

Chapter 10

An Improved Optical Tweezers Assay for Measuring the Force Generation of Single Kinesin Molecules

Matthew P. Nicholas, Lu Rao, and Arne Gennerich

Abstract

Numerous microtubule-associated molecular motors, including several kinesins and cytoplasmic dynein, produce opposing forces that regulate spindle and chromosome positioning during mitosis. The motility and force generation of these motors are therefore critical to normal cell division, and dysfunction of these processes may contribute to human disease. Optical tweezers provide a powerful method for studying the nanometer motility and piconewton force generation of single motor proteins in vitro. Using kinesin-1 as a prototype, we present a set of step-by-step, optimized protocols for expressing a kinesin construct (K560-GFP) in *Escherichia coli*, purifying it, and studying its force generation in an optical tweezers microscope. We also provide detailed instructions on proper alignment and calibration of an optical trapping microscope. These methods provide a foundation for a variety of similar experiments.

Key words Optical tweezers, Optical trapping, Optical trap alignment and calibration, Molecular motors, Kinesin, K560, Microtubules, Pyranose oxidase, Single-molecule assays, Force measurement, *Escherichia coli*, Protein purification

1 Introduction

Cytoskeletal filaments and their associated molecular motor proteins generate a multitude of forces that orchestrate the complex process of mitosis. While actin and myosin II drive the actual division of cells (the separation of the cytoplasm of the original cell into two new cells), it is primarily microtubules (MTs) and their associated motor proteins, cytoplasmic dynein and kinesins (composed of several distinct subfamilies) [1, 2], that govern the intricate rearrangements of the mitotic spindle preceding cytokinesis [3].

Numerous MT-associated motors exert forces on the MT mitotic spindle [4–8]. For example, during anaphase B, cytoplasmic dynein attached to the cell cortex walks toward the minus ends of astral MTs, pulling the spindle poles away from each other and toward the cell periphery [9]. The tetrameric, bipolar kinesin-5 (also called Eg5 or Kif11) likewise drives the spindle poles apart by

David J. Sharp (ed.), *Mitosis: Methods and Protocols*, Methods in Molecular Biology, vol. 1136,
DOI 10.1007/978-1-4939-0329-0_10, © Springer Science+Business Media New York 2014

simultaneously moving toward the plus ends of antiparallel interpolar MTs in the spindle midzone [10] (thereby sliding the oppositely oriented MTs away from one another). Interestingly, however, dynein appears to antagonize the function of kinesin-5 [11–13]. Kinesin-14 family proteins (e.g., HSET, Ncd)—which are homodimeric/unipolar, but nonetheless cross-link and slide antiparallel MTs—oppose the forces produced by kinesin-5 [14–17]. The kinesin-14 motor domain is near the C-terminus, and this motor walks in the opposite direction of most of its N-terminal kinesin relatives [1, 18, 19]. Thus, it pulls oppositely oriented MTs at the spindle midzone closer together as it walks toward the MT minus end.

During all of these processes, the MTs themselves are highly dynamic, regulated in part by members of the microtubule-destabilizing kinesin-13 family [20–22]. In addition, separate motors exert forces directly on chromosomes. The plus-end-directed kinesin-4 and kinesin-10 (the so-called chromokinesins) attach to chromosomes and carry them away from the spindle poles [1].

Despite growing knowledge, we are only beginning to understand how forces in mitosis are generated and regulated. As discussed, numerous different molecules are involved. In vitro single-molecule studies have proved useful in deciphering the functions of the various constituent proteins in isolation [13, 16, 23–25]. Among these techniques, optical trapping (also called optical tweezers; *see* refs. 26, 27 for excellent reviews) is particularly informative, because it can measure and exert forces on the piconewton (pN)-scale characteristic of molecular motors [28–32].

The basis for conventional optical trapping is the transfer of momentum from photons in a focused laser beam to a small dielectric particle. In the simple ray-optics picture (*see* **Note 1** for a more complete description), this occurs due to refraction of light by a microsphere ("bead") placed in a highly focused, near-infrared laser beam with a Gaussian intensity profile [26]. Refracted photons experience a change in momentum—i.e., a force—associated with their altered direction of propagation. Due to momentum conservation (or, alternatively, Newton's third law), the microsphere experiences an equal and opposite change in momentum (force) in the opposite direction as that experienced by the refracted photons. Near the laser focus, the geometry of the system (namely, the gradient of laser intensity surrounding the focus) causes the net force on the microsphere to pull it toward the center of the focus ("trap center") whenever it is displaced, thus "trapping" the particle in three dimensions. Conveniently, for small displacements, the microsphere position can be tracked with extremely high precision (a few nanometers or better), and the force exerted on it is linear with this displacement [26]. This is analogous to the force $F = -k\Delta x$ predicted by Hooke's law for an ideal spring extended a distance Δx, where the constant k is referred to as the

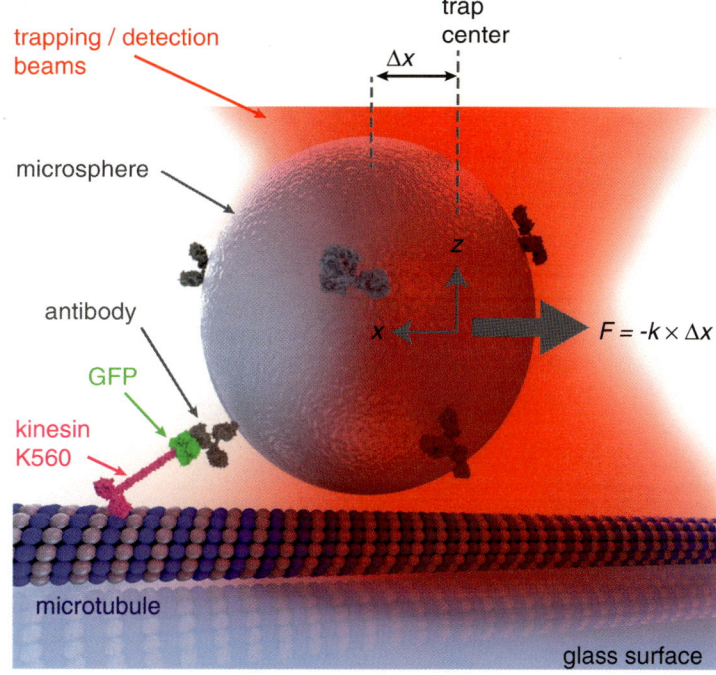

trapping / detection beams

trap center

Δx

microsphere

antibody

GFP

kinesin K560

$F = -k \times \Delta x$

microtubule

glass surface

Fig. 1 Optical tweezers assay for kinesin motility and force production (not to scale). A polystyrene microsphere covalently bound to anti-GFP antibodies binds a single kinesin K560-GFP dimer, and is trapped by a near-infrared optical trapping beam focused via a high numerical aperture microscope objective lens. The trap holds the microsphere directly above a MT that is covalently linked to the glass surface of the cover slip. When the kinesin binds to and moves along the MT in the presence of ATP, it pulls the attached microsphere with it. The trap resists this motion, exerting a force $F = -k \times \Delta x$ on the microsphere-motor complex, where k is the trap stiffness (force per displacement) and Δx is the distance from the trapping beam longitudinal (z) axis ("trap center") to the center of the microsphere. The detection beam—which overlaps the trapping beam, but does not contribute to trapping—is used to measure Δx via back focal plane interferometry. This figure was prepared with VMD [164] (using PDB entries 3KIN, 1GFL, and 1IGT) and the Persistence of Vision Raytracer (POV-Ray www.povray.org)

"spring constant" or "stiffness" (for optical traps, k is directly proportional to the laser intensity, but also depends on the diameter of the microsphere [33–35]). Therefore, by linking motor proteins to optically trapped microspheres, one can measure both the motility and force produced by these mechanoenzymes as they move along their cytoskeletal filaments and thereby displace the trapped bead from the trap center (Fig. 1). In fact, even some endogenous biological particles (e.g., endosomes) can be optically trapped, so that the technique can be applied in vivo [36–38] (though calibration and interpretation of results are more complicated).

Although optical trapping is still mainly a tool of specialists, its popularity and application have grown significantly since its first application to motor proteins [39], and it is feasible for nonexperts to build and use simple optical tweezers microscopes in their laboratories (some commercial options are also available [40]). While establishing microsphere trapping is relatively straightforward, accurately and precisely calibrating the instrument for force measurement presents some challenges. Although there is rich literature on technical aspects of optical trapping (*see*, e.g., [26, 27] and references therein), our own experience with building and calibrating such an instrument convinced us of the importance of a comprehensive, up-to-date resource addressing the many "hands-on" details and subtleties involved. Our goal is to help bridge the gap between "qualitative" optical trapping and research-quality force measurements.

In addition to details on instrumentation, we provide a simple system using optimized biochemical conditions, with which to establish a basic optical tweezers assay. Not long after its discovery over 25 years ago [41], kinesin-1 (also known as conventional kinesin) was the first molecular motor studied by optical trapping [39], and it has since become perhaps the best-studied MT-associated motor. As such, it serves as a well-characterized, predictable standard for establishing optical tweezers assays that measure the dynamic behavior of molecular motors. Here, we provide methods for isolating recombinant kinesin-1 and studying its motility and force production in vitro as it walks processively toward the plus ends of MTs. These optimized protocols provide a basis for a host of similar experiments.

Below, we describe the recombinant expression in *Escherichia coli* (*E. coli*) of a GFP- and polyhistidine-tagged construct (herein referred to as K560-GFP, or simply K560) containing the first 560 amino acids of human kinesin-1 [42] (*see* **Note 2**). We then provide methods for isolating functional kinesin motors to high purity by sequential steps of cell lysis, centrifugation, nickel-nitrilotriacetic acid (Ni-NTA), agarose affinity column purification (via the K560 polyhistidine-tag), and MT pulldown with ATP-induced release. These procedures yield functional K560 protein sufficient for a number of single-molecule experiments, including optical tweezers assays.

Next, we describe how to measure the movement and force generation of single K560 molecules as they walk along MTs in vitro. This includes methods for (1) preparing optical trapping beads (polystyrene microspheres) with covalently bound anti-GFP antibodies (*see* **Note 3**), (2) attaching K560-GFP to these beads, and (3) measuring K560 motility and force production as it pulls against the opposing load applied by an optical tweezers. We provide the pertinent details of the design and calibration of a combined optical trapping and total internal reflection fluorescence (TIRF) microscope capable of measuring nanometer- and

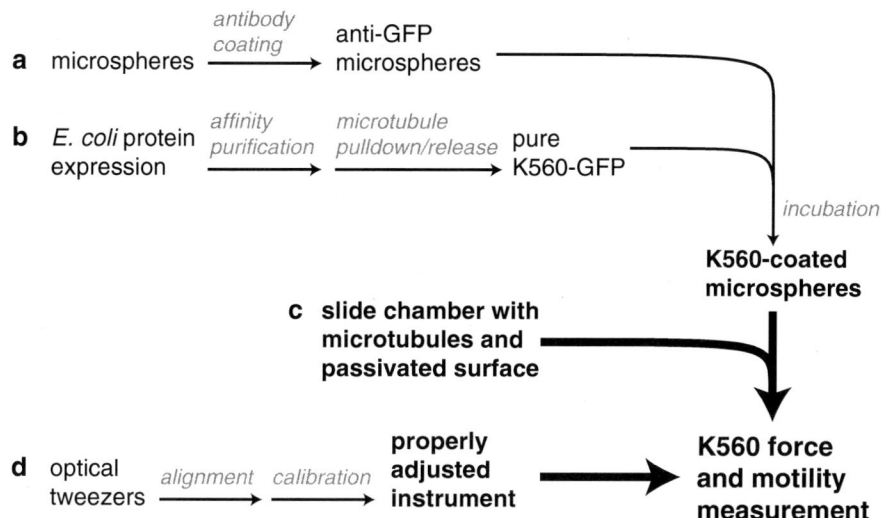

Fig. 2 Protocol summary. Each pathway (**a–d**) summarizes the major steps in preparing reagents and instrumentation for the final assay: (**a, b**) purifying kinesin and attaching it to optical trapping microspheres bound to antibodies; (**c**) preparation of a slide chamber with immobilized, Cy3-labeled MTs (*see* Chap. 9, this issue); (**d**) aligning and calibrating the optical tweezers instrument. These tools are combined to precisely measure force production and motility of single kinesin molecules

piconewton-scale displacements and forces. Figure 2 summarizes how the various procedures come together to enable measurements of force production by single kinesins.

In addition to the procedures and accompanying figures, we have made extensive use of notes to draw attention to important details and underlying principles. The reader is encouraged to consult this section thoroughly before performing the protocols. We also wish to direct the reader to our accompanying protocol for slide chamber preparation and MT fluorescence labeling, polymerization, and surface immobilization (*see* Chap. 9). The techniques in that protocol have wide applicability, but were specifically designed for use with the optical trapping methods presented here.

By implementing the methods described, researchers will be able to design and carry out a variety of assays for studying forces produced by MT-associated molecular motors.

2 Materials

2.1 Transformation of E. coli and Expression of the K560-GFP Construct

1. LB medium: suspend Difco™ LB Broth, Lennox (BD Diagnostic System, Cat. No. DF0402-07-0), in ddH$_2$O to 20 g/L. Autoclave and store at room temperature.

2. Carbenicillin (100 mg/mL): dissolve carbenicillin disodium (Sigma) in ddH$_2$O to 100 mg/mL. Sterilize by filtration

through a sterile 0.22 μm Millex® GS filter unit (Millipore). Store at –20 °C in the dark. The working concentration is 100 μg/mL in this protocol.

3. LB/carbenicillin agar plates: suspend Difco™ LB Agar, Lennox (BD Diagnostic System, Cat. No. DF0402-17-0), in ddH₂O to 35 g/L; then autoclave the solution. When the solution cools to ~55 °C, add carbenicillin. Mix well and pour into the plates. After solidification, store the plates at 4 °C.

4. BL21(DE3) competent *E. coli* cells for protein expression (New England Biolabs).

5. Isopropyl β-D-1-thiogalactopyranoside (IPTG) (1 M): dissolve appropriate amount of IPTG (Sigma) in ddH₂O to 1 M; then filter to sterilize. Store at –20 °C.

6. Plasmid DNA: K560-GFP-6×His (pET17b) (gift from the laboratory of R. Vale, University of California, San Francisco).

2.2 Ni-NTA Agarose-Based Protein Purification

For cell pellet from 1 L *E. coli* culture

1. Phenylmethanesulfonyl fluoride (PMSF) (100 mM): dissolve PMSF (Sigma) in isopropanol to 100 mM, and store at –20 °C (*see* **Note 4**).

2. β-mercaptoethanol (βME) (Sigma) (*see* **Note 5**).

3. Mg²⁺-adenosine 5′-triphosphate (Mg-ATP) (100 mM, pH ~7): dissolve ATP disodium salt (Sigma) in ddH₂O with equimolar MgSO₄, and use NaOH to adjust the pH to ~7 (*see* **Note 6**). Store at –20 °C in aliquots.

4. Imidazole (2 M, pH ~8): suspend imidazole (Sigma) in ddH₂O, and then adjust pH to ~8 with HCl. Store at –20 °C.

5. Lysozyme (50 mg/mL): dissolve lyophilized egg-white lysozyme powder (Sigma) in ddH₂O at room temperature to 50 mg/mL, and then keep on ice until use. Prepare fresh solution for each purification (*see* **Note 7**).

6. Lysis buffer (10 mL): 50 mM Tris, 300 mM NaCl, 5 mM MgCl₂, and 0.2 M sucrose, pH ~7.5.

7. Resin wash buffer (10 mL): 50 mM Tris, 300 mM NaCl, 5 mM MgCl₂, 0.2 M sucrose, and 10 mM imidazole, pH ~7.5.

8. Wash buffer (30 mL): 50 mM Tris, 300 mM NaCl, 5 mM MgCl₂, 0.2 M sucrose, and 20 mM imidazole, pH ~7.5.

9. Elution buffer (10 mL): 50 mM Tris, 300 mM NaCl, 5 mM MgCl₂, 0.2 M sucrose, and 250 mM imidazole, pH ~8.

10. Storage buffer (30 mL): 80 mM PIPES, 2 mM MgCl₂, 1 mM EGTA, and 0.2 M sucrose, pH ~7.

11. Nickel-nitrilotriacetic acid (Ni-NTA) agarose for purification of 6×His-tagged proteins by gravity-flow chromatography (Qiagen).

12. Poly-prep chromatography column (0.8×4 cm, BioRad).

13. Econo-Pac® 10DG Desalting Prepacked Gravity Flow Columns (BioRad).

14. Coomassie protein assay reagent (Thermo Scientific, based on Coomassie blue G-250).

15. 10 % SDS-PAGE gel (Amersham ECL™ or similar).

16. SDS running buffer: 25 mM Tris, 192 mM glycine, and 0.1 % SDS (pH 8.3).

17. Ultrasonic homogenizer (Fisher Scientific, model F550 Sonic Dismembrator).

2.3 MT Binding-and-Release Purification of K560

1. BRB80 buffer: 80 mM PIPES (Sigma P6757), 2 mM $MgCl_2$, and 1 mM EGTA, pH ~7.

2. Bovine brain tubulin (10 mg/mL): dissolve one vial of 1 mg lyophilized powder (Cytoskeleton) in 100 µL BRB80 buffer (final concentration 10 mg/mL). Aliquot and flash freeze. Store aliquots at –80 °C (*see* **Note 8**).

3. Paclitaxel (Taxol) (2 mM): dissolve one vial of lyophilized powder (0.2 µmol per vial, Cytoskeleton, Inc.) in 100 µL anhydrous DMSO. Store at –20 °C. The working concentration is 10–20 µM.

4. Mg^{2+}-guanosine 5′-triphosphate (Mg-GTP) (25 mM): dissolve GTP sodium salt (Sigma) with equimolar Mg^{2+} in ddH_2O. Aliquot and store at –80 °C.

5. Adenylylimidodiphosphate (AMP-PNP) (100 mM): dissolve AMP-PNP in ddH_2O to 100 mM and store at –80 °C (*see* **Note 9**).

6. Mg-ATP (100 mM): *see* Subheading 2.2, **step 3**.

7. Dithiothreitol (DTT) (1 M): dissolve DTT in ddH_2O and store at –20 °C. The working concentration is 1 mM.

8. Release buffer: 80 mM PIPES, 2 mM $MgCl_2$, 1 mM EGTA, and 300 mM KCl, pH ~7.

9. Glycerol cushion: 80 mM PIPES, 2 mM $MgCl_2$, 1 mM EGTA, and 60 % glycerol (v/v).

10. Sucrose (2 M): dissolve sucrose in the release buffer to 2 M.

11. Beckman Coulter Optima™ TLX Ultracentrifuge (120,000 maximum rpm).

12. Beckman TLA-120.1 fixed angle rotor (8×34 mm, 0.5-mL tube).

2.4 Coating Microspheres with Anti-GFP Antibodies

1. Tabletop centrifuge (e.g., Eppendorf 5430R) cooled to 4 °C with rotor for 1.5-mL microcentrifuge tubes.

2. Low-power bath sonicator (e.g., Branson B-3).

3. Carboxyl-modified polystyrene microspheres: ~1 μm diameter (*see* **Note 10**), and 100 mg/mL suspension (Bangs Laboratories). Store at 4 °C.

4. Activation buffer: 100 mM NaCl and 10 mM MES (2-(*N*-morpholino)ethanesulfonic acid). Adjust to pH 6.0 with NaOH.

5. Coupling buffer (100 mM sodium phosphate buffer): combine 77.4 mL 1 M Na_2HPO_4, 22.6 mL 1 M NaH_2PO_4, and 900 mL ddH$_2$O. Adjust to pH 7.4 with concentrated NaOH.

6. PBS rinse solution: 137 mM NaCl, 2.7 mM KCl, 4.3 mM Na_2HPO_4, and 1.47 mM KH_2PO_4. Adjust to pH 7.4 with concentrated KOH.

7. Quenching solution (30 mM hydroxylamine hydrochloride ($NH_2OH \bullet HCl$) in PBS): dissolve 0.42 g $NH_2OH \bullet HCl$ in 200 mL PBS rinse solution, and adjust to pH 8.0 with concentrated NaOH.

8. Water-soluble carbodiimide coupling reagent, 1-ethyl-3-(3-dimethylaminopropyl)carbodiimide, hydrochloride (EDAC, Life Technologies), and *N*-hydroxysuccinimidal stabilizing reagent, *N*-hydroxysulfosuccinimide, sodium salt (NHSS, Life Technologies). Store both reagents desiccated (or under argon) at –20 °C (*see* **Note 11**).

9. EDAC quencher: 14.3 M βME. Store at room temperature.

10. 100 mg/mL BSA solution. Store at –20 °C.

11. Anti-GFP antibody stock solution (1–4 mg/mL). Store at –20 °C.

2.5 Pyranose Oxidase/Catalase (POC) Oxygen Scavenger Preparation

1. Pyranose oxidase from *Coriolus* species, 250 U (Sigma Cat. No. P4234; *see* **Note 12** and Swoboda et al. [43] regarding additional steps for experiments involving double-stranded DNA). Store at –20 °C.

2. Catalase (Sigma, Cat. No. C40), store at 4 °C.

3. Tabletop centrifuge (e.g., Eppendorf 5430R) cooled to 4 °C with rotor for 1.5-mL microcentrifuge tubes.

4. Ultrafree Centrifugal Filters (Durapore PVDF 0.1 μm, Cat. No. UFC30VV00).

5. POC storage buffer: 40 mM Tris, 30 % (v/v) glycerol, and 5 mg/mL BSA, pH 7.4.

2.6 Optical Tweezers Setup, Alignment, and Calibration

1. Force-fluorescence microscope.

2. Immersion oil Type A (Nikon).

3. Slide chambers.

4. Trapping beads.

5. 25 mg/mL bovine β-casein (*see* Chap. 9 for preparation instructions).

6. TetraSpeck™ microspheres, 0.1 μm diameter (Life Technologies, Cat. No. T7279).

7. Analysis software. Typically, this software is custom written. However, published software packages are available [44–47], some of which provide libraries that can be used in other programs. Analysis methods to be implemented are described in Subheading 3.8.

8. *Optional*: magnification calibration standard (e.g., MRS-4.1, Geller MicroÅnalytical Laboratory, Inc.).

2.7 Sample Preparation for Optical Tweezers Assay

1. Slide chamber containing surface-immobilized, Cy3-labeled MTs. We suggest using the methods presented in Nicholas et al. (Chap. 9), but if another method is preferred, it can be substituted.

2. BRB80 buffer (*see* Subheading 2.3).

3. 25 mg/mL bovine β-casein (*see* Subheading 2.6).

4. Purified K560, store at –80 °C.

5. 10 mM paclitaxel (Sigma) in DMSO.

6. 100 mM ATP (*see* Subheading 2.2).

7. 1 M DTT (*see* Subheading 2.3).

8. Pyranose 2-oxidase/catalase (POC, *see* Subheading 3.5).

9. 1 M glucose (Fisher), aliquot in ~5 μL volumes and store at –80 °C.

10. Anti-GFP antibody-coated trapping beads (*see* Subheading 3.4).

11. Low-power bath sonicator.

12. Pieces of filter paper cut into strips ~2 in. long and 0.5 in. wide.

13. Vacuum grease (*see* **Note 13**) and cotton-tipped applicator.

2.8 Optical Tweezers Measurement of Motility and Force Generation

1. Force-fluorescence microscope and calibration software.

2. Immersion oil Type A (Nikon).

3. Slide chamber prepared according to Subheading 3.9.

2.9 Force-Fluorescence Microscope Instrumentation

Decisions regarding instrument design depend on many factors, including the desired precision, requirements for optical tweezers force feedback and beam steering, available space and budget, and the number of required fluorescence channels. As a result, instruments vary widely among laboratories, and we will therefore not attempt to give detailed instructions for the design and construction of the optical tweezers/TIRF fluorescence microscope. See, for example, Lee et al. [48] for a hands-on procedure for

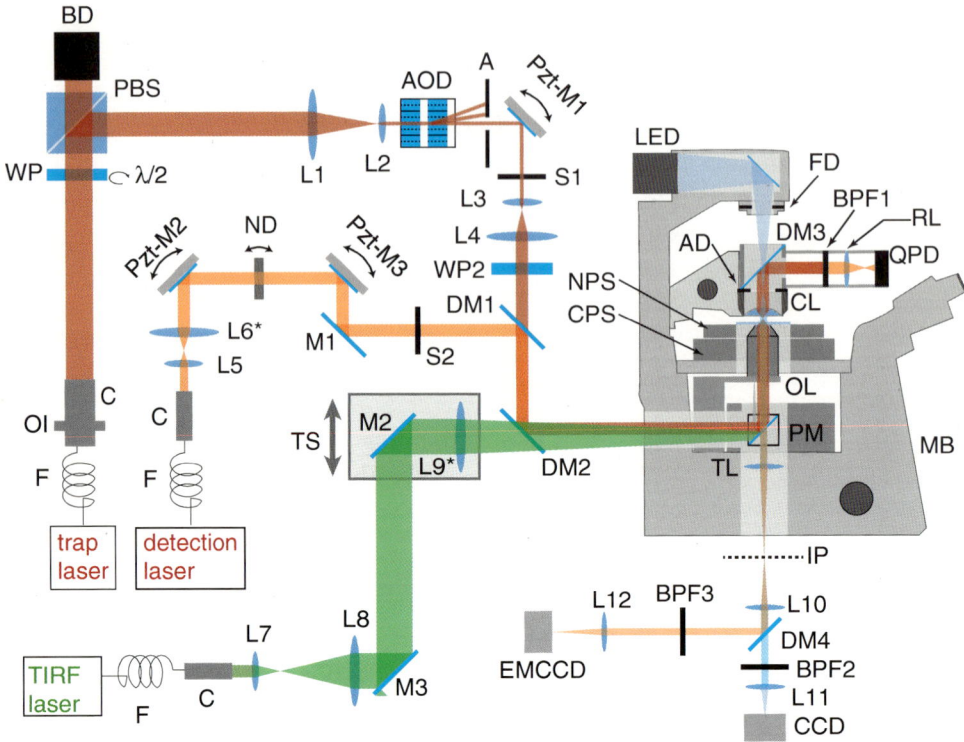

Fig. 3 Force-fluorescence microscope (see text for description). *A* aperture; *AD* aperture diaphragm; *AOD* acousto-optic deflector; *BD* beam dump; *BPF* band-pass filter; *C* beam collimator; *CCD* charge-coupled device (for bright-field detection); *CL* condenser/collection lens; *CPS* coarse-positioning stage; *DM* dichroic mirror; *EMCCD* electron-multiplying CCD (for fluorescence detection); *F* single-mode, polarization-maintaining optical fiber; *FD* field diaphragm; *IP* image plane; *L* lens; *L** lens mounted on a translation stage for fine focus adjustment; *LED* light-emitting diode; *M* mirror; *MB* microscope body; *ND* neutral density filter; *NPS* nanopositioning stage; *OI* optical isolator; *OL* objective lens; *PBS* polarizing beam splitter; *PM* polychroic mirror; *Pzt-M* piezo-driven mirror mount; *QPD* quadrant photodiode; *RL* relay lens; *S* shutter; *TL* tube lens; *TS* translation stage; *WP* half-wave plate

building a simple optical trap on an inverted fluorescence microscope, Selvin et al. [49] for a protocol for constructing a simple TIRF microscope, and Neuman and Block [26] and van Mameren et al. [50] for excellent discussions of many relevant design considerations for optical tweezers. Instead, we will outline a simplified version of the force-fluorescence microscope in our laboratory and make note of some practically important design and construction details common to all such microscopes. For the protocol instructions, we assume the optical tweezers setup in Fig. 3 and refer to components according to the labels therein.

The illumination pathways consist of beams for trapping, back focal plane detection (*see* **Note 14**), bright-field imaging, and TIRF microscopy. All optics should have antireflection coatings for the appropriate laser wavelengths. The *trapping laser* (1,064 nm) passes through a combined collimator and optical isolator (OI, to prevent back-reflection into the laser resonator via the glass fiber) and a variable beam splitter consisting of a computer-controlled wave plate (WP) and polarizing beam splitter (PBS), allowing precise control of the transmitted laser power (and therefore the trapping spring constant, k). The beam then passes through a two-channel acousto-optic deflector (AOD), the $(1, 1)$-order diffracted beam is selected using an aperture (A), and the beam is re-expanded to fill the rear entrance pupil of the microscope objective lens (OL). The "pivot point" within the AOD is conjugate telecentric to the back focal plane of the microscope objective, so that rotations of the diffracted beam originating from the AOD plane produce translations of the focus in the sample (front focal plane of the objective). The computer-controlled, piezo-electric-driven mirror mount following the AOD (Pzt-M1) is positioned such that when it rotates, it mostly translates the beam in the back aperture of the objective. Together, this mirror and the AOD allow precise control of the position and angle of the trapping beam as it enters the objective lens (note that this simplified diagram omits additional lenses required to achieve the aforementioned AOD and Pzt-M1 optical mappings). The trap is turned on and off by means of a simple mechanical shutter (S1). The half-wave plate (WP2) positioned directly before the microscope allows rotation of the laser beam polarization (*see* **Note 15**).

After beam expansion, the *detection beam* (830 nm) overlaps the trapping beam as it enters the microscope. Two beam steering mirrors (Pzt-M2 and Pzt-M3) work as a pair to precisely control the detection beam alignment. A neutral density filter (ND) on a computer-controlled, motorized mount allows switching of the detection beam from low power (~40 μW, normal operation for detection) to higher power (~3 mW, for trapping and visualization during alignment). Like the trapping beam, the detection beam is also shuttered (S2).

The *TIRF laser* (532 nm, for Cy3 excitation) is expanded in order to illuminate the entire visible sample plane and focused by lens L9* ($f=500$ mm) onto the back focal plane of the objective. M2 and L9* are mounted on a translation stage that allows the laser focus to be moved laterally (off-axis) in the rear aperture of the objective, thereby adjusting the angle of the beam exiting the objective for total internal reflection at the cover slip surface.

The *LED* (470 nm) provides bright-field illumination. The detection pathways consist of bright-field and fluorescence imaging (using a CCD and EMCCD, respectively) and back focal plane detection of the trapped bead position relative to the detection beam (using a quadrant photodiode, QPD). The *bright-field and fluorescence images* are separated using an appropriate dichroic mirror (DM4) and filtered with band-pass filters at 470 and 580 nm, respectively (primarily to block reflected laser light from the

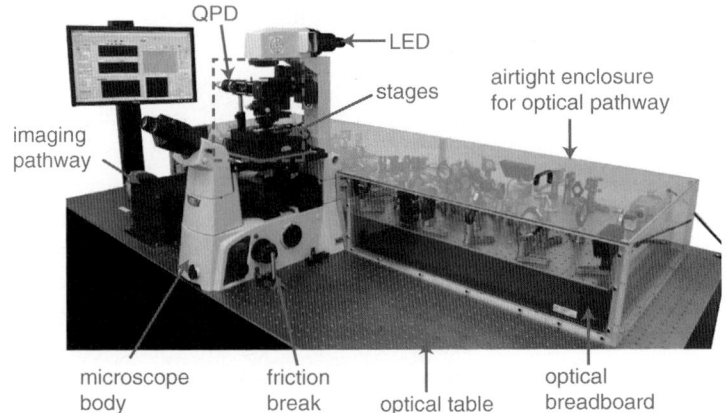

Fig. 4 Force-fluorescence trapping microscope. Note the optical pathway elevated on an optical breadboard and enclosed in an airtight box. The friction break [165] consists of an optical post or other rigid element bolted firmly to the table and pressed against the microscope fine-focus knob to prevent drift of the objective over time. The region demarcated by the red dashed line (back focal plane imaging arm) is shown in greater detail in Fig. 12. Refer also to Fig. 3

trapping beam). The images are then relayed onto the appropriate detectors and can be viewed simultaneously in the control software. Magnification should be chosen so that the effective pixel size of both imaging systems (physical pixel size divided by magnification) is close to equal. This allows the images to be easily registered and overlaid with minimal image processing.

The *back focal plane detection* arm collects the trapping and detection beams, filters out the trapping beam, and uses a relay lens (RL) to image the back focal plane of the condenser lens (CL) onto a QPD. The voltage signals from each quadrant are low-pass filtered at the Nyquist frequency (half the data acquisition frequency used by the software) to prevent aliasing.

Note that the polychroic mirror (PM) is a custom-designed element, which simultaneously reflects all laser beams, while transmitting the bright-field and fluorescence images (in our instrument, this mirror allows for simultaneous imaging of two fluorescence channels in addition to the one shown here).

Components of interest:

1. Foundational components: thermally and acoustically isolated microscope room (acoustic noise criteria, NC30 or 45; vibration criteria, VC-D or VC-E; temperature stability, ±0.2 °C or better), vibration isolation table (Technical Manufacturing Corp., 24 in. thick, Part No. 784-37397-01), and optical pathway enclosure (Fig. 4) with optical breadboard (Newport Corp., Part No. RG-26-4-ML) (*see* **Note 16**).

2. Inverted microscope body with illumination pillar (Nikon model Eclipse Ti-U). For stability, the rubber feet on the base of the microscope are removed, and the body is firmly bolted to the optical table using right-angle steel brackets.

3. 100× oil-immersion, high-numerical aperture (NA), apochromatic microscope objective lens (*see* **Note 17**), NA 1.49 (Nikon, model CFI Apo TIRF 100× Oil).

4. High-NA oil condenser lens, NA 1.4 (Nikon, model HNA-Oil).

5. Coarse-positioning stage (Physik Instrumente, Part No. M-686.D64; *see* **Note 18**).

6. Nanopositioning stage (Physik Instrumente, Part No. P-517.3CD).

7. 1,064-nm trapping laser: linearly polarized, diode-pumped CW Ytterbium laser with 10-W maximum output (IPG Photonics, Part No. YLR-10-1064-LP; *see* **Note 19**) with polarization-maintaining, single-mode optical fiber output and combined beam collimator/expander/optical isolator (Part No. ISO-1080-100).

8. 830-nm detection laser with polarization-maintaining, single-mode optical fiber and beam collimator (Qioptiq, Part No. iFLEX-P-10-830-0.7-50-NP).

9. Shutters and shutter controllers (Thorlabs, Part Nos. SH05 and SC10, respectively).

10. Picomotor™ piezo-controlled mirror mounts (New Focus/Newport Corp., model 8821; *see* **Note 20**).

11. High-capacity beam dump for trapping laser (Thorlabs, Part No. BT510).

12. Acousto-optic deflector (IntraAction Corp., Part No. DTD-274HD6; *see* **Note 21**).

13. Quadrant photodiode (QPD) and power supply (Electro-Optical Systems, Inc., Part Nos. S-078-QUAD-E4/1MHZ and PS1, respectively) with 1 MHz bandwidth and near-maximal output (~0.55 A/W) at 830 nm.

14. Two dual-channel low-pass filters (Stanford Research Systems, Inc., SR640), one channel for each QPD quadrant signal (*see* **Note 22**).

15. Digital acquisition board (National Instruments, Part No. 6281).

16. 532-nm laser for Cy3 TIRF excitation (MeshTel/INTELITE, Inc., Part No. GM32-100GSA-P10) with polarization-maintaining, single-mode optical fiber and integrated collimator (Qioptiq).

17. Achromatic doublet lens with antireflection coating for visible wavelengths, for TIRF illumination (f=500 mm; diameter=50.8 mm; Newport Corp., Part No PAC091AR.14).

18. kineMATIX fiber manipulators (Qioptiq) for laser collimator alignment (*see* **Note 23**).

19. Polychroic mirror (Chroma Technologies, custom Part No. zt488/532/633/830/1064rpc; *see* **Note 24**).

20. Dichroic mirrors designed for laser beam combination (Chroma Technologies or Semrock, Inc.; *see* **Note 25**).

21. Piezo-driven rotation stage (Newport Corp, Part No. AG-PR100) for half-wave plate rotation.

22. LED (Thorlabs, Part No. M470L2) for bright-field imaging.

23. CCD for bright-field imaging (The Imaging Source, Part No. DMx 31BU03).

24. EMCCD for MT visualization (Andor Technology, model Luca S 658; *see* **Note 26**).

25. Trap control and analysis software (programmed in-house using a combination of National Instruments LabView and MathWorks MATLAB; *see* **Note 27**).

3   Methods

3.1 Transformation of BL21(DE3) Competent E. coli Cells and Expression of the K560 Construct

This protocol describes *E. coli*-based expression of K560-GFP [42]. **Steps 1–7** describe BL21(DE3) competent *E. coli* cell transformation with the K560-GFP plasmid, and subsequent steps describe cell growth for protein expression.

1. Take out two plates per transformation from 4 °C storage, and set them on the bench for 0.5 h to allow them to rise to room temperature. Place plates upside-down to prevent condensation of water on the agar, and label them.

2. Thaw one vial (50 μL) of *E. coli* competent BL21(DE3) cells on ice for 10 min (*see* **Note 28**).

3. Add 1–5 μL plasmid DNA (≥1 ng) into the tube; gently flick the solution 4–5 times to mix cells and plasmid (do not vortex).

4. Incubate on ice for 30 min (do not mix) (*see* **Note 29**).

5. Heat shock the cells at 42 °C in water bath for exactly 10 s; do not mix (*see* **Note 30**).

6. Put the tube on ice for 5 min (do not mix).

7. Spread 2 μL cells with 98 μL LB on one plate evenly, and the rest onto another plate. Seal the plates with Parafilm and incubate them at 37 °C overnight. Colonies should appear after 12 h.

8. Pick a single colony and inoculate it in 5 mL LB/carbenicillin in a test tube. Shake vigorously at 37 °C overnight.

9. Check the optical density of the overnight culture at wavelength of 595 nm (OD_{595}). Inoculate an appropriate amount

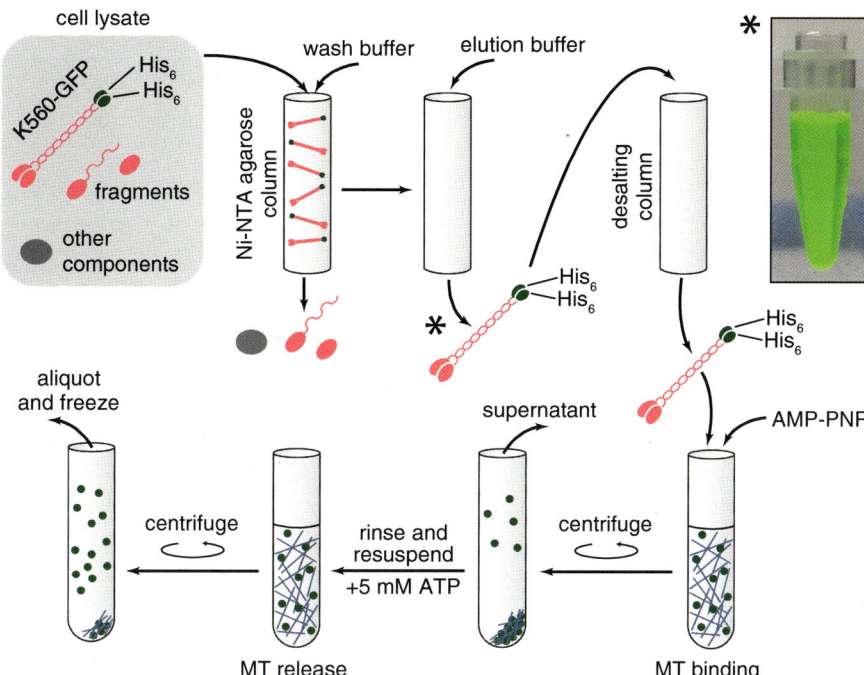

Fig. 5 K560 purification. The full-length motor is first purified from other proteins and incomplete fragments by nickel-nitrilotriacetic acid (Ni-NTA) chromatography, isolating only those proteins with the polyhistidine (His$_6$) tag (marked by the *asterisk*; note the inset photograph showing the elution with intense *green color* due to GFP). Next, His$_6$-containing fragments (degradation products) and motors incompetent to bind MTs are removed by MT binding and sedimentation in the presence of AMP-PNP. Motors unable to release MTs in response to saturating concentrations of ATP are then removed, and the purified, functional K560 is aliquotted, flash-frozen in liquid nitrogen, and stored

into 50 mL LB/carbenicillin such that the starting OD$_{595}$ ≈ 0.1. Incubate culture at 37 °C for ~2–3 h with vigorous shaking. OD$_{595}$ should reach ~0.6–1.

10. Inoculate the 50-mL culture into 1 L LB/carbenicillin for ~2–3 h until OD$_{595}$ reaches ~0.8–1 (*see* **Note 31**).

11. Cool culture on ice until temperature is <20 °C. Reserve 10 μL for gel. Add IPTG to final concentration of 0.1–0.2 mM. Shake vigorously overnight at 18 °C (*see* **Note 32**).

12. Harvest cell culture by centrifugation at 6,000 rpm (4,000 rcf) for 15 min at 4 °C. Discard the supernatant. Store the cell pellet in a 50-mL Falcon tube at –80 °C (*see* **Note 33**).

3.2 Ni-NTA Agarose-Based Protein Purification

This is the first step in purifying the K560-GFP construct expressed above. It utilizes the polyhistidine tag included in the construct and is based on methods given by Bornhorst and Falke [51] and Stock and Hackney [52] (*see* Fig. 5).

1. Chill all the buffers on ice; add βME (final 5 mM) and Mg-ATP (final 50 μM) to all the buffers right before purification. Add protease inhibitor PMSF into lysis buffer to 0.5–1 mM (*see* **Note 34**). Add dissolved lysozyme into lysis buffer to 1–2 mg/mL.

2. Keep the pellet at room temperature until it is almost completely melted. Add 10 mL lysis buffer to the pellet, and gently invert the tube several times until the solution is homogenous. Incubate on ice for 30 min. The solution will become viscous due to the release of DNA from the cells.

3. To lyse the cells, sonicate the solution at 35 % power for 20 cycles of 30 s pulsing with 2.5-min rests between pulses (*see* **Note 35**). Keep the container holding the solution in contact with ice all the time.

4. Supply the lysate with 10 mM imidazole and 0.5 mM PMSF after sonication.

5. Clear lysate by centrifugation at 30,000 rcf for 30 min at 4 °C.

6. Wash 2 mL Ni-NTA agarose with 10 mL resin wash buffer in a BioRad column, and transfer the agarose to a prechilled 50-mL Falcon tube.

7. Add cleared lysate (reserve 2 μL for SDS-PAGE gel analysis) to Ni-NTA agarose, and nutate the mixture at 4 °C for 1 h in the dark.

8. In a 4 °C cold room, pour the mixture into a BioRad column; reserve 2-μL flow-through for gel.

9. Wash the resin with 3×10 mL wash buffer; collect 2 μL from each wash fraction for gel.

10. When the solution is almost drained, cap the column tightly. Add 1 mL elution buffer, resuspend the resin, and incubate for 5 min. Elute 0.5 mL into a chilled 1.5-mL microcentrifuge tube. Hereafter each time add 0.5 mL elution buffer, incubate for 5 min, and then elute 0.5 mL.

11. Estimate eluted protein concentration on a 96-well plate. Basically, add 1–2 μL eluent from each fraction to 200 μL Coomassie blue reagent, mix well, and identify the wells with the most intense blue color.

12. Wash a BioRad desalting column with 2×10 mL storage buffer.

13. Combine fractions that are most concentrated (typically ~4–6 fractions), and load the solution onto the desalting column.

14. Allow the solution to sink into the desalting column, and discard the first 2-mL flow-through.

15. Each time add 0.5 mL storage buffer and collect 0.5 mL. Estimate protein concentration on the 96-well plate, and

combine concentrated fractions. Determine bulk protein concentration using Bradford assay (*see* **Note 36**). Run a SDS-PAGE gel to estimate purity and yield.

16. Aliquot and flash freeze the solution. Store at −80 °C. For further purification via MT binding and release, 100 μL aliquots are recommended.

3.3 MT Binding-and-Release Purification of K560

Ni-NTA agarose-based purification isolates all proteins with a polyhistidine tag and native proteins with intrinsic high affinity for Ni. Here, we further purify this fraction to isolate "full-length," functional (non-degraded/cleaved) kinesins (*see* **Note 36**). **Steps 1–7** describe how to polymerize tubulin to form MTs, while the subsequent steps describe kinesin purification via MT-mediated pulldown and nucleotide-induced release (Fig. 5).

1. Calculate the amount of tubulin needed. Three- to fourfold excess molar amount of tubulin over kinesin is recommended. For example, for 100 μL of 1 mg/mL kinesin solution, use 400 μg tubulin.

2. Add 0.4 μL of 100 mM Mg-GTP (1 mM) and 0.4 μL of 0.1 M DTT (1 mM) to 40 μL of 10 mg/mL tubulin. Incubate at 37 °C for 15 min.

3. Add 4 μL of 0.2 mM paclitaxel (final 20 μM paclitaxel and 10 % DMSO), and incubate at 37 °C for 15 min (*see* **Note 37**).

4. Add 3 μL of 2 mM paclitaxel (final 20 μM) and 0.3 μL of 1 M DTT (final 1 mM) to 300 μL glycerol cushion and to 300 μL BRB80 buffer. Pipette up and down gently to mix well. Transfer the glycerol cushion to a clean 0.5-mL TLA-120.1 tube. Lay the polymerized tubulin solution on top of the glycerol cushion.

5. Use TLA-120.1 rotor to spin the MTs through the glycerol cushion for 10 min at 80,000 rpm (227,000 rcf average; k-factor 18.3) at room temperature.

6. Rinse top of the solution with 100 μL of BRB80 buffer. Carefully take off the supernatant, and then wash the pellet with 100 μL of BRB80.

7. Gently resuspend the pellet in 40 μL of BRB80 to render ~10 mg/mL MTs.

8. Add 1 μL of 0.1 M AMP-PNP (final 1 mM) and 1 μL of 2 mM paclitaxel (final 20 μM) to 100 μL of motor solution. Incubate on ice in the dark for 5 min.

9. Warm the motor solution to room temperature; then add 40 μL of prepared MTs. Incubate at room temperature for 15 min in the dark (reserve 1 μL for gel).

10. Add DTT and paclitaxel to 300 μL glycerol cushion, 200 μL BRB80, and 100 μL release buffer (final 1 mM and 20 μM, respectively).

11. Add the glycerol cushion to TLA120.1 tube, carefully lay the mixture on top, and spin at 80,000 rpm (227,000 rcf average; k-factor 18.3) for 12 min at room temperature (reserve 1 μL of supernatant for gel).

12. Wash top of the cushion with 100 μL BRB80, remove supernatant, and then wash pellet with 100 μL BRB80 (be careful not to disturb the pellet).

13. Resuspend pellet in 100 μL release buffer with 5 mM Mg-ATP (reserve 1 μL for gel), and incubate at room temperature for 5 min.

14. Spin at 80,000 rpm (227,000 rcf average; k-factor 18.3) for 10 min at room temperature.

15. Transfer supernatant (which contains released kinesin) to a prechilled 1.5-mL microcentrifuge tube on ice, and add 33 μL prechilled 2 M sucrose (reserve 1 μL for gel).

16. Aliquot the solution into small tubes in 3 μL volumes (or appropriate amount for future experiments), flash freeze, and store at −80 °C.

17. Run a 10 % SDS-PAGE gel to determine the released fraction (*see* **Note 38**).

3.4 Coating Microspheres with Anti-GFP Antibodies

In this procedure, microspheres are coated with anti-GFP antibodies and BSA (Fig. 6). These coated microspheres will bind K560-GFP in the optical trapping assay. The protocol given here is similar to the one we reported previously [53]. *See* ref. [54] for a detailed discussion of the relevant chemistry.

1. Briefly vortex microsphere stock suspension to evenly distribute the microspheres. Pipette 100 μL of the microsphere stock into a 1.5-mL microcentrifuge tube. Add 900 μL activation buffer to the microspheres and mix well using the pipette.

2. Centrifuge at 1,200 rcf for 15 min at 4 °C. Carefully remove and discard the supernatant (do not remove any microspheres), and resuspend with 1 mL of activation buffer. Vortex and sonicate for 10–20 s in the low-power bath sonicator to ensure there is no aggregation (*see* **Note 39**).

3. Repeat **step 2** two more times. During the last centrifugation, obtain the EDAC and NHSS from the freezer. Allow them to equilibrate to room temperature before opening (*see* **Note 11**).

4. In a separate 1.5-mL microcentrifuge tube, weigh 5 mg of EDAC and 10 mg of NHSS and immediately add the 1 mL bead suspension to activate the beads. Mix well and sonicate for 10 s. Incubate for 30 min at room temperature while gently mixing on a rocking platform.

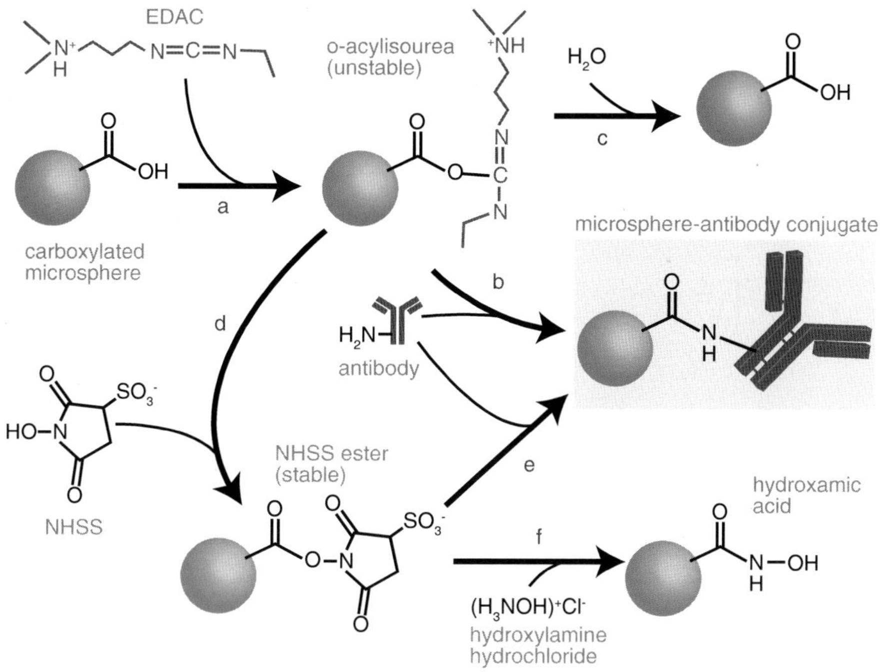

Fig. 6 Coupling antibodies to optical trapping microspheres. First, a microsphere bearing surface carboxyl groups reacts (**a**) with the carbodiimide EDAC (1-ethyl-3-(3-dimethylaminopropyl)carbodiimide) to form an unstable *O*-acylisourea intermediate. This short-lived species is reactive toward primary amines and thus capable of forming stable, covalent amide bonds with lysyl residues on the antibody surface (**b**) and yielding an isourea byproduct (not shown). However, this *O*-acylisourea is also readily and rapidly hydrolyzed, yielding the original carboxyl group (**c**). Adding NHSS (3-sulfo-*N*-hydroxysulfosuccinimide) to the reaction greatly enhances microsphere–antibody coupling efficiency by forming a relatively water-stable, but amine-reactive NHSS ester (**d**) that can likewise bind antibodies (**e**). BSA (not shown) is added with the antibody and also binds via its surface lysines, thus blocking the remainder of the microsphere surface. After antibodies and BSA are bound, any residual NHSS esters are removed by adding hydroxylamine (**f**), which reacts with the ester to form hydroxamic acid

5. *Optional*: add 1.4 µL of 14.3 M βME (20 mM final concentration) to quench the EDAC reaction (*see* **Notes 5** and **40**). Invert gently back and forth by hand for ~30 s.

6. Repeat **step 2** three times, using coupling buffer in place of activation buffer (*see* **Note 41**) and using 250 µL for the final resuspension. During the final centrifugation, thaw the BSA and antibody solutions and place them on ice.

7. In a separate tube, mix 55 µL of 100 mg/mL BSA with a volume of antibody corresponding to ~0.2–0.5 mg (~6 mg total protein). Add coupling buffer to bring the total volume to 500 µL. Mix well. Then add the microsphere mixture to the protein solution, mixing rapidly, but gently with the pipette (*see* **Note 42**). Sonicate briefly (3–5 s).

8. Allow the mixture to react for 2–4 h at room temperature (or overnight at 4 °C) with constant nutation.

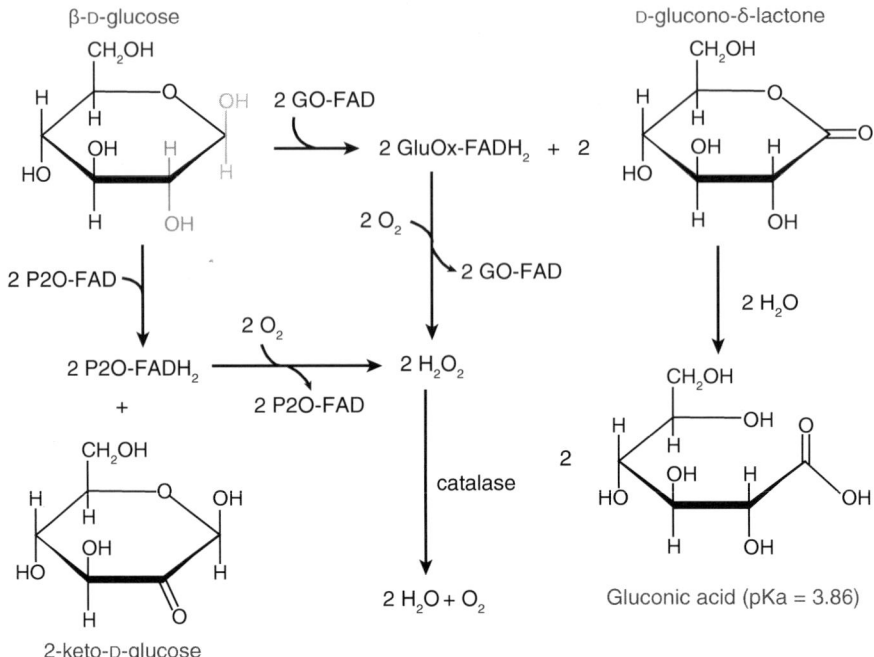

Fig. 7 Pyranose 2-oxidase (P2O) oxygen scavenging system, with comparison to glucose oxidase (GO). Both flavoenzymes contain flavin adenine dinucleotide (FAD) prosthetic groups that are reduced by glucose, thereby generating FADH$_2$ and removing two hydrogens from the sugar. Whereas GO removes hydrogens from the C1 position (*green*), P2O acts on C2 (*magenta*), yielding D-glucono-δ-lactone and 2-keto-D-glucose, respectively. Whereas 2-keto-D-glucose is stable, in the presence of water, D-glucono-δ-lactone hydrolyzes to gluconic acid, thereby acidifying the reaction solution. The reduced enzymes (FADH$_2$ forms) are oxidized by O$_2$, regenerating the original enzymes and producing H$_2$O$_2$, which is converted by catalase to H$_2$O and O$_2$. In the net reaction, for each O$_2$ removed, two glucose molecules are consumed, and for GO, two gluconic acid molecules are generated. Although β-D-glucose, the substrate for GO, is shown, P2O has no anomeric preference and can also catalyze the oxidation of α-D-glucose [43, 111]

9. Repeat **step 2** twice using PBS rinse solution in place of activation buffer. After the final centrifugation, resuspend in 1 mL of quenching solution.

10. Incubate in quenching solution for 30 min at room temperature (or overnight at 4 °C; *see* **Note 43**).

11. Repeat **step 2** three times using PBS rinse solution in place of activation buffer. After the final centrifugation, resuspend in 100 μL of PBS or assay buffer. Add 0.5 μL of 100 mg/mL BSA, mix, and store at 4 °C (*see* **Note 44**).

3.5 Pyranose Oxidase/Catalase (POC) Oxygen Scavenger Preparation

The procedure below yields a solution of approximately 600 U/mL pyranose oxidase and 18 kU/mL catalase, which work together to deplete oxygen from the sample chamber in the trapping assay (Figure 7 demonstrates the basis for oxygen removal and compares POC to the conventional system using glucose oxidase; *see* **Note 12**).

This is diluted 200× in the trapping assay, to yield 3 and 90 U/mL pyranose oxidase and catalase, respectively.

1. Remove catalase from refrigerator, invert bottle several times to suspend contents evenly, and transfer 75 μL to a 1.5-mL microcentrifuge tube on ice.

2. Centrifuge at 20,000 rcf for 5 min at 4 °C. Remove the supernatant (which contains antimicrobial thymol preservative), estimate its volume with the pipette, and discard. Resuspend the pellet in 200 μL ddH$_2$O.

3. Centrifuge again at 20,000 rcf for 5 min at 4 °C. Discard the supernatant. Resuspend the pellet in POC storage buffer in a volume equal to that of the supernatant removed in **step 2**. Keep on ice.

4. Remove the pyranose oxidase from the freezer and place the vial on ice. Add 350 μL of POC storage buffer to dissolve the pyranose oxidase. Pipette up and down gently, avoiding bubbles and dissolving all protein in the vial. The solution will have a bright yellow color owing to the flavin adenine dinucleotide (FAD) prosthetic group in the enzyme.

5. Transfer the pyranose oxidase solution to a clean 1.5-mL microcentrifuge tube, and add 22 μL of the catalase solution. Measure the volume with the pipette.

6. Using a sufficient volume of POC storage buffer to bring the POC solution to 415 μL total, rinse the pyranose oxidase vial, and add the rinse to the POC solution.

7. Spin filter the solution two times using the same centrifuge settings as above (split the solution between two different spin columns for the first centrifugation to avoid clogging the filter).

8. Aliquot in 3 μL volumes in the smallest plastic tubes available, snap freeze in liquid nitrogen, and store at −80 °C.

3.6 Optical Tweezers Setup

Here we describe the initial adjustments and calibrations necessary to perform quantitative experiments with the optical tweezers. These procedures need only be performed once during initial instrument setup, but it is good practice to check the adjustments periodically. The aims of these steps are to (a) align the axes of the AOD, CCD, and nanopositioning stage, (b) determine the AOD beam deflection in response to applied voltage, and (c) align the bright-field and fluorescence images. As a starting point, we assume that the trapping laser is coupled into the microscope and coarsely aligned along the optical axis and that TIRF imaging is established. The lateral position of the trapping beam focus should be recorded and displayed in real time on the CCD image as a point of reference (e.g., as a crosshair superimposed on the image).

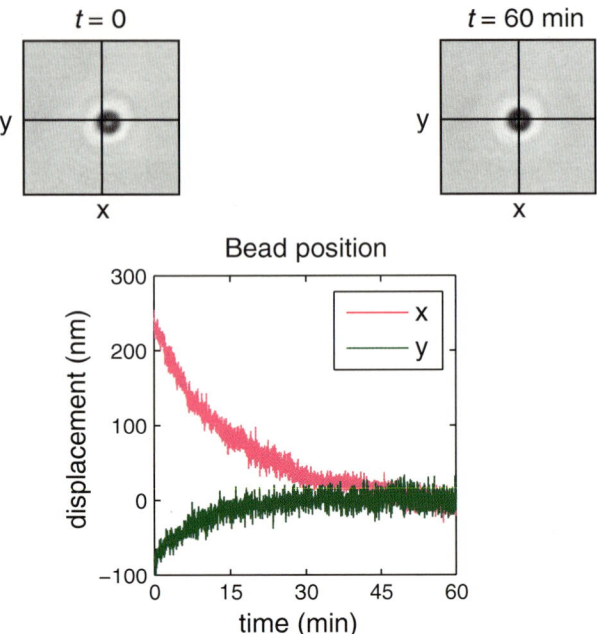

Fig. 8 Thermal equilibration of the optical tweezers laser pathway. When the optical trapping laser is initially powered on, the trap center is typically offset from its previous stable position (*left-hand image*; the cross to the *upper left* of the bead is the stable position to which the bead moves after the optical elements in the pathway expand due to heating by the trapping laser). The new position is typically very close (tens of nanometers) to the stable position during the preceding use of the instrument (*right-hand image*). The graph shows a representative example of microsphere movement over time as the optical pathway thermally equilibrates. The position traces were obtained by tracking the trapped microsphere in images acquired every second for 60 min, using a simple centroid-finding algorithm. The paths shown are followed fairly consistently each time the instrument is powered on before experiments. The instrument requires approximately 45–60 min to fully stabilize

Steps 1–4 describe the thermal equilibration of the instrument and initial focusing, as required whenever the instrument is used. For these adjustments, use a high trap power (~100 mW entering the back aperture of the objective) in order to reduce bead diffusion.

1. Turn on the trapping and detection lasers and all associated electronics, QPD last. Set the output power of the trapping laser to 100 % (or whichever setting exhibits the best power stability). Open shutters S1 and S2, and allow the trap to thermally equilibrate for at least 30–45 min (*see* Fig. 8 and **Note 45**).

2. Place a drop of immersion oil on the objective lens. Fix the slide chamber securely on the nanopositioning stage, cover slip side down, and raise the objective to contact the center of the cover slip. On the top of the slide chamber, apply approximately 0.2 mL of immersion oil, and lower the condenser to make contact with the oil (*see* **Note 46**).

3. Turn on the bright-field LED and observe the CCD image. Raise the objective until microspheres come into focus. Then use the fine focus to move the objective down until new beads stop coming into focus, and the visible beads appear white. These beads are diffusing into the cover slip surface, just slightly above the focal plane. Move the objective upward just until these beads appear dark.

4. Close the field diaphragm almost completely. Adjust the condenser height until the edges of the diaphragm come into sharp focus. This achieves Köhler illumination. Lock the condenser position and open the field diaphragm to just beyond the field of view.

*The following steps align the CCD, nanopositioning stage, and AOD axes (see **Note 47**).*

5. Prepare a slide chamber and flow in 15 µL of trapping beads diluted ~1:1,000 from stock in BRB80. Incubate for 15 min to allow the beads to bind the surface, and flow in 20 µL of 1 mg/mL casein in BRB80. Incubate 5 min, and flow in 15 µL of trapping beads diluted ~1:1,000 from stock in the 1 mg/mL casein solution. Seal the chamber ends with vacuum grease, and place the chamber on the microscope (follow **steps 2–4** above).

6. Focus on the stuck beads so they appear dark, and move one to the marked focal position of the trapping beam (this should be the center of the CCD field of view). Move the stage in a 3 µm × 3 µm grid pattern with 200-nm steps, centered at the trapping beam mark. Record an image with the CCD at each position (this process should be automated).

7. Create a minimum projection image for the stack of images acquired in the previous step (either in the trap software itself or using an analysis program such as NIH ImageJ 55, 56). This will form a dark square.

8. Draw a box around the square to determine the rotation of the CCD relative to the stage. Carefully rotate the CCD to correct this rotation.

9. Repeat **steps 6–8** until there is no observable rotation of the CCD relative to the stage.

10. Trap a bead, and move it close to the cover slip surface (~100 nm above the point of bead–cover slip contact, as judged by eye). Apply appropriate voltages to the AOD driver frequency modulation inputs to step the bead in a similar grid to that used in **step 6**, again recording images at each position.

11. Create a minimum projection image, as in **step 7**. Overlay this image with the final image formed after adjusting the CCD (**step 9**). Draw lines along the edges of each square to determine the angle of rotation θ (in radians) of the AOD axes relative to the stage axes (*see* **Note 48**). The sign of θ is positive if

the AOD is rotated *clockwise* relative to the stage and negative otherwise.

12. Apply a counterclockwise rotation transformation (in the software) to the voltages sent to the AOD, of the form

$$A_{x,r} = A_{x,0} + (A_x - A_{x,0})\cos(\theta) - (A_y - A_{y,0})\sin(\theta) \text{ and}$$
$$A_{r,y} = A_{y,0} + (A_y - A_{y,0})\cos(\theta) + (A_x - A_{x,0})\sin(\theta),$$

where A_x and A_y are the original voltages in the x and y directions, respectively, $A_{x,0}$ and $A_{y,0}$ are constant offsets (if any) applied when the beam is in its center position, and $A_{r,x}$ and $A_{r,y}$ are the corresponding voltages after rotation (*see* **Note 49**).

13. Recheck the rotation of the AOD axes relative to the stage, following **steps 10** and **11**, while applying the rotation in **step 12** to each point. If the rotation is corrected, the transformation in **step 12** should be applied henceforth to all beam steering by the AOD.

The following steps determine the distance moved by the trap in response to voltage applied to the AOD:

14. Determine the effective pixel size of the CCD. Using the same sample chamber as in the previous steps, move the bead by several known distances, saving an image at each position (*see* **Note 50**). Track the bead position, and determine the separation in pixels D_{px} between subsequent bead positions. Divide the known distance moved D_{nm} by D_{px} to find the effective pixel size in nanometers $P_{nm} = D_{nm}/D_{px}$. Repeat this measurement several times and take the average. *Optional*: repeat the computation of P_{nm} using a magnification calibration standard such as the MRS-4.1. The answers from both methods should agree to within ~2 nm/pixel or less.

15. Trap a bead. Apply known voltages to the AOD driver frequency modulation inputs (while applying the rotation transformation determined above), in order to step the bead in various lines, saving an image at each position. Track the bead centroid position (in pixels) in each image, and plot the x and y pixel positions vs. the applied voltage. Fit a line to these data to determine the distances in pixels moved per volt applied (*see* **Note 51**), i.e., the line slopes $L_{x,px}$ and $L_{y,px}$ for x and y, respectively.

16. Calculate the conversion factor W needed to determine the voltage required to move the trapping beam by a given distance: $W_x = P_{nm} \times L_{x,px}$ (and identically for the y direction). For a desired displacement Δx, a voltage $A_x = \Delta x / W_x$ must be applied to the AOD. Save these conversion factors and apply them permanently in the trap control software.

The following steps align the CCD and EMCCD:

17. Prepare a slide chamber (using a clean, but non-silanized cover slip) with fluorescently labeled 100-nm TetraSpeck beads in 1 M

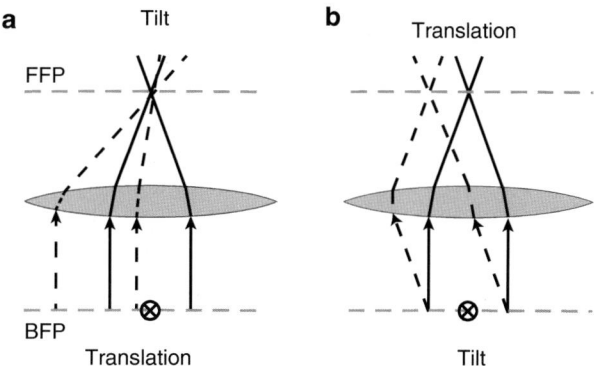

Fig. 9 Important optical relationships for focusing a collimated beam. BFP, back focal plane; FFP, front focal plane. Solid rays represent a beam perfectly aligned with and centered on the optical axis. (**a**) Pure translation of the beam in the back aperture leads to tilting of the beam as it focuses in the FFP, but does not change the position of the focus. This induces asymmetry in intensity pattern of the retroreflected beam. (**b**) Pure tilting of the beam in the BFP leads to translation of the focus in the FFP, essentially without affecting the beam angle as it approaches the FFP. This shifts the position of the intensity pattern of the retroreflected beam, with minimal effects on the intensity distribution

NaCl solution (this will cause the beads to stick to the glass surface) and place it on the microscope (follow **steps 2–4** above).

18. Bring the beads into sharp focus in the bright-field image, and move a bead to the center of the CCD image. Turn on the 532-nm laser and observe the fluorescence channel. Identify the fluorescent spot corresponding to the bead in the center of the CCD field of view, and carefully adjust the position of the image on the EMCCD (by moving the EMCCD itself and/or using the appropriate mirrors in the imaging pathway) so that the spot is also centered on the EMCCD.

19. Move to a region where there are several beads visible on both cameras. Overlay the CCD and EMMCD images in software, applying any scaling or translation necessary for the images of the beads to overlay perfectly. Rotate the EMCCD slightly, if necessary. Focus on the central region of the field of view, ignoring any slight misalignments in the peripheral regions.

3.7 Optical Tweezers Alignment

These procedures comprise a systematic, reproducible method for properly aligning the trapping and detection beams. This procedure is typically done once at the beginning of a series of experiments. After initial adjustment, for a trap that is in frequent use, adjustment on subsequent days typically requires only minute corrections, and in practice, the coarse adjustment can often be skipped. During these procedures it is useful to recall the simple optical diagrams in Fig. 9, which demonstrate the consequences of tilting vs. translating the trapping beam entering the objective lens. Refer to Fig. 3 regarding the location of optical components in the pathway.

The following steps coarsely align the trapping and detection lasers:

1. Prepare a sample slide chamber as described in Nicholas et al. (Chap. 9), but using a clean cover slip (non-aminosilanized). Fill with 1 mg/mL of β-casein in BRB80, and incubate 1 min to block the cover slip surface. Dilute trapping beads ~1:1,000 in 30 μL of the 1 mg/mL β-casein solution. Flow the bead suspension into the chamber and seal it with vacuum grease.

2. Trap a bead. Raise the nanopositioning stage to bring the cover slip into contact with the bead, and continue raising it until the bead starts to be pushed out of the trap (the bead will start to appear white).

3. Turn off the LED and increase the exposure time on the CCD. Remove filter BPF2 in order to view the retro-reflected trapping beam on the CCD (Fig. 10; *see* **Note 52**). Adjust the focus, if needed, in order to view the cross shape clearly at the center of the beam.

4. Zoom in to the center of the pattern on the image display. If the focal spot is not already centered on the marked position of the trap center, move it there using the AOD, and save this position as the new AOD center position ($A_{x,0}$ and $A_{y,0}$ in Subheading 3.6, **step 12**).

5. Zoom out on the image display so that the whole pattern is visible, and observe whether it is symmetrical. It is useful to move the nanopositioning stage up and down while observing the pattern (Fig. 10b). If the pattern is not symmetrical, use mirror Pzt-M1 to correct this.

6. After **step 5**, the center position of the pattern may have shifted slightly. If so, repeat **steps 4** and **5** until the trapping beam pattern is centered and symmetrical (Fig. 10c). The step size of the movements will need to be decreased as the target position is approached.

7. Close shutter S1 to block the trapping beam. Open shutter S2 and move filter ND out of the path to increase the detection beam power.

8. Observe the center position and symmetry of the detection beam. Use mirrors Pzt-M2 and Pzt-M3 to correct the center position and symmetry of the pattern as done for the trapping beam. Begin by using mirror Pzt-M2 to adjust the symmetry of the pattern (important: *see* **Note 53**), and then use Pzt-M3 to center the pattern. Repeat this process iteratively until the detection beam is centered and symmetrical (Fig. 10d). The trapping and detection beams should now be fairly precisely overlapped (Fig. 10e).

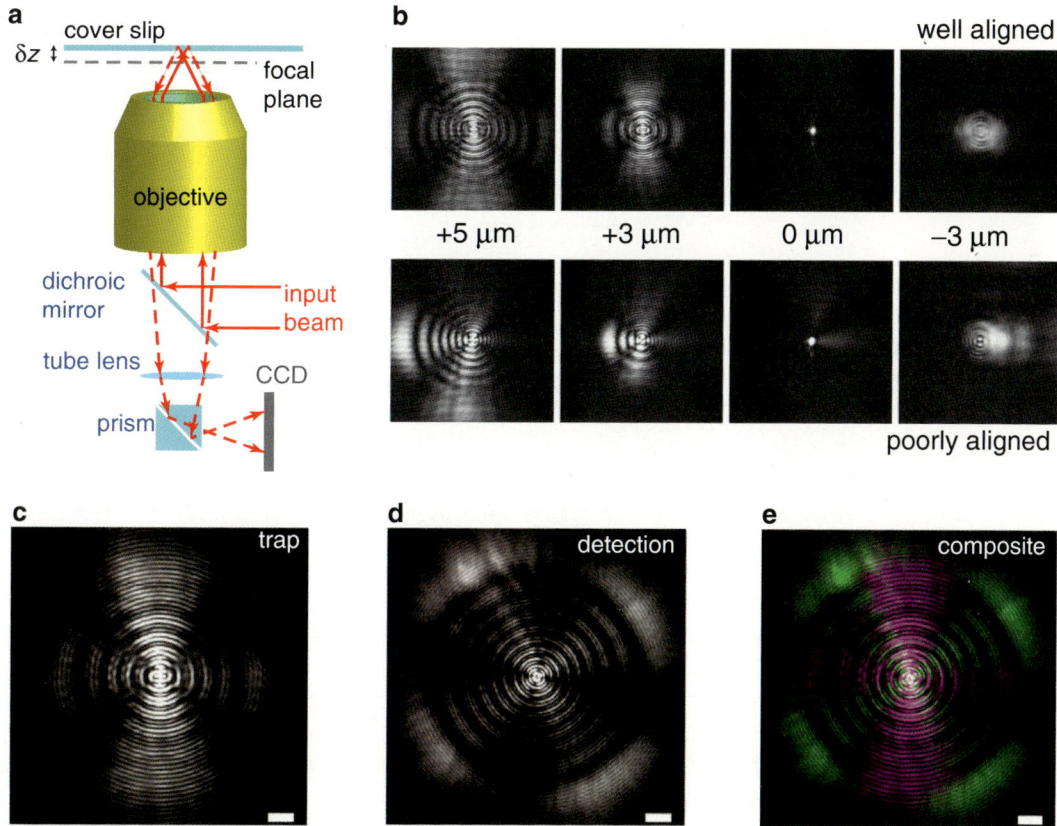

Fig. 10 Coarse alignment using the retroreflected trapping and detection beams. (**a**) Diagram of retroreflected beam detection. The input beam (trapping or detection, *solid rays* in the diagram) reflects off a dichroic mirror with high reflectivity at the laser wavelength and passes through the microscope objective lens to be focused on the cover slip. When the focused beam strikes the interface between the cover slip and the aqueous solution of the sample chamber, a small proportion of the intensity is reflected back into the objective, traveling a reverse path through the system (*dashed rays*). Despite the high reflectivity of the dichroic mirror, a small fraction of this light is transmitted to the imaging optics to form an intensity pattern on the CCD. The size and shape of the pattern depend on the distance δz between the glass–solution interface and the focal plane of the objective (δz is negative when the focal plane is below the interface). (**b**) Images of the retroreflected trapping beam intensity pattern for various values of δz (adjusted by moving the nanopositioning stage holding the slide chamber), for well-aligned (*top row*) and poorly aligned (*bottom row*) beams. The well-aligned beam passes directly through the center of the objective, parallel with the longitudinal axis of the lens (optical axis), forming a symmetrical pattern at each stage position. The poorly aligned beam, which may be displaced, tilted, or both relative to the well aligned beam, forms an asymmetrical pattern that changes (and may shift position in the image) depending on δz. Note that when $\delta z = 0$ (near perfect focusing on the glass–solution interface), both patterns look very similar, and it is only when the beams are "defocused" that the differences become apparent. After proper alignment of the (**c**) trapping beam and (**d**) detection beam, the two patterns appear much more symmetrical and are concentric with each other (**e**, composite image: detection beam pseudocolored *green*, and trapping beam *magenta*). Scale bars in (**c–e**) are 2.25 µm

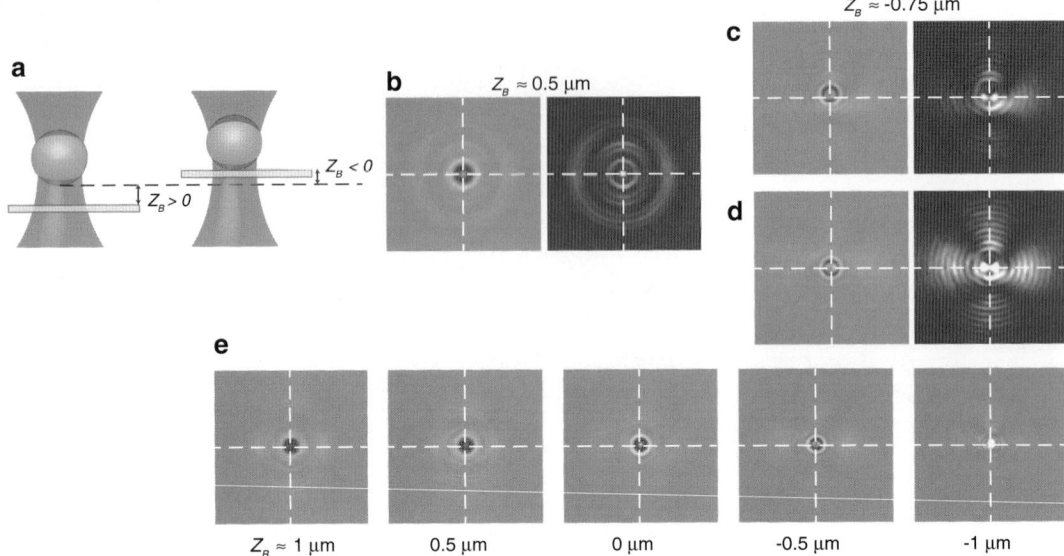

Fig. 11 Refinement of trapping beam alignment. (**a**) The trap holds the microsphere above the cover slip with a distance Z_B between the glass and the surface of the microsphere (*left*, not to scale; Z_B is measured as the axial stage position minus the height of the bottom surface of the bead at its normal trapped position). As the nanopositioning stage moves the cover slip upward, the magnitude of Z_B progressively decreases. Once the cover slip moves high enough, it displaces the bead axially from its equilibrium trapped position (*right*, at this point, Z_B is negative). (**b**) Following coarse adjustment using retroreflections, with $Z_B \approx 0.5$ μm, the microsphere is well centered, and the reflected laser light forms a fairly symmetrical concentric pattern around it (*left*). With the bright-field illumination turned off and the CCD gain increased, it is easier to view the reflected light (*right*). (**c**) After moving the stage upward ($Z_B \approx -0.75$ μm), the microsphere position deviates laterally (*left*) and the reflected light pattern is asymmetrical and off-center (*right*), indicating a slight misalignment of the laser beam. (**d**) After readjustment, the microsphere is centered (left) and the reflection pattern is more symmetrical and centered. (**e**) Following multiple rounds of minor adjustments, the microsphere remains well centered and the retroreflection symmetrical, over a wide range of stage displacements

The following steps refine the alignment of the trapping beam:

9. Close shutter S2. Turn the LED on and adjust the CCD exposure for bright-field imaging. Focus on the beads near the cover slip surface and trap one. Turn down the LED power until the laser back-reflection off the bead becomes visible (with the bead still visible in bright field, Fig. 11b). By turning the LED off momentarily, this back-reflection will appear clearer (Fig. 11b). If the preceding adjustment was done correctly, it will appear quite symmetrical (*see* **Note 54**).

10. Move the stage up (+z direction) slowly, approximately 0.75 μm past the position at which the bead begins to appear bright (Fig. 11a). The bead will likely be displaced radially (Fig. 11c), and the back-reflection will become asymmetrical, with intensity concentrated on one side of the bead.

11. Use mirror Pzt-M1 to adjust the beam so that the back-reflection becomes symmetrical again, and the bead returns to the trap center position (Fig. 11d). Adjust in small increments (*see* **Note 55**). It is useful to switch the LED off periodically to examine the back-reflection more carefully.

12. Move the nonpositioning stage upward by 50–100 nm, and repeat **step 11**. If no adjustment is needed, move an additional 50–100 nm upward.

13. Move the cover slip away from the bead so it is not in contact anymore, and observe the bead position (the adjustments using Pzt-M1 may have small unintended effects on the position). If it is no longer centered at the trap center marked on the CCD, move it back to the center using the AOD, and save the new center positions. Repeat this 2–3 times. When finished, the bead should remain centered with a symmetrical back-reflection, even when the stage presses against it and displaces it axially from the trap center (Fig. 11e).

14. Replace filter BPF2.

The following steps refine the alignment of the detection beam, using back focal plane (BFP) detection signals calculated from the QPD voltages. Figure 12 presents BFP detection and the associated equations for the QPD x and y normalized voltage signals:

15. Close shutters S1 and S2, and completely block the QPD from the light. Set the acquisition frequency for the QPD data to 3 kHz and the low-pass filter frequency to 1.5 kHz. For each preamplifier (preamp) gain setting on the low-pass filters, acquire a few seconds of data, take the average for each quadrant, and save these values in the software as offsets to be subtracted automatically from the corresponding quadrant when using the specified preamp gain (*see* **Note 56**).

16. Repeat **step 15** for 65,536-Hz sampling rate (and any other sampling rates to be used).

17. Open shutter S2, close shutter S1, and remove filter ND. Set the preamp gain on the filters to 0 dB. Trap a bead using the detection laser. If it is not aligned with the marked center position, adjust its position using mirror Pzt-M3. Alternate between trapping with the trapping beam and the detection mean during the adjustment. Adjust Pzt-M3 so that the bead stays in essentially the same position by eye when switching between the two lasers.

Steps 18 and **19** *can often be skipped if the instrument is already in fairly good alignment:*

18. Close both shutters S1 and S2. Adjust the QPD positioner and the QPD relay lens (RL) to positions near the center of their ranges. Move the cover slip well above the focal plane of the

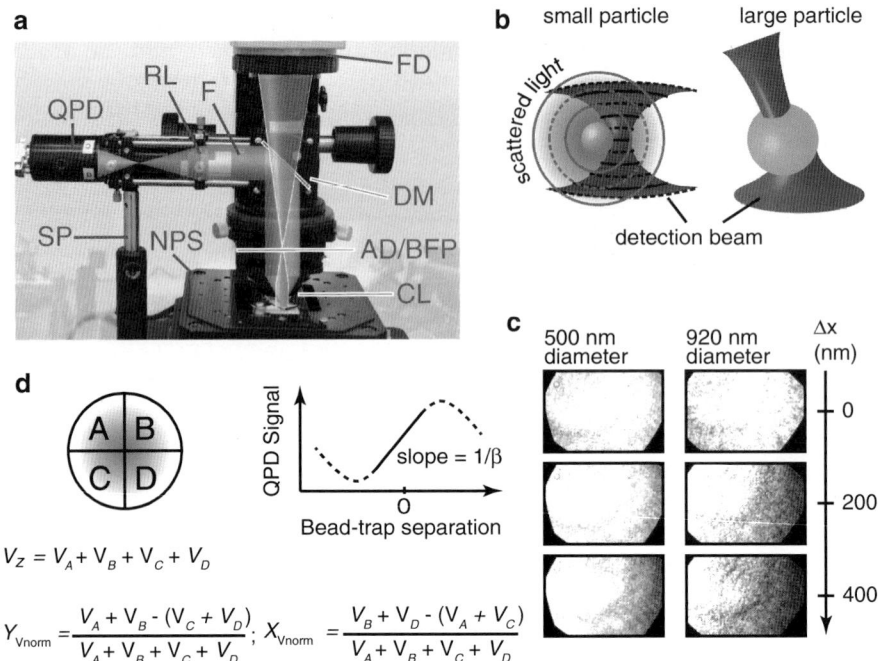

$$V_Z = V_A + V_B + V_C + V_D$$

$$Y_{Vnorm} = \frac{V_A + V_B - (V_C + V_D)}{V_A + V_B + V_C + V_D}; \quad X_{Vnorm} = \frac{V_B + V_D - (V_A + V_C)}{V_A + V_B + V_C + V_D}$$

Fig. 12 Back focal plane detection of microsphere position. (**a**) Optical configuration for back focal plane detection. The bright-field illumination (*gray*, exiting the field diaphragm (FD) from above) and detection beam (*yellow*) propagate in opposite directions through the system. After exiting the objective lens on the inverted microscope, the detection beam enters the slide chamber fixed to the nanopositioning stage (NPS), interacts with the trapped microsphere, and is collected by the condenser lens (CL), which is confocal with the objective. The detection beam is then redirected toward the quadrant photodiode (QPD) detector by a short-pass dichroic mirror (DM) that reflects the 830-nm detection beam, but transmits the 470-nm bright-field illumination. The trapping beam, which follows the same path, is blocked by an 830-nm band-pass filter (F). The relay lens (RL) is positioned such that it images the aperture diaphragm/back focal plane (AD/BFP) of the condenser lens onto the QPD (in this case, the lens is placed three focal lengths from the AD/BFP and 1.5 focal lengths from the QPD, respectively, to achieve a magnification of ½). A support post (SP) with vibration-dampening foam on top helps stabilize the QPD detection arm and eliminate unwanted movements. (**b**) The nature of the interaction of the detection beam with the bead depends somewhat on its size. Very small (<1 μm) particles act as scattering point sources. In this case, the pattern in the back focal plane arises due to interference between the unscattered portion of the detection beam (which essentially propagates without interacting with the particle) and the light scattered by the particle (*left*, *solid* and *dashed lines* represent optical wave fronts). Larger particles may significantly alter the path of the detection beam via refraction of the light as it passes through the microsphere (*right*), causing the entire pattern in the back focal plane to shift. (**c**) Examples of the patterns observed by replacing the QPD with a CCD camera. The dark regions with sharp edges in each corner are images of the aperture diaphragm (located at the back focal plane of the condenser and visible here because it has been partially closed). The *left* and *right* columns correspond to 500- and 920-nm-diameter beads, respectively. Approximate bead–trap separations for each set of images are given on the right side. Note that the larger particle has a more pronounced effect on the overall beam position. (**d**) The voltage signals from the four quadrants of the QPD (labeled V_A, V_B, V_C, and V_D, respectively) are used to calculate response signals in three dimensions (normalized by the total voltage for the x and y directions). The response signals each have similar shapes and are linear with displacement near the center of the detection beam (solid black line, slope $= 1/\beta$). In this region, the QPD response X_{Vnorm} can be directly converted to displacement $\Delta X = \beta_x X_{Vnorm}$ (and identically for ΔY)

objective, and open S1 so that the trapping beam scatters microspheres away from the optical axis (this minimizes fluctuations in the QPD signals).

19. Open shutter S2. Adjust lens RL so that both the x and y signals ($V_{\mathrm{norm},x}$ and $V_{\mathrm{norm},y}$, respectively) are zero. This centers the detection beam on the QPD.

20. Close shutter S1. Move the stage back downward so the beads come into focus and trap one with the detection laser. Adjust the position of the QPD so that $V_{\mathrm{norm},x}$ and $V_{\mathrm{norm},y}$ fluctuate around zero (often no adjustment is needed at this step).

21. Reinsert filter ND to reduce the power of the detection beam. Open shutter S1 and trap a bead within roughly 50–100 nm of the cover slip surface. Observe the normalized QPD signals and use mirror Pzt-M3 to center them at zero (*see* **Note 57**).

22. Using the AOD, sweep the trapped bead in a triangle-wave pattern along the x axis with amplitude ~1.2 μm peak-to-peak (i.e., ±600 nm) and a period of ~2 s (Fig. 13).

23. Observe the QPD response signals, focusing on the axis along which the bead is moving. The objective is to make this response symmetrical and centered about the origin (Fig. 13b, panel **4**). For our instrument, this corresponds to a maximum deflection of ±0.4 V_{norm} in each channel (*see* **Note 58**). Start by using Pzt-M3 to move the overall signal in the direction required to center it about zero (Fig. 13b, panel **2**). This will significantly perturb the symmetry of the response waveform, which is then reestablished using Pzt-M2 (Fig. 13b, panel **3**), moving it in the same direction as Pzt-M3 (up, down, right, or left buttons, configured as described in **Note 53**). Repeat this process until the response waveform is symmetrical and centered at zero (Fig. 13b, panel **4**).

24. Stop the sweeping of the bead, and re-zero both QPD signals using Pzt-M3.

25. Repeat steps **22–24** for the y axis. Adjustments to one axis may initially perturb the response in the other axis. Iterate between the two axes until the response in both axes is symmetrical and zero centered.

26. During the early rounds of iterating **steps 22–24** for the two axes, there may be considerable "crosstalk" between the x and y QPD signals (displacements in the signal of the non-sweeping axis correlated with the displacements in the sweeping axis). After multiple rounds of adjustment, the crosstalk should be negligible. If not, it may be symptomatic of an improperly rotated QPD (*see* Fig. 14 *and associated legend*). Correct this by sweeping the bead along one of the axes and carefully rotating the QPD around the optical axis until the crosstalk is minimized.

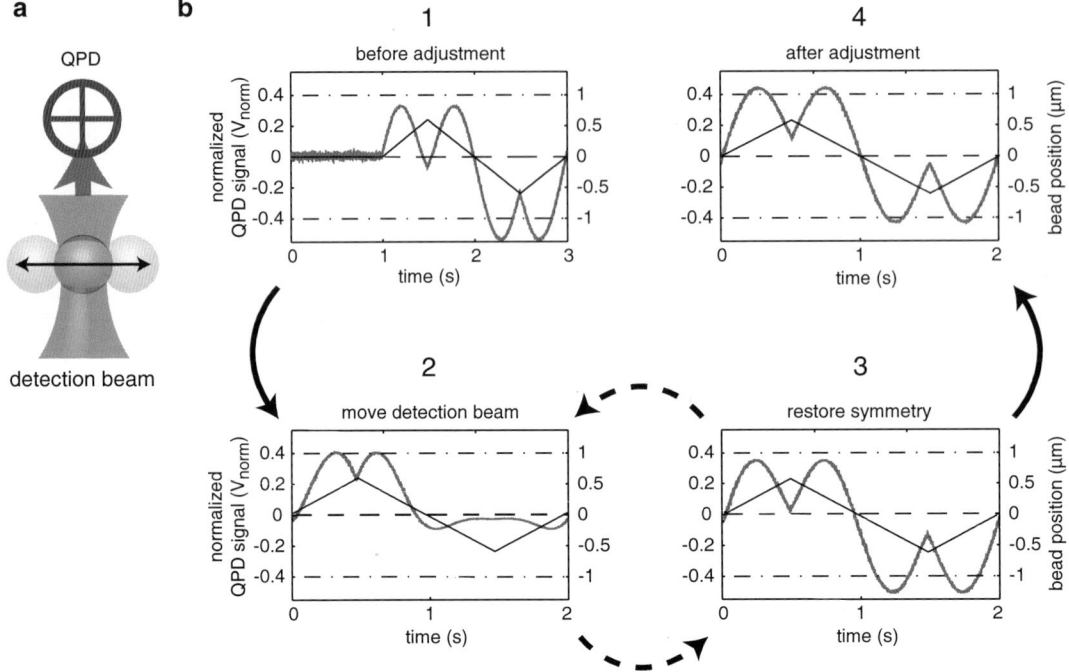

Fig. 13 Alignment of the trapping and detection beams by sweeping the trapping beam in a triangle-wave pattern. (**a**) A trapped bead is swept transversely across the detection beam (along either the *x* or *y* axis) by using the AOD to move the trap position in a triangle-wave pattern. This modulates the normalized QPD position signals, X_{Vnorm} and Y_{Vnorm}. The objective of adjustment is to position the detection beam so that the response in each channel is symmetrical about the origin. (**b**) After coarse alignment of the trap and detection beams, the QPD position signal (*red* trace, V_{norm}) for each channel is very nearly zero. However, when the trap position is modulated in a triangle-wave pattern (*solid black line*), it is clear that the response is not symmetrical about the origin (but rather around another position, in this case one slightly negative), indicating the detection beam and trapping beam do not overlap (step 1, the *dashed lines* through 0 V_{norm} and ±0.4 V_{norm} can be used as landmarks to help guide the eye). The center position of the sweep pattern is moved toward the origin using mirror M1 (*see* Fig. 3), with the consequence that the symmetry of the response is lost (step 2). Adjusting M2 restores the symmetry, while reversing some of the movement toward the origin (step 3). However, the net result is that the response is symmetrical and that the center of the pattern is closer to the origin. Performing steps 2 and 3 repeatedly yields the desired response (step 4). This process is repeated iteratively for each channel, until both exhibit symmetrical responses centered at the origin

3.8 Optical Tweezers Calibration

BFP detection (Fig. 12) [57–61] measures the separation between the detection beam and a trapped bead to within nanometers at a rate of many kilohertz. If the spring constant for the trap is known, these displacements are converted easily into forces. Thus, two calibrations must be performed: one to characterize the QPD detector response to bead displacements ("position calibration") and another to determine the spring constant ("force calibration"). Both calibrations involve measurements of microsphere displacements, either known displacements induced by the experimenter or natural fluctuations due to Brownian motion. Several types of calibration should be done when the optical trap is initially

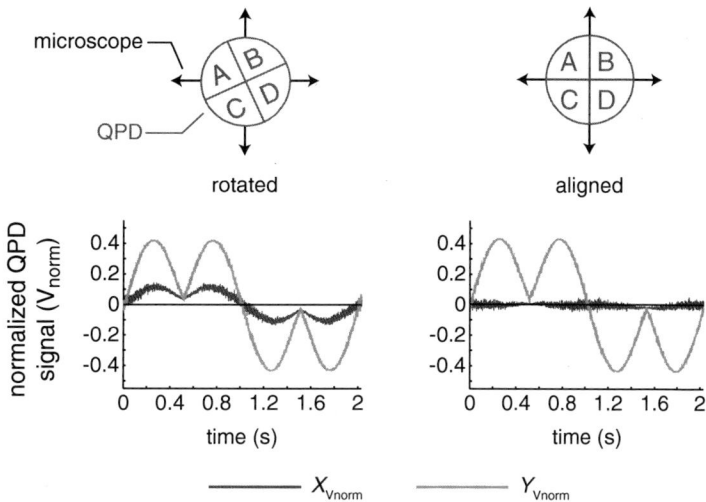

Fig. 14 Identification and correction of erroneous QPD rotation. After aligning the instrument as precisely as possible, when the trap sweeps a bead back and forth across the detection beam in the in one axis (*y* in this example), the QPD signal (X_{Vnorm}) for the non-sweeping axis exhibits substantial displacements if the QPD is rotated relative to the axes of the microscope (*left*). This example is exaggerated for clarity, and in practice, the "crosstalk" between channels is usually much subtler. Correcting the QPD rotation decreases excursion in the non-sweeping axis, and when the axes are aligned (*right*), the signal for the non-sweeping axis is virtually unaffected even by large displacements in the orthogonal direction. For counterclockwise (as judged when looking toward the face of the QPD detector) rotations relative to the microscope axes, *y* sweeping produces deflections of the same sign in X_{Vnorm} and Y_{Vnorm} (as shown in the figure), whereas *x* sweeping produces deflections of the opposite sign in X_{Vnorm} and Y_{Vnorm}. For clockwise rotations, the situation is reversed

constructed and checked for a reasonable degree of consistency before conducting experiments. Detailed, automated calibration analysis should also be integrated into the trap control software, since it can be very useful in assessing instrument performance during daily use and when diagnosing problems.

Because both force and position calibrations depend on the separation between the bead and the cover slip (see below), the first step is to determine the bead–cover slip distance. This is done by moving the nanopositioning stage in the axial (*z*) direction, finding the point of bead–cover slip contact (Fig. 15), and correcting the measured distance for the effective focal shift [62, 63] (Fig. 16).

For position calibration, the most straightforward method is to simply sweep the bead across the detection beam focus in known increments (Fig. 17), by either steering the trap or, with the nanopositioning stage, using a bead stuck to the cover slip

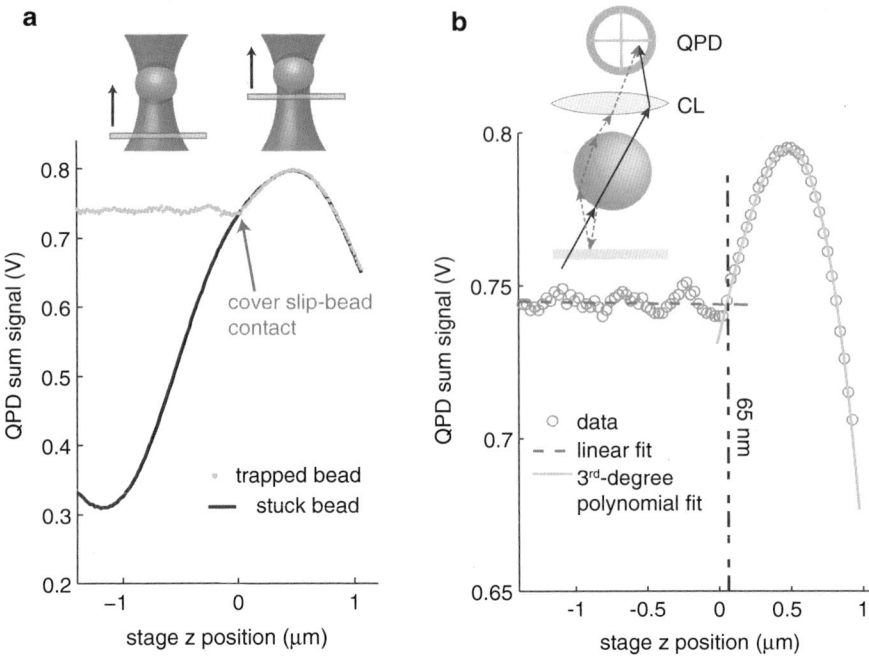

Fig. 15 Determination of bead–cover slip separation. (**a**) The stage is swept over a ±1,200 nm range in the axial (*z*) direction. First, a bead strongly stuck to the cover slip is carefully centered on the detection beam so that the *x* and *y* QPD signals are zero. Sweeping the bead through the trapping beam induces the characteristic response in the QPD sum signal (*solid black line*). If the same experiment is performed with a trapped bead (*green dots*), essentially no response is produced until the cover slip makes contact with the bead surface and pushes it out of the trap (at which point it follows the same trajectory and induces the same QPD signal as the stuck bead). Note that while the position (in nm) at which the contact point occurs depends on the distance between the cover slip and bead, the QPD voltage at which this happens depends on the position of the detection beam focus relative to the trapped bead (i.e., the trapping beam focus). In this example, the focus of the detection beam is slightly below the trapped bead, since the two curves overlap beyond the central voltage value for the stuck bead. (**b**) To find the bead–cover slip separation during an experiment, the position of the inflection point (where the sum signal for the trapped bead turns upward and follows the path of the signal for the stuck bead) is determined by finding the intersection of a line fit to the initial segment and a third-order polynomial fit to the peak, in this case 65 nm. Note that this value must be corrected to account for the focal shift (*see* Fig. 16), so that the true initial separation was actually ~0.82 × 65 = 53 nm. The periodic modulation observed in the sum signal as the cover slip comes very close to the bead is expected behavior and arises because the surfaces of the bead and cover slip essentially form a miniature Fabry–Pérot interferometer [62], as shown in the inset. Light backscattered/reflected from the bead (*red dashed rays*) reflects off the cover slip and interferes with forward-propagating (unscattered/forward-scattered) light (*solid black rays*) at the detector. The constructive/destructive nature of the interaction depends on the bead–cover slip distance and oscillates with a spatial frequency dependent on the wavelength of the detection beam

(*see* **Note 14**). In the central region (±~100 nm), the calculated QPD response signals ($X_{V\text{norm}}$ and $\Upsilon_{V\text{norm}}$) are linear with bead displacement and may be converted to physical positions by multiplying with the inverse slope: $x = X_{\text{nm}} = \beta_x \, X_{V\text{norm}}$ (*see also* Fig. 12). Beyond this central region, the response function is nonlinear, but the useful range can be extended to the entire region for which the

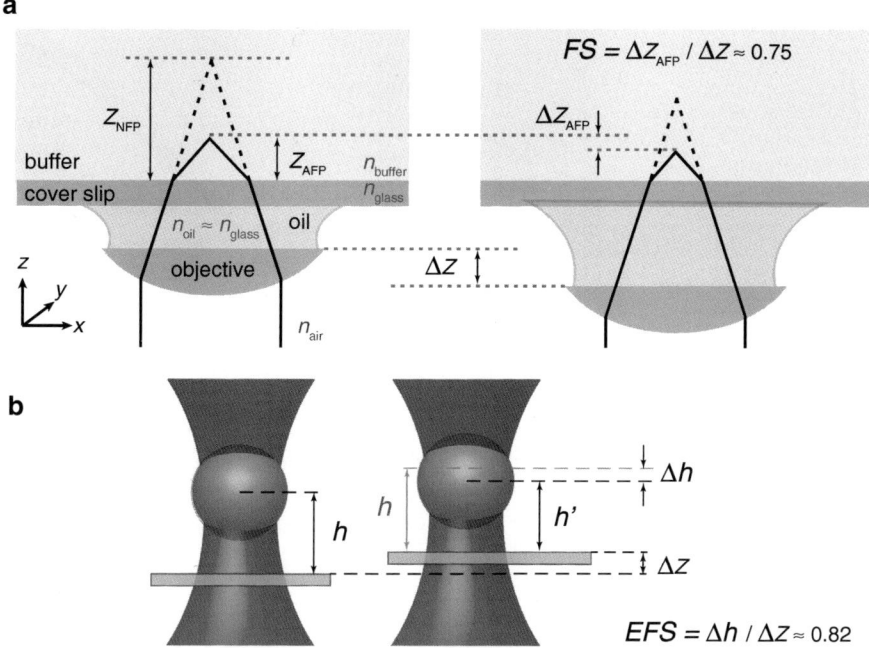

Fig. 16 The focal shift and its effects on bead–cover slip separation. (**a**) Ray diagram for a collimated beam focused by an oil immersion microscope objective. Although the indices of refraction are closely matched between the objective, immersion oil, and cover slip ($n_{oil} \approx n_{glass} \approx 1.51$), the refractive index of the sample buffer is significantly smaller ($n_{buffer} \approx n_{water} = 1.33$). This causes the rays to refract at the interface, bringing the actual focus position (Z_{AFP}, measured relative to the cover slip surface) closer to the objective than the nominal focus position (Z_{NFP}) expected if the refractive indices were equal (dashed rays). Z_{AFP} depends on the indices of refraction and on the distance the beam travels in each medium (glass/oil vs. buffer), which is determined by the separation between the cover slip and objective. A change ΔZ in this separation (due to movement of the objective or cover slip) induces a corresponding change ΔZ_{AFP} in the actual focus position, which is linear with ΔZ over several micrometers. The proportionality constant relating ΔZ_{AFP} to ΔZ is referred to as the focal shift, *FS* (which is defined differently by some authors, as the difference [$Z_{NFP} - Z_{AFP}$]). Using high-NA oil-immersion objectives and aqueous buffer, a full electromagnetic treatment of the problem gives $FS \approx 0.75$. (**b**) In optical tweezers, the figure of interest is not the focal shift per se, but rather the effective focal shift (*EFS*), defined analogously, though in reference to the trapped bead position, *h*, rather than to the focus itself. $EFS \approx 0.82$ is slightly greater than *FS* for aqueous solutions, but still significantly different from unity. This must be accounted for when moving the objective or the nanopositioning stage holding the cover slip in the axial direction. Moving the stage toward the trapped bead by 100 nm decreases the separation not by 100 nm, but rather by ~82 nm. Truly reducing the bead–cover slip separation by 100 nm requires moving the stage by ~122 nm instead

response function is single-valued by fitting it with a third-degree polynomial, rather than a line (Fig. 17; *see* **Note 59**). Some less direct methods employing spectral analysis (see below) or rapid sweeping of the AOD [64] have also been used (*see* **Note 60**).

Force calibration is usually done after position calibration, either by analyzing the Brownian motion of a trapped bead or by applying a known force to the bead and measuring how far it is displaced from the trap center as a result. The simplest method is

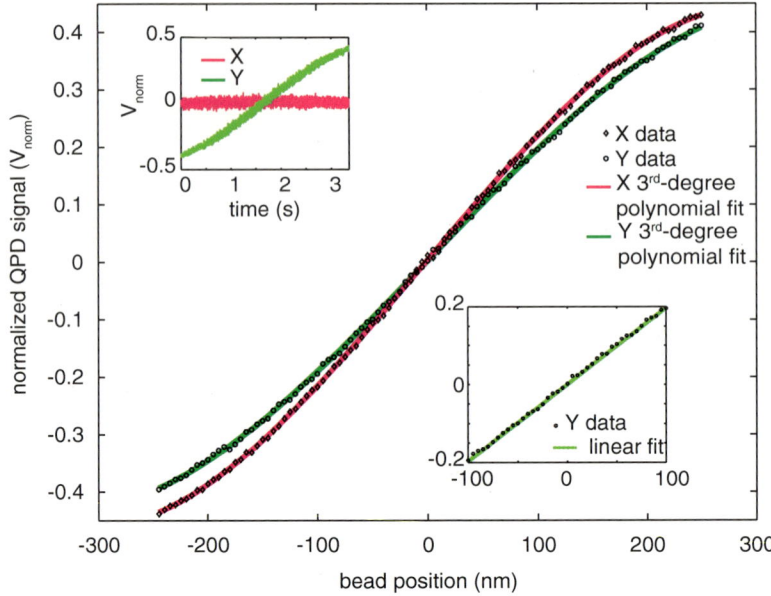

Fig. 17 QPD position calibration. The trap is used to step a bead across the detection beam separately in each axis by a known distance, in this case ±250 nm (the inset in the *upper left* shows the sweep in the *y* axis). For the central region (approximately ±100 nm), the response is linear (the *lower right inset* shows this region for the *y* axis), but exhibits curvature for larger displacements (*see* Fig. 13). At each step (every 5 nm in this example), many samples of the QPD response signal are recorded and averaged (weaker traps require more samples because the bead is less confined; in this example, with $k \approx 0.06$ pN/nm, 100 samples were averaged at each point). These values are then plotted against the position and fit with a third-degree polynomial. The coefficients from this fit are then used to determine the bead position from the QPD signal during experiments

equipartition analysis (Fig. 18a and c), in which the spring constant is calculated as $k_x = k_B T / <x^2>$ (and similarly for k_y) [26], where k_B is the Boltzmann constant, T is the absolute temperature (in Kelvin), and $<x^2>$ is the variance of the distribution of bead positions over time (*see* **Note 61**). The advantage of this method is that essentially no specific information about the system (e.g., bead size, solution viscosity, distance of the bead from the cover slip) is required, and it is very easy to compute k. However, the value of the variance is inflated by any drift or other artifacts (e.g., electronic noise) in the optics or detection system (since the variance depends on the square of the displacement, even if artifacts average to zero, they increase the computed value). Therefore, the equipartition method tends to underestimate k.

A related method for force calibration, also employing a statistical mechanical analysis of trapped microsphere fluctuations, directly computes the trapping potential $U(x)$ [65]. From the Boltzmann law, $P(x) = \exp(-U(x)/k_B T)/Z_p$ (*see* **Note 61** for

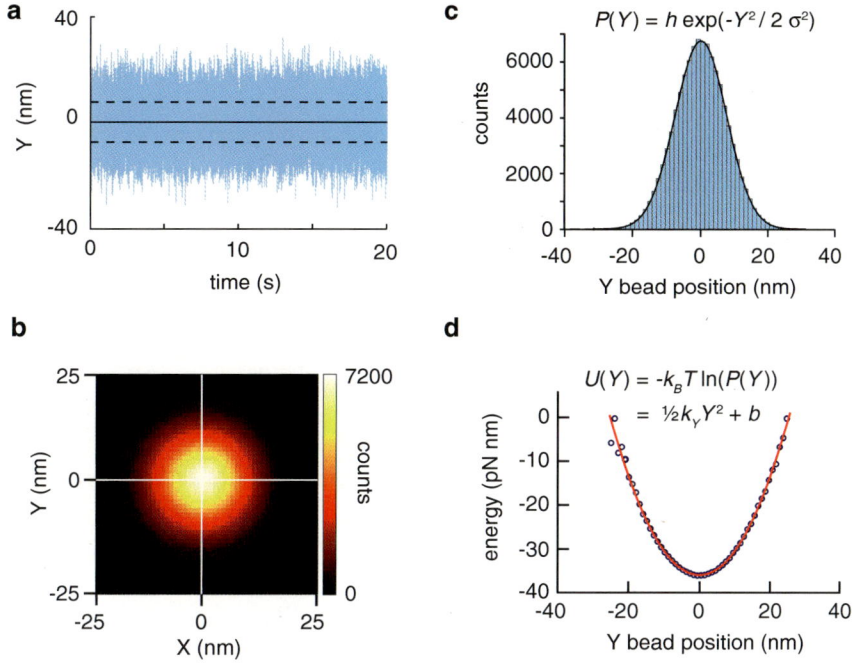

Fig. 18 Mapping of bead diffusion and determination of the trap spring constant. (**a**) Y position trace for a trapped bead over a period of 20 s (*dashed lines* are ±standard deviation, $\sigma = 7.6$ nm). The variance of the positions is $\sigma^2 = 57.4$ nm^2, which for a value of $k_BT = 4.116$ pN nm ($T = 25$ °C) yields $k_y = 0.072$ pN/nm by the equipartition method. (**b**) For the same experiment as in (**a**), plotting a two-dimensional histogram of bead positions provides a useful method for determining the shape of the trapping potential (limited to the central region in which the bead can diffuse) and identifying any irregularities or asymmetries. (**c**) The data from (**a**) are plotted in a histogram (50 equally spaced bins of ~1.3 nm width) and fit with a Gaussian function $P(Y)$, also yielding $\sigma = 7.6$ nm (agreement between σ calculated "blind" and by fitting shows that the data are Gaussian distributed and do not require fitting to determine σ). (**d**) Given the (non-normalized) probability distribution in (**c**), it is possible to calculate the potential $U(Y)$, which, assuming a harmonic potential, can be fit by the parabola $\frac{1}{2}k_y\,Y^2 + b$, where b is a constant. This procedure yields $k_y = 0.071$ pN/nm, in good agreement with the equipartition method. The good fit of the parabola to the calculated potential shows that the trap potential is indeed harmonic (like a Hookean spring) near the center

symbol definitions). We can compute the potential as $U(x) = -k_B T \ln(P(x)\,Z_p) = -k_B T \ln(P(x)) - k_B T \ln(Z_p)$, where the last term is a constant offset that is ignored. $P(x)\,Z_p$ is simply the (non-normalized) histogram of position fluctuations of the trapped microsphere (Fig. 18c). Assuming that $U(x) = \frac{1}{2}k_x x^2 + b$ (where b is a constant offset), this model can be fit to the potential calculated from the histogram in order to find k_x (Fig. 18d). This method has the added benefit of creating a map of the trapping potential, even if it is not in fact harmonic. However, only a fairly limited range near the trap center is mapped, and like the equipartition method, this analysis can be hindered by unwanted noise sources.

Spectral analysis offers a useful alternative to statistical analysis (Gittes and Schmidt [66] provide a very accessible introduction), whereby the Fourier transform is used to evaluate the power spectral density $S(f)$ of the bead motion (*see* **Note 62**), where f is frequency. Considering the thermal motion of the bead in the trap as that of a damped harmonic oscillator driven by a (random) Langevin force [67], the power spectral density takes the form of a Lorentzian, $S_{xx}(f) = k_B T/[\pi^2 \gamma \ (f_c^2 + f^2)]$, where γ is the hydrodynamic drag for the bead, and the "corner" or "roll-off" frequency is $f_c = k/2\pi\gamma$ (frequency at which the power spectral density reaches half the maximum value of $4\gamma k_B T/k^2$). Thus, by computing the power spectral density and fitting this function with a Lorentzian, the spring constant can be determined from the corner frequency found by the fit (Fig. 19a), i.e., $k = 2\pi\gamma f_c$ (*see* **Note 63**).

The power spectral density can also be used for position calibration. It can be shown [26, 58, 68] that the product $S_{xx}(f) f^2$ asymptotically approaches the limit $C = k_B T/\pi^2 \gamma$ for $f \gg f_c$ (Fig. 19c). Assuming a linear conversion β_x between detector response (V_{norm}) and displacement (nm), then $S_{xx}(f) = \beta_x^2 S_{VV}(f)$, and $\beta_x = S_{xx}(f) f^2/S_{VxVx}(f) f^2 = [k_B T/(\pi^2 \gamma \ S_{VxVx}(f))]^{1/2}$ for $f \gg f_c$. Alternatively, if β_x is known, this method can be used to determine γ, i.e., $\gamma = k_B T/(\pi^2 \beta_x^2 \ S_{VxVx}(f))$.

Force calibration via power spectral density analysis has the advantage that low-frequency drift and other noise sources are easily visible in the spectrum and can be disregarded during the fitting procedure. It is complicated, however, by the need to know γ. In bulk solution, for a sphere of diameter d, the drag coefficient is $\gamma = 3\pi d\eta$, where η is the viscosity of the solution. There is usually some degree of uncertainty in the bead diameter, and η depends on the composition of the solution and the temperature. More importantly, γ is strongly influenced by the proximity of the bead to the cover slip surface, according to Faxén's law (Fig. 19d) [67], $\gamma/\gamma_0 = \gamma_0/[1 - 9/16R + 1/8R^3 - 45/256R^4 - 1/16R^5]$. Here, γ_0 is the drag in bulk solution ($3\pi d\eta$), $R = d/(d + 2Z_B)$, and Z_B is the distance from the surface of the cover slip to the surface of the bead (*see* **Note 64**). The value used for γ must therefore be scaled according to the bead-surface separation. Our approach is to estimate γ from the power spectrum (Fig. 19c, and see above) and also by directly measuring the bead–cover slip separation (Figs. 15 and 16) and computing γ via Faxén's law. We then use these values as initial guesses in the Lorentzian fit to the power spectral density and fit with both γ and f_c as free parameters (Fig. 19a).

A more conceptually straightforward method for force calibration relies on Stokes's law for the viscous drag force exerted on a sphere subjected to continuous fluid flow. In this method, fluid flow past the trapped microsphere is created by moving the slide chamber with the nanopositioning stage. The drag force on the bead is $F_{drag} = \gamma v_{stage}$, where v_{stage} is the velocity at which the stage moves (Fig. 20a; *see* **Note 65**). This force displaces the bead by a distance $\gamma v_{stage}/k$ from the trap center. Thus, assuming v_{stage} and γ are known, k may be calculated directly from the observed

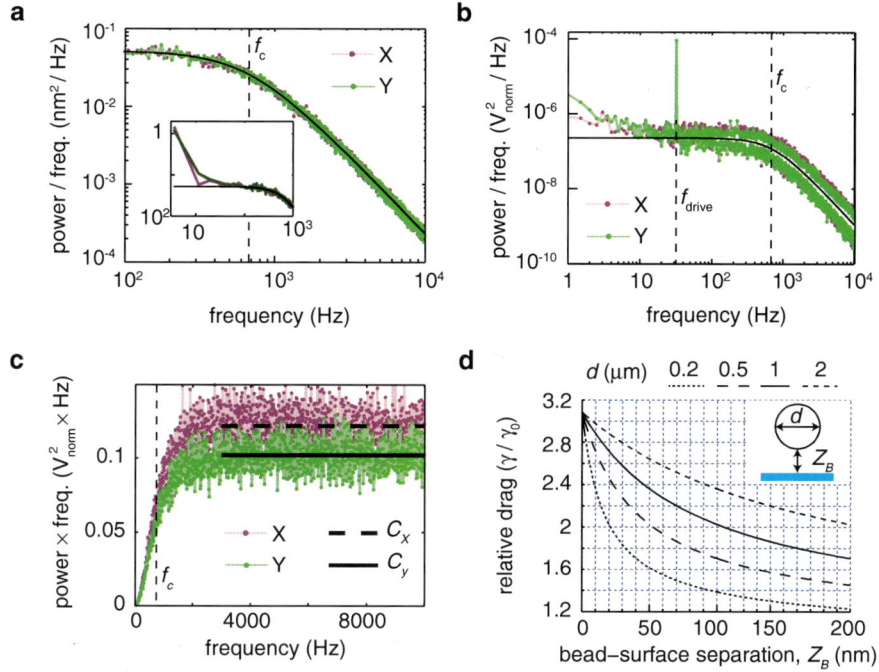

Fig. 19 Power spectrum analysis for optical trap calibration. For (**a**–**c**) the separation Z_B between the bead and cover slip is ~70 nm, and the sample rate for the data is 65,536 Hz (2^{16} Hz). (**a**) Log–log plot of the x and y power spectral densities for a trapped bead, with Lorentzian fit to the frequencies above 100 Hz (x, *dashed line*; y, *solid line*; note that the x power spectrum and dashed-line fit are only partially visible because they almost exactly overlap the y spectrum). The inset shows the entire spectrum, which exhibits non-Lorentzian character at very low frequencies (this is attributable to slow, long-term drift and is minimized by enclosing the optical pathway and otherwise stabilizing the instrument). The parameters extracted from the fit (for x and y, respectively) are corner frequencies (f_c) 683 and 685 Hz and effective drag coefficients $2.23 \times \gamma_{water}$ and $2.22 \times \gamma_{water}$, yielding spring constants $k_x = k_y = 0.074$ pN/nm (compared to $k_x = 0.074$ pN/nm and $k_y = 0.072$ pN/nm calculated by the equipartition method). (**b**) Power spectrum analysis for forced oscillations (sinusoidal nanopositioning stage movement at $f_{drive} = 32$ Hz) of the same bead as in (**a**). In this example, the stage oscillates in the y direction with 98 nm measured amplitude (*see* **Note 75**), giving rise to a sharp peak at f_{drive} only in the y channel (note that the spatial units here are V^2_{norm} rather than nm^2). The amplitude of this peak is used to calculate β (the V_{norm}-to-nm conversion factor for the linear region of the QPD response signal). A Lorentzian fit (*black line*) to the block-averaged y data above 100 Hz (*white trace*) yields $f_c = 696$ Hz, $\beta_y = 474$ nm/V_{norm}, $\gamma_y = 2.22 \times \gamma_{water}$, and $k_y = 0.073$ pN/nm, in very good agreement with the standard power spectrum analysis in (**a**) and with the AOD-based QPD calibration ($\beta_y = 478$ nm/V_{norm}; not shown). The x and y power spectra apparently overlap somewhat less than in (a); however, this is attributable to the small difference in β between the two channels ($\beta_x = 434$ nm/V_{norm}), rather than a difference in physical motion. (**c**) For frequencies much greater than f_c, the product of frequency squared times the power spectral density approaches a constant value C, such that $\gamma = k_B T/[C (\pi \beta)^2]$. Thus, if γ is known, C can be used to find β and vice versa. Given β from the AOD-based calibration of the QPD response, we use this phenomenon to generate an initial guess for γ in our Lorentzian fits of the power spectra in (**a**) and (**b**). The data shown here (from the same underlying position data as the power spectrum in (**a**)) give $C_x = 0.122 V^2_{norm}$ Hz and $C_y = 0.102 V^2_{norm}$ Hz, yielding $\gamma_x = 2.31 \times \gamma_{water}$ and $\gamma_y = 2.22 \times \gamma_{water}$, in relatively good agreement with the values used in the final fit. (**d**) The drag γ experienced by a sphere in close proximity to a surface is significantly larger than that experienced in bulk solution, γ_0, and is approximated by Faxén's law. The ratio γ/γ_0 is plotted for different bead diameters as a function of the bead-surface–cover slip surface separation, Z_B

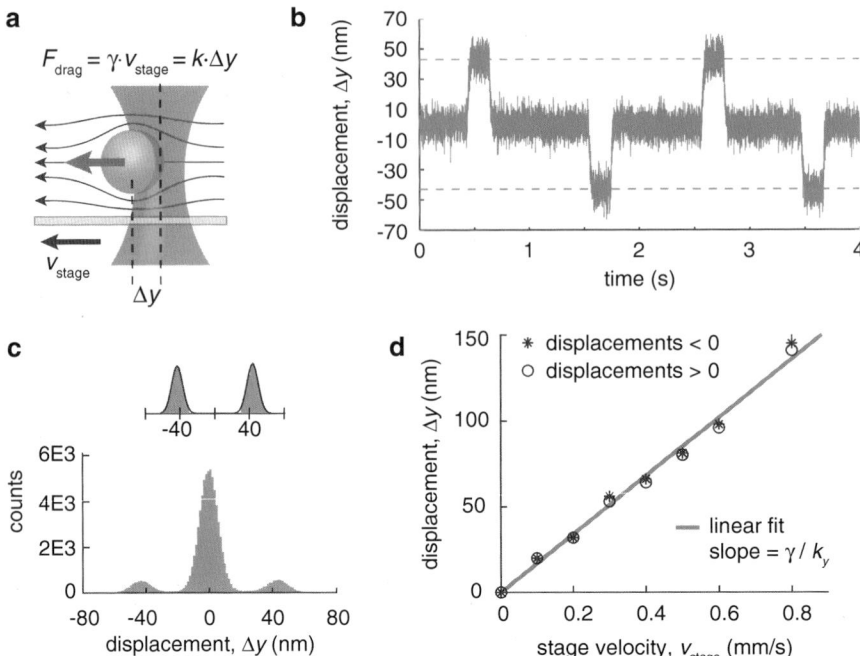

Fig. 20 Viscous drag determination of trap stiffness and mapping of trap potential. (**a**) With a bead in the trap, the stage is moved in the y direction at a constant velocity, v_{stage}. This pulls the solution in the chamber past the bead, thus inducing a viscous drag force γv_{stage} that pulls the bead away from the trap center and is balanced by the restoring force of the trap, $k_y \Delta y$. (**b**) Example data, showing bead displacements during constant-velocity stage movement, interspersed with periods of pausing. For this experiment, the 920-nm-diameter bead was approximately 500 nm above the cover slip, the stage velocity was 0.3 mm/s, and k_y was measured to be 0.076 pN/nm via equipartition and power spectrum, with $\gamma \approx 1.5 \gamma_{water}$. The predicted Δy for these parameters is ~46 nm, in very good agreement with the 43 nm measured (*red dashed lines*; see (**c**)). (**c**) Measurement of Δy. The positions from the data in (**b**) are plotted in a histogram (100 equally spaced bins), with a major peak at 0 nm due to the pauses between movement. The inset shows fits to the two side peaks, with data from ±20 nm discarded, yielding means of ±43 nm. (**d**) Repeating these processes at many different velocities allows the mapping of the trapping potential in regions far from the center, in order to test the validity of trap linearity in these regions, and provides a more robust estimation for k_y, as determined from the slope of a line fit to the data. During the course of the experiment, k_y was measured to be ~0.070 pN/nm on average (via equipartition and power spectrum), with $\gamma \approx 1.5 \gamma_{water}$. The fit is consistent with a linear force–displacement relationship for displacements of at least 150 nm, and the calculated slope of 1.7×10^{-4} s yields $k_y = 0.068$ pN/nm, in reasonably good agreement with the standard methods

bead displacement. Moreover, the bead displacement should be linearly related to v_{stage}, with slope γ/k (Fig. 20d), allowing a more precise calculation of k than using a single value of v_{stage}. Any nonlinearity (e.g., at bead displacements more than 100–200 nm) indicates deviation from a harmonic potential. The ability of this technique to detect such nonlinearity and map the trapping potential over a wide range of displacements represents a major advantage compared to other techniques. However, similar to the power spectrum method, γ must be known precisely and is likely to change as the stage is moved, since the cover slip surface is never perfectly perpendicular to the optical axis (*see* **Note 66**). Likewise, v_{stage} must be known to high precision, and measurements should only include data for which the stage has achieved a constant, steady velocity. This calibration method can also be cumbersome since it is relatively time-consuming, and multiple beads can often be accidentally caught in the trap as the stage moves over relatively long distances.

A "hybrid" method of calibration [63, 69] (Fig. 19b), combining the power spectrum and viscous drag analyses, allows simultaneous position and force calibration, without the need to know the drag coefficient (in fact, this method has been used to *measure* the drag coefficient and solution viscosity [63, 70]). During the data acquisition, the nanopositioning stage oscillates in a sinusoidal pattern with known amplitude and frequency, f_{drive}, adding a small (almost imperceptible) driven oscillation to the bead motion, in addition to the thermal "noise." This gives rise to an isolated "spike" in the power spectrum (Fig. 19b), the total power of which (i.e., area under the curve) is determined by β, the conversion factor used to convert the QPD response to distance in the linear region of the response (Fig. 12d). This feature is used to directly determine the position calibration, while a Lorentzian fit (as described above) is used to find f_c and γ in order to calculate k (theoretical details of this method are presented in detail in ref. 69). This method is of particular utility because it does not require an AOD or other beam steering device for position calibration and also does not require a separate detection beam. Note that the position calibration is only valid within the linear response range of the QPD. This method also takes somewhat longer than the aforementioned power spectral density analysis, since the x and y spring constants must be determined by separate experiments. Like the simple viscous drag calibration method, this hybrid method can in principle be used to map the trapping potential by using very large amplitude oscillations [69].

Below we give instructions for calibrating via each of the methods discussed, assuming a well-aligned instrument as the starting point. The vast majority of these steps should be automated.

The following steps determine the "height" Z_B of the trapped bead (bead–cover slip separation) and adjust it to the desired position for calibration and experiment, Z_{target}:

1. Set the QPD sampling rate to 3 kHz.

2. Trap a microsphere, and move the cover slip upward using the nanopositioning stage until the bead is clearly in contact (bead will appear white in the image and may also be displaced).

3. Move the nanopositioning stage downward by ~150 nm more than the desired Z_B.

4. Step the nanopositioning stage axially over a large range (at least $5Z_B$ in each direction around the current position) in 1–10-nm steps. At each position, record the average value of the QPD sum signal over 50–200 samples.

5. Plot the averaged sum signal vs. stage position. Fit a line to the data on the left side of the curve (stage positions before the inflection point; *see* Fig. 15b), and a third-degree polynomial to the data on the right side (stage positions after the inflection point; *see* **Note 67**).

6. Find the position (in nm) of the intersection of the fit line and polynomial. This is the nominal position of the bead-surface contact, Z_s. Due to the effective focal shift (Fig. 16b), this must be corrected to find the true $Z_B = \text{EFS} \times Z_s$, where for an oil–water interface, $\text{EFS} \approx 0.82$ [62, 63], i.e., $Z_B \approx 0.82 \times Z_s$.

7. If $Z_B \neq Z_{\text{target}}$, move the stage by $(Z_{\text{target}} - Z_B)/0.82$, and repeat **steps 1–7** until the height of the bead is within the desired range of Z_{target}.

The following steps calibrate the position detection (Fig. 17):

8. Use the AOD to sweep a trapped bead across the detection beam focus in the x direction, with a range of ±250 nm (or whatever range over which the response signal $X_{V\text{norm}}$ is single-valued) and a step size of 5 nm. At each position, record the average of 100–200 values of $X_{V\text{norm}}$. Repeat for the y direction (Fig. 17, top left inset).

9. Fit a line to the central region of each curve (±~100 nm) and determine the slope (Fig. 17, lower right inset, and Fig. 12d). The inverse of this slope, β, can be used to quickly convert the QPD response signal to physical bead position: $X_{\text{nm}} = \beta_x X_{V\text{norm}}$ (and identically for y). This also serves as a useful check for comparison to β obtained by other calibration methods.

10. For each recorded curve, fit a third-degree polynomial of the form $p(x) = a_0 + a_1 x + a_2 x^2 + a_3 x^3$ (and identically for y) to the entire range. This considerably extends the useable detection range (*important: see* **Note 68**). To convert $X_{V\text{norm}}$ to X_{nm}, compute the following:

$M_1 = (a_0 - X_{V\text{norm}})/(2a_3)$.

$M_2 = a_2^3/(27a_3^3)$.

$M_3 = (a_1 a_2)/(6a_3^2)$.

$M_4 = a_1/(3a_3) - a_2^2/(9a_3^2)$.

$M_5 = M_1 + M_2 - M_3$.

$M_6 = [(M_4^3 + M_5^2)^{1/2} - M_5]^{1/3}$.

$X_{\text{nm}} = \text{Re}\{M_4/2\,M_6 - a_2/3a_3 - M_6/2 + (3^{1/2})i/2(M_6 + M_4/M_6)\}$.

where $\text{Re}\{\dots\}$ indicates the real part of a complex number and i is the square root of -1.

The following steps calibrate the force via the equipartition method:

11. Set the QPD sampling rate to 65,536 kHz or higher (*see* **Note 69**).

12. Collect data for several seconds. Decimate the data by an appropriate amount so that each point is separated by at least $3\gamma/k$ (*see* **Note 70**).

13. *Optional*: subtract low-frequency drift from the position data (*see* **Note 71**).

14. Convert the raw QPD data into a time series of position data using the calibration procedures above (Fig. 18a). Compute the variance $<x^2>$ of these data in nm^2. *Optional*: plot the data in a histogram (Fig. 18c), fit a Gaussian function in order to determine the standard deviation, and square this quantity to find $<x^2>$. The two answers should agree very closely for a large data set.

15. Calculate $k_x = k_B T / <x^2>$, where $k_B T = 0.013807 \times (T_C + 273.15)$ pN nm; T_C is the magnitude of the temperature in °C. At 25 °C, $k_B T = 4.116$ pN nm.

The following steps describe force calibration by computing the trapping potential:

16. Complete **steps 11–13** above (including the optional part of **step 13**).

17. From the histogram $P(x)$, compute the potential $U(x) = -k_B T \ln(P(x))$, and fit this with the equation $U(x) = \frac{1}{2} k_x x^2 + b$, where b is an arbitrary constant (Fig. 18d). Use the value calculated for k_x via the equipartition method as an initial guess. Repeat for the y direction.

18. *Optional*: create a two-dimensional histogram of positions, and plot the data as an image (Fig. 18b). This image gives a reasonable measure of the symmetry and homogeneity of the trap potential (at least over the narrow region sampled).

The following steps describe force calibration via power spectral density analysis. Analysis steps given for the x direction are identical for the y direction:

19. Complete **steps 11–13** above.

20. Split the time series data into several series of equal length. Compute the one-sided power spectral density $S_{xx}(f)$ (in $nm^2/$Hz; Fig. 19a) for each series, and average the results to compute the final spectrum (*see* **Note 72**).

21. Use Faxén's law to estimate the drag coefficient, γ, based on the bead-surface separation:

$$R = d/(d + 2Z_B),$$

$$\gamma_0 = 3\pi d\eta \ (\text{\textit{see} } \textbf{Note 73}),$$

$$(\gamma/\gamma_0) = \gamma_0/[1 - 9/16R + 1/8R^3 - 45/256R^4 - 1/16R^5],$$
$$\gamma = \gamma_0 \times (\gamma/\gamma_0).$$

22. *Optional*: estimate γ from the power spectrum itself. Compute $S_{xx}(f)\,f^2$ (in nm^2 Hz) and fit a flat line to the high-frequency region in which the curve approaches a constant value C_x (Fig. 19c). Then compute $\gamma = k_B T/[C_x \pi^2]$ (note that in Fig. 19c, the curve is calculated in V^2_{norm} Hz, and thus the conversion formula in the legend includes the conversion factor β).

23. Fit the equation $S_{xx}(f) = k_B T/[\pi^2 \gamma\,(f_c^2 + f^2)]$ to the power spectral density function (Fig 19a). Use the estimate of γ as an initial guess and $f_c = k/2\pi\gamma$ as the guess for f_c, using the value of k determined from the equipartition method (*important: see* **Note 74**). We typically fit the region from $f = 50$–100 Hz and above, excluding the region containing low-frequency drift (Fig. 19a).

24. Compute the spring constant $k_x = 2\pi\gamma f_c$. Check that the value for γ is reasonable given Z_B.

The following steps calibrate the trapping force using the "hybrid" method [69] combining viscous drag and power spectral density analysis:

25. With a trapped bead, oscillate the stage with a known amplitude A_{drive} (we use $A_{drive} = 100$ nm) in a sinusoidal pattern at $f_{drive} = 32$ Hz (*see* **Note 75**), with a QPD sampling frequency of 65,536 Hz.

26. Record the QPD response signals X_{Vnorm} and Y_{Vnorm}, and compute (and average) the power spectral density $S_{VxVx}(f)$ for each (in V^2_{norm}/Hz) similarly to **step 20** above (do not block average, as this would alter the height of the spike). Whereas the power spectral density for the axis orthogonal to stage oscillation is essentially unaffected, a clear spike at f_{drive} will be visible in the power spectral density for the axis parallel to the oscillation (Fig. 19b).

27. Estimate γ as done in **steps 21** and **22** above (in this case, C_x will be in V^2_{norm} Hz, and thus $\gamma = k_B T/[C_x\,(\pi\,\beta_x)^2]$; β_x can be estimated from previous calibrations).

28. Determine f_c and γ^{Vnorm} via fitting a Lorentzian to the power spectral density, as done in **step 23** (Fig. 19b). If the fitted region includes f_{drive}, remove the data constituting the spike when fitting the curve. Note that the unit of length for γ^{Vnorm} is V_{norm} rather than nanometers.

29. Compute the quantity $W_{th} = A^2_{drive}/[2\,(1 + f_c^2/f^2_{drive})]$, the theoretical value of the power in the spike (area under the curve).

30. Compute the quantity $W_{ex} = [S_{VxVx}(f_{drive}) - S^{fit}_{VxVx}(f_{drive})]/t_{msr}$, the experimentally determined value for the power in the spike (*see* **Note 76**), where $S_{VxVx}(f_{drive})$ is the value of the computed power spectral density at f_{drive} (i.e., the absolute amplitude of the spike), $S^{fit}_{VxVx}(f_{drive})$ is the value of the fit at f_{drive} (i.e., the "thermal background" to which the spike is added), and t_{msr} is the measurement time for the data set from which the power spectral density was computed.

31. Compute $\beta_x = (W_{th}/W_{ex})^{1/2}$ and $\gamma = \gamma^{Vnorm}/\beta_x^2$. Check that both values are reasonable based on the alternate calibration methods and previous calibrations.

32. Compute $k_x = 2\pi\gamma f_c$.

33. Repeat the calibration for the y direction.

The following procedure uses the viscous drag method to calibrate the trapping force:

34. With a bead in the trap, move the stage at a known velocity v_{stage} over its entire range ($\pm\sim50$ μm). The periods of constant velocity will produce "steps" in the position trace for the bead (Fig. 20b). Move many times in both directions (positive and negative) along the given axis.

35. Repeat **step 34** for a number of different stage velocities, recording the data for each.

36. For each stage velocity, plot the position data in a histogram (Fig. 20c). Remove the central peak and fit a Gaussian (Fig. 20c, inset) to the two remaining peaks to determine their center positions (i.e., bead displacement for the given stage velocity).

37. Plot the bead displacement vs. the stage velocity, and fit a line to the data (the fit should be forced through the origin).

38. Estimate γ based on Z_B (**step 21** above) or one of the other calibration methods.

39. Divide γ by the slope of the line fit in **step 37** to determine k.

40. Observe any major deviations from the linear fit at large displacements. This indicates a nonlinear force–displacement relationship at these positions. Poor overlay of points acquired via sweeping in one direction vs. the other along the given axis may indicate an asymmetrical trapping potential.

The following steps establish the linearity of trap strength with laser power and provide a useful calibration for adjusting trap strength during experiments:

41. For a range of laser powers from ~2 to 60 mW (measured at the back aperture of the objective lens), measure the spring

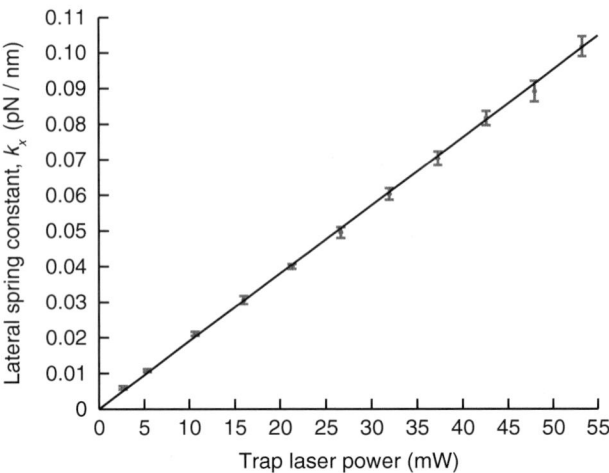

Fig. 21 Linear dependence of lateral trap stiffness (spring constant) on the trapping laser power demonstrated for the *x* dimension. Laser power was measured at the entrance pupil of the objective lens, using a weakly focusing lens to ensure complete collection of the entire beam. Each measurement of the spring constant was obtained as an average between the equipartition and power spectrum analyses. Each point represents the mean of five different measurements taken over several minutes (error bars are ± standard deviation), and the line is a least-squares fit (weighted by the inverse variance of each point) through the origin; slope = $(1.90 \pm 0.02) \times 10^3$ pN nm^{-1} mW^{-1}. Identical analysis for the *y* dimension (not shown) essentially overlaps with the data here, with the fit yielding an identical slope

constant at the bead–cover slip separation typically used for experiments.

42. Plot the spring constant vs. the input laser power, and check that the data are well described by a line (Fig. 21). This graph serves as a useful measure of trap performance and can be used to precisely adjust laser power for the desired spring constant in future experiments.

3.9 Sample Preparation for the Optical Tweezers Assay

The procedure below describes how to prepare a sample for measurement of K560 motion and force generation in the optical trap, starting with a slide chamber with immobilized Cy3-labeled MTs on the surface (*see* Chap. 9 for preparation instructions). Trapping microspheres are incubated with K560, which binds to the antibodies on the microsphere surfaces (Fig. 1). The beads are then introduced to the MT-containing, surface-passivated chamber in the presence of ATP. The slide chamber is then sealed and taken to the instrument. Final assay concentrations are as follows: 1 mM ATP, 10 μM paclitaxel (*see* **Note 77**), 1 mg/mL β-casein, 10 mM DTT, 3 U/mL pyranose oxidase, 90 U/mL catalase, and 22.5 mM glucose.

Prepare the following freshly at the beginning of a set of experiments:

1. "BRB/CS": 96 μL of BRB80 and 4 μL of 25 mg/mL β-casein. Keep on ice.

2. "POC/CS": thaw a POC aliquot freshly and mix 16 μL of β-casein, 2 μL of POC, and 2 μL of BRB80. Keep on ice (*see* **Note 78**).

3. "BRB/Tx": 350 μL of BRB80 and 0.5 μL of 10 mM paclitaxel. Keep at room temperature.

4. Motility mix: 141.5 μL of BRB/Tx, 4.5 μL of 1 M glucose, 2 μL of 100 mM ATP, and 2 μL of 1 M DTT. Keep on ice.

5. Trapping beads: dilute 1 μL trapping bead stock solution in 50 μL BRB80. Mix thoroughly and sonicate in the low-power bath sonicator for 5–10 s (never sonicate the main stock solution). Keep on ice. These trapping beads can be stored at 4 °C for days to months without significant loss of activity.

Do the following for each experiment:

6. Prepare a flow chamber by incubating fluorescence-labeled MTs in a glutaraldehyde-functionalized flow chamber (*see* Chap. 9).

7. During the MT incubation, prepare the K560 and trapping microspheres. Thaw an aliquot of K560 and pre-dilute it in BRB/CS as needed (typically 10,000–50,000×; *see* **Note 79**).

8. Mix 4 μL of pre-diluted motor with 4 μL of trapping beads and incubate for 10 min, so that the MT incubation in the slide chamber and motor-bead incubation end at the same time.

9. About 1 min before the end of the incubations, flush the flow chamber with 40 μL of casein blocking solution (*see* Chap. 9).

10. When the incubations end, add 2 μL of POC/CS to the motor-bead mixture, followed by 30 μL of motility mix. Pipette up and down gently to mix.

11. Immediately flow the entire assay solution into the flow chamber. Dry the ends with a Kimwipe, wiping away from the center of the chamber and taking care not to suck solution out of the chamber itself. Seal the chamber with vacuum grease as described by Nicholas et al. (Chap. 9).

3.10 Optical Tweezers Measurement of Motility and Force Generation

1. Turn on the optical tweezers microscope and follow the protocol in Subheadings 3.7 and 3.8 to align and calibrate it using a test slide.

2. Prepare a sealed sample chamber containing surface-bound MTs and K560 bound to trapping beads according to Subheading 3.9.

3. Mount the slide chamber on the microscope and adjust for Köhler illumination, as done during the alignment procedure.

4. Turn on the 532-nm fluorescence excitation laser, and observe the MTs on the EMCCD. This may require changing focus with the nanopositioning stage. Use the minimal excitation power that still permits visualizing the MTs (use EM gain on the camera).

5. Find an MT oriented perfectly along the y axis (use a vertical line drawn on the EMCCD display to check), preferably separated from other MTs by a few micrometers. Move the nanopositioner so that the MT is directly under the cross that marks the trap center on the display. Save the xy position of the nanopositioner in the software, and then move to a spot on the cover slip with no MTs. Turn the 532-nm laser and EMCCD imaging off.

6. Follow the steps in the calibration protocol (Subheading 3.8) to obtain a bead in the trap separated from the cover slip surface by ~50–70 nm (*see* **Note 80**), and measure the spring constant. For K560, an appropriate spring constant is approximately 0.05–0.07 pN/nm (adjust if necessary). Record the spring constant and polynomial coefficients determined for the QPD position response. Save the z height of the nanopositioner in the software. Set QPD data acquisition rate to 3 kHz and the low-pass filters to 1.5 kHz with 20 dB preamp gain.

7. Move the stage back to the xy position saved in **step 5**, keeping the z position the same as in **step 6**.

8. Observe the microsphere for 2–3 min as it diffuses in the trap above the MT (*see* **Note 81**). Watch both the x and y signals and the QPD sum signal (z signal) simultaneously. When a motor binds the MT, the fluctuations in all three signals will decrease. Often, the change in the z signal will be the most noticeable initially, though K560 moves fast enough in the presence of 1 mM ATP that there is usually no noticeable delay between binding and movement of active motors. If the bead moves, record the data and score it as a motile bead. If the bead does not move, score it as immotile (*see* **Note 79**), and trap a new bead.

9. Move the nanopositioner to a spot on the cover slip with no MTs, and repeat **steps 6–8** for at least ten beads. If the fraction of motile beads is significantly larger than 30 %, repeat the experiment using a higher motor dilution during the sample chamber preparation. This ensures that the motility observed is due to single molecules [71].

10. For each raw data set, convert the QPD responses to bead–trap separations (in nm) using the recorded polynomial coefficients. Next, to obtain the corresponding force data (in pN), multiply the position data by the spring constant ($F = -kx$). Figure 22 presents typical data obtained from this assay for K560.

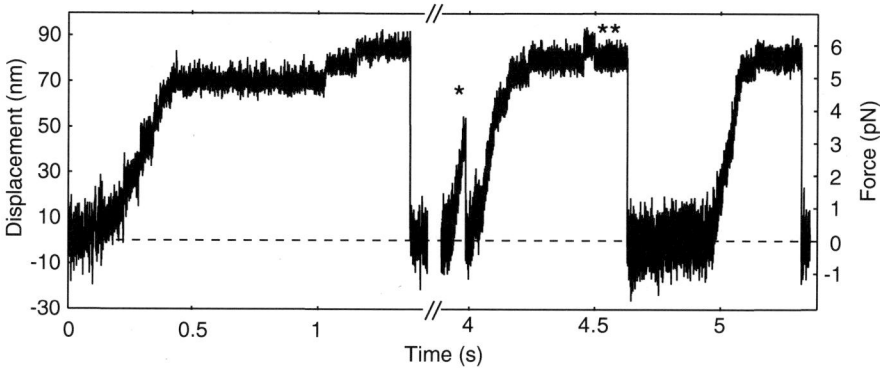

Fig. 22 K560 motility and force generation in the optical tweezers. The force/motion trace shows repeated force generation events in the trap in the *y* direction ($k_y = 0.07$ pN/nm). The motor repeatedly stalls at ~6 pN backward load. As it approaches stalling, the velocity slows considerably, and ~8-nm steps are clearly resolvable. Occasionally, the motor detaches from the MT before stalling (*asterisk*). Although the bead displacement is almost entirely monotonically increasing, the motor occasionally takes backward steps when it stalls (*double asterisk*)

4 Notes

1. The geometrical/ray-optics description offers an intuitive and reasonably accurate description of optical trapping for large particles, for which the microsphere can be considered to be essentially a ball lens that refracts the incident light. However, this model applies poorly to microspheres with diameters *d* much smaller than the wavelength λ of the trapping beam (note that this regime is defined by the wavelength of the laser *in the sample medium*; for a 1,064-nm laser trapping a particle in aqueous buffer, this corresponds to $\lambda/n_{\text{water}} = 1{,}064$ nm/1.33 = 800 nm). Electromagnetic scattering theory provides a better description for this case. In this formulation, the small dielectric particle in the trap is considered to be a point dipole (with a dipole moment induced by the electromagnetic field of the trapping beam) that scatters light according to Rayleigh scattering theory. In the electromagnetic field of the trapping beam (due to the intensity gradient imparted by focusing of the beam and its inherent Gaussian intensity profile), this dipole experiences a Lorentz force (called the gradient force) that draws it toward the trap center. This balances the net momentum transfer in the axial direction of the trap that results from Rayleigh scattering (since scattering is isotropic, net momentum is only transferred in the direction of beam propagation, which tends to push the particle out of the trap; this is called the scattering force). The net result is a stable trap in three dimensions (note: due to the scattering force, the position of the trapped bead is slightly shifted beyond the highest intensity of the laser focus).

While refractive models apply well to particles with $d \gg \lambda$, and Rayleigh scattering/Lorentz force models describe trapping of particles $d \ll \lambda$, neither theory applies well to the situation typically encountered in single-molecule optical tweezers studies, in which $d \approx \lambda$ [72]. Over the last several years, a great deal of progress has been made in this area by employing the full Mie solution to Maxwell's equations (also called the Lorenz–Mie, Mie–Debye, or Lorenz–Mie–Debye solution, or Mie scattering), which fully describes the scattering of light by dielectric spheres of any size. This complex theory converges appropriately with the simpler Rayleigh scattering (for very small particles) and geometrical optics (for large particles) descriptions. Nieminen et al. provide an informative review [73] with useful references therein. For any such theory to be of practical use, it must also encompass the effects of nonideal optical aberrations on optical trapping [74], particularly the spherical aberration and astigmatism. Recently, Dutra et al. [75] extended the so-called MDSA (Mie–Debye spherical aberration) theory [76] to include these effects, finding good agreement with experimental results.

2. The kinesin-1 motor domain is N-terminal, so that the truncation in K560 occurs just after the first coiled-coil stalk. The second (truncated) coiled coil is not required for heavy chain dimerization, or to produce motion [42] or force (see, e.g., supplemental information of Gennerich et al. [31]). The polyhistidine (6×His) and GFP tags are on the C-terminus of the construct, immediately following amino acid 560.

3. In the optical trapping field, there continues to be some variability in the surface-coupling techniques used to attach molecular motors to microspheres. Techniques can be categorized broadly according to whether they employ specific vs. nonspecific and covalent vs. non-covalent binding strategies. We advocate the use of specific binding to a known structure on the motor (preferably the tail), if possible, so that there is a high degree of confidence regarding the orientation of the motor on the bead. For K560-GFP, the GFP tag on the tail provides a convenient and specific target for coupling to the bead via high-affinity anti-GFP antibodies. We link the antibodies to beads covalently to ensure strong association and block the remaining surface with covalently bound bovine serum albumin (BSA).

4. Alternatively, PMSF can be substituted by the combination of Pefabloc SC, leupeptin, and pepstatin (Roche), or the cOmplete ULTRA EDTA-free Tablet (Roche), which are less toxic and more effective in long-term protease inhibition.

5. βME is toxic and flammable and should only be opened in a fume hood.

6. Kinesin requires divalent metal cations in order to hydrolyze ATP. Mg^{2+} is the most effective [77]. When preparing the ATP solution, keep it on ice to limit ATP hydrolysis. For assays in which the ATP concentration is non-saturating and needs to be known with precision, the weight of the ATP powder weighed should be corrected for water and residual solvent content (found by examining the certificate of analysis from the supplier).

7. Lysozyme dissolves slowly at 4 °C; hence, it is best to dissolve it at room temperature and then keep the solution on ice. It is preferred to add the imidazole after the lysis by lysozyme, since imidazole inhibits lysozyme activity [78]. Alternatively, other highly specific, active lysozyme can be applied [79].

8. The vial should always be kept on ice, since the tubulin will polymerize at higher temperature.

9. AMP-PNP does not hydrolyze in aqueous solution. However, it becomes unstable at low pH.

10. Microspheres of different diameter can be substituted with only slight modifications to this protocol. It is important that the trapping beads be as perfectly spherical as possible and have a uniform and precise diameter, since irregularly shaped particles or deviations from the expected diameter may lead to errors in trap stiffness determination and prevent the microspheres from tumbling uniformly (therefore preventing certain portions of the surface area from contacting the MTs in the motor trapping assay).

11. EDAC and NHSS are hygroscopic and sensitive to hydrolysis. Avoid any contact of the stocks with moisture or humidity. If the powder forms clumps, it has absorbed water and should be replaced (if in doubt, always replace old stocks with fresh reagents to ensure reactivity). When opening vials, allow them to warm to room temperature in a desiccator beforehand, in order to avoid condensation of moisture from the air.

12. The POC "oxygen scavenging" system is used to eliminate molecular oxygen from solution. The problems caused by molecular oxygen, and our choice of POC to prevent them, warrant discussion. In this assay, the presence of O_2 leads the formation of a variety of reactive species, particularly singlet oxygen ($^1O_2$) and molecular oxygen in the first ($O_2(^1\Delta_g)$) or second ($O_2(^1\Sigma_g^+)$) excited states above the ground state ($O_2(^3\Sigma_g^-)$; *see* ref. 80 for a comprehensive review). These reactive oxygen species rapidly oxidize nearby molecules, leading to protein damage [81, 82] and loss of organic dye fluorescence (simultaneous photobleaching and physical destruction of fluorescently labeled MTs, and prevention of these phenomena by oxygen scavengers [83, 84], provide a dramatic demonstration of both effects). The initial formation of $^1O_2$ generally

occurs as a photochemical reaction in which a "sensitizer" molecule enters an excited triplet state and transfers its energy to $O_2(^3\Sigma_g^-)$, yielding $O_2(^1\Delta_g)$ directly, or $O_2(^1\Sigma_g^+)$, which rapidly converts to $O_2(^1\Delta_g)$ [80] (the $O_2(^3\Sigma_g^-)$ triplet ground state thus quenches the excited triplet state of the sensitizer). In single-molecule experiments, the sensitizer is usually an organic dye excited by a laser, but a vast range of other molecules can participate as well [85], and even polystyrene microspheres in optical trapping beams can act as sensitizers [86].

The typical approach for removing O_2 is to add an oxygen "scavenger" (an enzymatic system that consumes oxygen) while sealing the sample chamber to prevent additional O_2 from entering. The traditional oxygen scavenger system is glucose oxidase/catalase (GOC) in the presence of glucose, for both fluorescence [87–94] and optical trapping assays [31, 53, 95–100]. Glucose oxidase (β-D-glucose:oxygen-1-oxidoreductase, EC 1.1.3.4) [101–103] is a flavoenzyme that catalyzes the oxidation of glucose by O_2, yielding hydrogen peroxide (H_2O_2, which is removed by the catalase) and D-glucono-δ-lactone (which is converted to gluconic acid). The more recent protocatechuic acid (PCA)/protocatechuate-3,4-dioxygenase (PCD) [104, 105] employs a single-step reaction in which PCD catalyzes the cleavage of the PCA extradiol ring, incorporating one O_2 molecule and yielding a muconic acid. This system has several advantages, including that no reactive H_2O_2 is produced. However, both GOC and PCA/PCD produce acid, significantly lowering the pH of the assay solution [43, 90, 106] unless strong buffers are used. With GOC, we observe widespread casein precipitation over long periods (>1 h), correlated with increased acidity and directly attributable to the presence of glucose oxidase (casein itself is acidic and is highly insoluble near its isoelectric point of ~4.6 [107, 108]). This acidification also correlates, over shorter periods, with unwanted and nonspecific sticking of trapping microspheres to glass and cytoskeletal filaments. Swoboda et al. [43] recently introduced the pyranose oxidase/catalase (POC) system [109] to single-molecule fluorescence studies, replacing glucose oxidase with pyranose 2-oxidase (pyranose:oxygen-2-oxidoreductase, EC 1.1.3.10) [110–113], which yields a stable 2-keto-D-glucose that does not convert to an acid (*see* Fig. 7). This system behaves at least as well as PCA/PCD in single-molecule assays, without acidification [43]. We have not rigorously compared the performance of oxygen scavenging by POC vs. GOC in the optical trapping assay, but have observed no negative effects with POC, and qualitatively identical effects on fluorescence. Based on this and the extensive characterization by Swoboda et al. [43], we employ POC in all our current assays.

It should be noted that for relatively high-power trapping beams, photobleaching is likely caused by multiphoton excitation and subsequent dye ionization, rather than $^1O_2$ formation [114]. While elimination of $^1O_2$ is still important (e.g., to prevent damage to K560 motors), photobleaching in this setting is limited by addition of an antioxidant rather than an oxygen scavenger [114]. Antioxidants are also known to inhibit triplet-state-induced fluorophore blinking under anaerobic conditions ($O_2(^3\Sigma_g^-)$ is a triplet quencher, as mentioned above, so its absence enhances population of the fluorophore triplet state). DTT serves this purpose in our assay (in addition to its primary role in preventing unwanted protein disulfide bond formation). Other reducing agents such as βME and tris(2-carboxyethyl)phosphine (TCEP) may also be beneficial, though ascorbic acid and the vitamin E analog Trolox® (6-hydroxy-2,5,7,8-tetramethylchroman-2-carboxylic acid) exhibit superior performance in preventing blinking via triplet quenching [104].

13. Slide chamber ends are sealed to prevent evaporation, ensure an anaerobic environment, and provide physical closure of the chamber (assumed by the theory used for optical trap calibration employing periodic stage oscillation, Fig. 19b, although the theory can be modified for a chamber with open ends [69]). As noted in Nicholas et al. (Chap. 9), vacuum grease is highly preferred over nail polish or other organic solvent-based adhesives. Using these substances to seal the chamber can lead to protein precipitation and nonspecific binding of trapping beads to glass and MTs, even in the absence of attached motors.

14. In principle, back focal plane detection of the microsphere position can be done with the same beam used to do the trapping itself. However, this presents some limitations and challenges. The major limitation is that it is not possible to perform force-clamp (also called force-feedback) experiments in which a constant force is applied by steering the trapping beam in real time to maintain a constant bead–trap separation, since this would prevent detection of bead movement by the attached motor. This technique therefore requires a separate detection beam. In addition, the separate detection beam allows convenient in situ position calibration of the detection system by sweeping the trapped bead across the detection beam using the trap. This allows easy calibration for each trapped bead. If only one beam is available, this sweeping must be done instead via the nanopositioning stage, using a bead stuck to the cover slip. Not only is this more tedious, but the resulting calibration is likely to be less accurate (e.g., due to different proximity of the bead to the cover slip, variations in bead sizes, and imprecision in positioning the stuck bead relative to the

true trap center of the laser beam). This problem can be circumvented using a technique (discussed in the Methods section) that calibrates the detector by analyzing the response to oscillatory movement of the nanopositioning stage [69], but this technique confines the detection to the range in which the detector response is linear. Finally, the high-power lasers typically used for trapping generally have wavelengths >~850 nm. The poor absorption of light of these wavelengths by most silicon photodiodes can cause unwanted low-pass filtering of the position data (called "parasitic filtering") [115, 116] that complicates trap stiffness determination.

15. Although the geometry of the optical trap is radially symmetrical, the linear polarization of the laser introduces an inherent asymmetry that can lead to asymmetry in the trap spring constants [34, 117, 118]. The basis for this effect can be understood fairly simply in the ray-optics regime [119] by considering that the reflection and refraction of light at a dielectric interface (i.e., the Fresnel coefficients at the bead surface) depend on polarization. In practice, the asymmetry can be largely removed by using a quarter wave plate oriented at 45° to the laser polarization to produce circularly polarized light [34], and this effect is also minimized when $d \approx \lambda$ [34, 117]. It is important to understand, however, that it is not entirely straightforward to control the polarization in the focal plane of the objective. First, mechanical stress on any of the optical elements in the pathway can partially orient the otherwise amorphous structure of glass, thereby imparting a slight birefringence (the degree of which is usually spatially inhomogeneous) [120]. This effect, known as photoelasticity or stress birefringence, may be significant in a long pathway containing many lenses and mirrors, due to cumulative effects. Moreover, the dielectric layers on dichroic mirrors can be birefringent, in addition to enhancing or suppressing S vs. P polarization modes depending on the orientation relative to the incident beam [121–123]. Finally, simply focusing light through a high-NA objective can significantly alter the polarization state in the focal plane [122, 124, 125]. These effects can be compensated for in order to achieve the desired polarization in the sample [123, 126]. In practice, we have found that the slight asymmetry in our setup can be removed using the half-wave plate (WP2), by simply applying adjustments, checking the spring constant calibration, and choosing the wave plate orientation at which $k_x = k_y$ within reasonable precision.

16. Because the MTs along which the kinesin walks are attached to the cover slip, the trapping bead is (indirectly) mechanically coupled to the microscope and optical table during experiments. Even miniscule mechanical vibrations will therefore

affect the bead position and produce unwanted noise and artifacts in the trapping data. Therefore, trapping microscopes should be built in the most acoustically and vibrationally isolated environment available. Where possible, all components that generate vibrations (e.g., any equipment with a cooling fan) should be placed in a separate room, with laser beams and electrical signals transmitted via appropriate optical fibers and cables, respectively. In addition, the optical pathway the trapping beam follows should be as mechanically stable as possible, to prevent the trapping beam from drifting (and therefore moving the beam focus/trap center). Optical posts should be thick (we use 1-in.-diameter posts) and of fixed height (not adjustable). We also use an optical breadboard on top of our main table for the majority of our pathway, in order to limit the length of the posts needed to raise the input beams to the height of the microscope's rear port. Temperature fluctuations should also be minimized, since they will cause mechanical elements to expand and contract. Ideally, even the person working with the instrument should be isolated from it during experiments. On our instrument, we have two identical monitor displays and keyboards inside and outside the microscope room, so that after the initial setup, experiments can be conducted from outside.

Air currents within the room should be minimized. In our customized microscope room (approximately ±0.01 °C maximum fluctuations), we use a valance cooling system (essentially a system of chilled-water heat sinks along the perimeter of the room) to control the temperature without the need for air flow. Even in this environment, we have found that small air currents can lead to relatively substantial instabilities in the optical trap position (approximately ±15-nm drift over a few seconds), especially if a person is moving anywhere in the room. This is attributable to (a) beam scattering off any particulate matter moving through the air (i.e., dust) and (b) beam refraction due to air turbulence that causes local, dynamic fluctuations in air density and, therefore, the index of refraction. Both effects reduce the pointing stability of the trapping laser as it travels along the optical pathway leading to the microscope. These problems are minimized by building an acrylic glass (poly(methylmethacrylate), PMMA) airtight enclosure around the optical pathway (this enclosure also adds to thermal stability). In our experience, such an enclosure is absolutely necessary in order to obtain short- and long-term trap stability of better than ~10 nm (even better stability can be attained under these conditions by systematically eliminating sources of mechanical instability in the microscope body, e.g., by replacing the filter turret and objective nosepiece with solid blocks, or even replacing the entire commercial microscope body completely). A more economical version of this enclosure can be made using a combination of optical posts and air-filled plastic packing material ("bubble wrap") as the walls. Other groups have improved stability further by replacing the atmosphere in the enclosure with helium gas (its index of refraction is very nearly that of vacuum, so that air density fluctuations have smaller effects) [127] or tracking cover slip fluctuations and correcting for them in real time using the nanopositioning stage [128, 129]. Controlling temperature fluctuations of the objective (which are caused in part by the incident laser and are therefore unavoidable) with a feedback control system can also significantly reduce long-term drift of the trap, especially in the axial direction [130].

17. To overcome the scattering force and trap the microsphere in the axial direction requires a tightly focused laser beam and therefore a microscope objective of high NA (this is also a requirement for the objective-based TIRF illumination). To obtain the steepest gradient (tightest focus), the incoming beam must take advantage of the entire NA, i.e., illuminate the entire rear entrance pupil of the lens. Since Gaussian beams have very low intensity toward the edges, it is common practice to "overfill" the objective (expand the input beam to be slightly larger than the entrance pupil). However, it should be noted that the marginal rays in high-NA illumination do not contribute to trapping, due to total internal reflection at the glass–buffer interface (the same phenomenon used for TIRF illumination), and therefore represent wasted energy. This energy (and the wasted energy blocked by the back aperture of the objective) can serve only to produce unwanted heating of the objective and slide chamber. Mahamdeh et al. [35] provide a detailed analysis that concludes that *under-filling* the objective to obtain an effective NA of ~1.25 leads to the most efficient use of laser power (lowest input power to achieve a given spring constant). This should be taken into account when designing the illumination pathway leading to the objective.

18. In principle, a computer-controlled coarse positioning stage is not required. However, most standard stages with manual micrometer-based positioning are highly unstable on the nanometer scale and thus undermine the stability of the entire instrument. We use the M-686.D64 stage because, although it has a very long range of movement that allows exploration of the entire sample chamber, it is highly mechanically stabile when not in motion.

19. To achieve practically useful spring constants requires only ~10–50 mW to enter the microscope objective. However, a much greater laser power is required due to losses in the optical fiber and in diffraction by the AOD. In addition, maximal

power output often decreases over the lifetime of a laser, so that having "reserve" power is desirable. The extra power can also be used to create additional, separately controlled traps.

20. Computer-controlled mirror mounts for beam positioning are more or less required since the optical pathway is enclosed in the isolation box. This also saves a tremendous amount of time during adjustment and allows precise mirror movements that are not possible by hand. It is important that the positioning system be stable when not in motion and also that it allows very small steps in positioning. The Picomotor™ system is ideal because it has excellent resolution and uses a piezo element to turn an adjustment knob that is otherwise stationary when no voltage is applied.

21. Depending on the application, electro-optic deflectors (EODs) may be preferable to AODs given their superior transmittance (which is also essentially independent of beam position), deflection-angle precision [131], and speed. EODs can be substituted for this protocol. Advantages of AODs over EODs are the larger deflection range, ease with which the power of the transmitted beam can be modulated, and the generally cheaper price.

22. High-quality electronic digital filters are expensive. Although they are very convenient, low-pass filtering can also be done in software (i.e., after data acquisition), if necessary. This is done by acquiring data at a sample rate well above the final desired one, low-pass filtering, and then decimating the data to the desired sample rate (no more than twice that of the filter frequency).

23. Particularly for the detection and trapping beams, stable mounting of the fiber output collimator is crucial. The kine-MATIX manipulators are particularly convenient because they allow very precise adjustment in combination with highly stable locking of the final position. This also allows the removal of additional mirrors from the pathway that would otherwise be needed to adjust the beam angle and position.

24. The custom polychroic mirror, the spectrum of which is available from Chroma Technologies, will need to be fabricated on request, which can take a few months. For applications not requiring multiple fluorescence channels, a simpler mirror can be substituted. Using a polychroic mirror in the microscope body (PM in Fig. 3) removes the need to switch elements in the microscope filter turret (in fact, the filter turret can be completely removed and the mirror rigidly mounted for added stability). This allows simultaneous trapping and fluorescence imaging and prevents slight misalignments that occur due to the filter turret not returning to precisely the same position each time it is rotated.

25. When purchasing dichroic mirrors, it is crucial to check that the substrate on which the dielectric coating is applied be

designed for use with lasers. Dichroic mirrors for routine fluorescence microscopy have insufficiently flat surfaces for use with coherent laser light and will give rise to interference that significantly degrades the Gaussian beam profiles of reflected lasers. In addition, it is imperative that the dichroic mirrors be mounted with no stress or torque whatsoever on the glass (the resultant bending will also lead to interference effects that ruin the beam profile). This precludes the use of virtually all mechanical means of mounting and even the use of most glues (which contract as they dry). We apply 732 RTV multipurpose silicone sealant (Dow Corning) with a syringe in a small bead along the perimeter of the mirror (*not* on the optical surface, but along the edge). This holds the glass very securely to the mount without deformations, but remains pliable enough to be removed if necessary.

26. The Luca series EMCCDs are not ideal for single-molecule imaging, but are relatively cheap and provide a very wide field of view for MT visualization. For instruments intended for combined trapping and single-molecule imaging, a higher-quality detector will be needed (e.g., the Andor Technology iXon series EMCCD), either in addition to or in place of the Luca EMCCD.

27. We highly recommend custom-writing the software to control the instrument, as this is virtually the only way to integrate control of all components into a single user interface and affords the ability to alter/add capabilities as needed. We have found that this initial investment of time was well worth the enhanced efficiency of a customized program. While LabView offers rapid interface development and contains built-in support for controlling a wide array of cameras and other hardware, we find it to be somewhat cumbersome for data analysis (e.g., trap spring constant calculation) compared to MATLAB. However, the two programs can share data in real time.

28. It is not recommended to refreeze competent cells, which decreases the competent efficiency of the cells. If several transformations are to be performed, prechill several 2-mL microcentrifuge tubes on ice, and divide cells into tubes. 50 μL of cells can be used for several transformations.

29. If time is limited, this step can be omitted. However, expect twofold decrease in efficiency for every 10-min reduction in incubation time (*see* NEB website, http://www.neb.com).

30. Different competent cells require different heat shocking times—adjust accordingly. The 10-s heat shock time here is for BL21(DE3) competent *E. coli* cells from NEB.

31. Check OD_{595} frequently after 2-h growth. After the lag phase, *E. coli* replicates every 20 min, which can easily lead to an overgrown culture.

32. Long kinesin constructs such as K560 tend not to fold properly and form inclusion bodies at 37 °C induction temperature [52]. Stability of the GFP in the K560 construct also decreases dramatically above room temperature [52, 132]. Therefore, an induction temperature <20 °C is recommended for this expression.

33. With the GFP tag, the cell pellet should be bright green. It is recommended to subject the pellet to a cycle of freezing and thawing to assist cell lysis, even if cell harvest and protein purification are performed in the same day [133].

34. ATP, βME, and PMSF should be as fresh as possible, since at pH 7, ATP is prone to hydrolysis, βME can be oxidized and form disulfide bonds in solution when exposed to air, and PMSF has a short half-life (110 min) in water solution at 25 °C [134].

35. The settings given here are appropriate for the Fisher Scientific model F550. Other ultrasonic homogenizers may require different settings to achieve the same results.

36. This is the bulk protein concentration, which overestimates the real concentration of full length K560, since there are fractions of truncated protein and other contaminants. Detailed methods in determining protein concentration are described in [135].

37. 6–12 % DMSO enhances tubulin polymerization [136].

38. Especially after the first K560 purification, it is advisable to perform standard assays for single-molecule function in addition to optical trapping and to rule out protein aggregation, in order to ensure that the motor is "well behaved" and functions as expected. Brouhard [137] provides a very useful approach to quality control for motor proteins in single-molecule studies.

39. The purpose of rinsing the microspheres is to remove surfactants (included to prevent aggregation during storage), soluble polystyrene-carboxyl synthesis byproducts (which could react with cross-linkers), and any other contaminants present in the stock solution. MES activation buffer is used because it does not contain amines or carboxyls that would interfere with the reaction. The slightly acidic pH is chosen to increase the fraction of protonated carboxyl groups (i.e., –COOH rather than –COO⁻) that will react with EDAC.

40. Quenching of the EDAC reaction is done to eliminate any free cross-linking reagents that could react with protein in subsequent steps. However, since (a) the microspheres are subsequently washed and (b) the reactive intermediates hydrolyze fairly quickly to regenerate the carboxyl groups on the microsphere surface, this quenching step can usually be omitted.

41. Coupling is done in a non-amine-containing buffer and at alkaline pH to promote deprotonation of the primary amines in lysine (i.e., $-NH_2$ rather than $-NH_3^+$; note that the pK_a for this amine is ~10).

42. The amount of total protein (BSA plus antibody) used here is significantly higher than the theoretical amount required to produce a monolayer on the microsphere surface (approximately 0.2 mg). However, we have found that at lower protein concentrations, the microspheres are more likely to aggregate due to multiple microspheres binding the same protein. This is also avoided by adding the beads to the protein, and not vice versa, because it effectively increases the ratio of protein-to-microsphere concentrations. If problems with aggregation occur, possible solutions are (a) to add a small amount of non-ionic surfactant and (b) to use a lower ionic strength coupling buffer (ions can shield the negative charge of the carboxylate ions on the microsphere surface, thereby minimizing ionic repulsion between the microspheres, and allowing attractive hydrophobic forces to dominate). The Bangs Laboratories website (www.bangslabs.com) has extensive technical notes on this subject.

43. The amine-containing quenching solution reacts with any remaining reactive groups on the microsphere surface. Because these groups are short-lived, this incubation can be shortened (or perhaps even skipped).

44. Microspheres retain GFP-binding activity for months when stored at 4 °C (never freeze the microspheres). Store in the smallest available tube to prevent evaporation over time, and wrap the mouth of the tube in Parafilm. When choosing the storage buffer, keep in mind that this buffer will enter the final assay in a roughly 1:20 dilution. For example, if studying enzymatic effects of phosphate in the final assay, PBS may be a poor choice. The protein-coated microspheres are stable over time in a variety of buffers.

45. It is important that the instrument be given sufficient time to equilibrate to a stable configuration. The instrument typically "settles" at a configuration very close to that of its previous use (Fig. 8), so that this "warm-up" period constitutes a type of self-alignment of the instrument. If adjustments are made as soon as the instrument is powered on, the system will continue to drift, thus misaligning the instrument again. The equilibration time will depend on the instrument, and an analysis similar to that shown in Fig. 8 is advisable.

46. Avoid bubbles, which will interfere with bright-field imaging and back focal plane detection. Spread the oil evenly in a thin pool approximately the diameter of the condenser. This helps prevent bubble formation when the condenser contacts the oil.

47. When building the instrument, take care to position all components carefully so that the microscope body and stages are aligned with the optical axis of the illumination pathway. In practice, it is difficult to align the stages with the pathway to better than about 1–2°. Since the stage rotation cannot be finely adjusted, we take the approach of using it as a reference to which the CCD and AOD are then aligned.

48. The most precise way to align the two sets of axes is to track the bead centroid in each image, thereby creating a grid of precise bead positions. Lines can then be fit to various regions of each grid to determine the exact degree of rotation between the two sets of axes. This procedure will also reveal any non-linearity in the AOD or stage movements.

49. For small angles ($\theta < \sim 0.07$ rad, or $\sim 4°$), the approximations $\sin(\theta) \approx \theta$ and $\cos(\theta) \approx 1 - \theta^2/2$ may be used. This reduces computational requirements with essentially no effect on the calculated positions.

50. Use distances of $\sim 2–10$ μm, near the center of the field of view. Distances that are too small will be more affected by errors in bead localization, while very large distances that involve placing the bead far off-axis are more susceptible to errors arising from any aberrations in the imaging pathway (e.g., curvature of field).

51. The overall displacement of the bead is highly linear with voltage applied to the AOD, so that simply measuring the distances between pairs of points is theoretically acceptable. However, because AODs exhibit "wiggles" in their position response on the order of a several nanometers [131], fitting a line to the overall data is more reliable.

52. The aberrated "cross" shape of the retroreflected beam pattern on the CCD may be somewhat unexpected (e.g., as opposed to a small circular spot or concentric rings). The pattern arises as a consequence of the laser beam polarization and the combined effects of the high NA of the objective and the lower index of refraction of water than of oil, which lead to total internal reflection of the marginal rays at the glass/water interface. This causes a phase shift of the wave components undergoing total internal reflection, and as a result, the apparent reflection point is displaced, leading to the aberrated reflection pattern. Novotny et al. have analyzed this phenomenon in detail [138]. Note that the orientation of the pattern depends on the angle at which the beam is linearly polarized (and can thus be changed by rotating WP2).

53. The mirror farther from the microscope should be used to adjust the pattern symmetry, since the major effect of this mirror is to move the beam in the back aperture of the objective, thereby changing the angle of the output beam (Fig. 9a),

while tilting the nearer mirror has a greater effect on the angle of the input beam, and therefore the position of the focal spot (Fig. 9b). By using the mirrors in combination, any position/angle can be achieved, even if there are drifts in the overall system over time. On our instrument, we have the four controls for each mirror (up, down, right, left) configured so that adjustments to Pzt-M2 translate the beam in the *opposite* direction of the button pressed (e.g., pressing the down button shifts the beam in the +y direction in the image). Conveniently, this means that the intensity within the pattern shifts in the *same* direction as the button pressed, so that when adjusting the pattern symmetry, one needs only to press the buttons in the direction in which intensity needs to be shifted to make the pattern symmetrical. Pzt-M3, which is used to adjust the focus position, is configured oppositely, so that the beam moves in the same direction as the button pressed. This opposite configuration of the two mirrors is convenient because it means that when aligning the detection beam position/symmetry, the adjustments for both mirrors are always made in the same direction (both for adjustments of using back-reflections and with the QPD signal, later in the protocol).

54. If after repeated attempts to adjust the back-reflection symmetry and position, the trapped bead position is not on the center position marked on the CCD, or the retroreflection from the bead does not appear symmetrical, it is likely that the marked position on the CCD is not truly on the optical axis. In practice, there is a range of positions for which the retroreflected beam pattern can look acceptable, but some will result in trapping at a different position in the x–y plane. If this occurs, experiment with centering the back-reflection at a different position (and readjusting the symmetry) and then trapping a bead and observing its position relative to the new center position. If the alignment is worse, move the back-reflection in the opposite direction. If it is better, move the back-reflection further and repeat. By iterating through this process, a position can be found such that the back-reflection center and trapped bead position coincide in the x–y plane. This position should then be saved as the new reference point for aligning the beam.

55. If the adjustments to Pzt-M1 are too large, the asymmetry will "flip" and the bead will be displaced from the trap center in the opposite direction. If this happens, move in smaller increments in the opposite direction.

56. All switching of filter frequencies and gains should be done automatically via RS-232 communication between the trap software and the low-pass filters. The purpose of this step is to account for offsets caused by the dark current of the photodiode

and/or voltage offsets introduced by other components. In practice, the offsets are different for each setting on the filters and also vary somewhat from day to day (presumably due to temperature changes at the QPD itself, which alter the dark currents). It is important to correct for these offsets mostly because the trap calibrations are not done at the same filter settings as data acquisition. Without subtracting the offsets, the normalized voltages calculated for the same bead displacements may be different during calibration vs. data acquisition, leading to subtle errors in position detection and/or spring constant measurement. Note that preamplifier gain should be used rather than output gain, because the filter adds some very low-voltage artifacts to the high-frequency components of the input signal. If the input signal is preamplified, these artifacts are negligible relative to the true signal, whereas if the amplification occurs at the output, they are not and will hinder power spectrum analysis during calibration.

57. Using the control configurations mentioned in **Note 53** for Pzt-M3, up/down = +/−y and right/left = +/−x.

58. In the optical trapping literature, the QPD response signals are often said to be in units of volts. While this is conceptually useful (e.g., when considering conversion of the QPD signal to nanometers), the x and y QPD signals are in fact normalized by the total QPD voltage and are therefore unitless. As a compromise, we refer to the units of the QPD signal in "normalized volts," V_{norm}.

59. The useful range of the QPD response function can be extended even further, for example, by taking into account the unavoidable crosstalk that occurs between the lateral and axial response functions when the bead is significantly displaced [139], or by limiting the effective NA of the objective for the detection beam [140], in order to create a larger focal spot. In addition, for experiments in which accurate two-dimensional position detection is needed over a wide range, the full two-dimensional detection area can be scanned and fit with a two-dimensional, fifth-degree polynomial [141]. This is generally unnecessary for experiments in which the motion is directed along an MT that is well aligned with one of the microscope axes.

60. Calibration in the axial direction is also possible [26, 60–62, 142, 143], but the QPD sum signal is not normalized, so laser power fluctuations cannot be easily distinguished from axial displacements. In this protocol, measurement of axial position is far less important than that of lateral position, especially since the bead-surface separation is adjusted with fairly good precision at the beginning of each experiment (*see* Fig. 15).

There is generally a trade-off between axial and lateral position detection sensitivity, depending on the effective NA of the condenser lens (adjustable using the aperture diaphragm): higher NA increases lateral position detection sensitivity, with the opposite effect for axial position detection [60, 61]. We generally use the maximum NA (aperture diaphragm fully open), though for very large or small beads, one should be aware that high NA of the detection system can produce unintended effects [61].

61. Assuming the trap behaves as a spring that applies force $F_x = -k_x x$, then the associated potential energy is $U(x) = \int k_x x \, dx = \frac{1}{2} k_x x^2$. The average value is $\langle U(x) \rangle = \langle \frac{1}{2} k_x x^2 \rangle = \frac{1}{2} k_x \langle x^2 \rangle$, where the brackets ("$\langle \ldots \rangle$") denote the average and $\langle x^2 \rangle$ is thus the variance of the distribution of bead positions over time. In classical statistical mechanics, the so-called equipartition theorem states that, for each degree of freedom for which an associated energy has quadratic dependence, the average of that associated energy at equilibrium is $\frac{1}{2} k_B T$ [144]. Thus, $\langle U(x) \rangle = \frac{1}{2} k_x \langle x^2 \rangle = \frac{1}{2} k_B T$, i.e., $k_x = k_B T / \langle x^2 \rangle$. It is worth noting, however, that the more generalized theorem [145] simply states that the energy is equally partitioned between the degrees of freedom. Boltzmann's law gives the probability of a particle having a particular energy/position: $P(x) = \exp(-U(x)/k_B T)/Z_p$, where Z_p is a normalization constant known as the "partition function." Computing $\langle U(x) \rangle = \int P(x) U(x) \, dx$ gives the associated average energy, which is only $\frac{1}{2} k_B T$ if $U(x)$ is quadratic in x, i.e., if the potential is harmonic. Thus, the equality $k_x = k_B T / \langle x^2 \rangle$ does not apply if the trap potential is not harmonic in the region in question.

62. In the literature, the power spectral density is often referred to simply as the "power spectrum." Although the two are related, they are not precisely the same: the power spectral *density* describes the power *per unit frequency* (units of displacement squared/Hz), while the power spectrum simply measures power (units of displacement squared). This simply amounts to a difference in scaling, but using the power spectrum instead of the power spectral density may lead to errors in the parameters extracted from fits. A good way to tell if the spectrum has been scaled properly is to check that the following is true (Parseval's theorem): the integral of the power spectral density (i.e., total power) should be equal to the variance of the original position data. Both MATLAB (via the function "pwelch") and LabView (via the "FFT Power Spectral Density" virtual instrument) have functions to directly calculate the appropriate quantity, and fast Fourier transform (FFT) functions that can also be used for explicit calculation. In either case, it is the single-sided (frequencies greater than zero) spectrum that

should be calculated, multiplied by 2 in order to maintain the total power in the original spectrum.

63. Note that larger k implies larger f_c. As a result, measurement of spring constants for traps with increasing stiffness requires increasing the bandwidth of the measured spectrum to include higher frequencies. Note also that f_c decreases for larger γ (larger bead size or greater solution viscosity), so that the bandwidth requirements also depend on the experimental conditions. Larger k and γ both decrease the overall amplitude of the power spectral density, though for constant k, the total area under the curve is constant (and equal to the variance in position data; *see* **Note 62**), since decreases in γ (thus reducing the amplitude) are balanced by corresponding increases in f_c (increasing the area under the curve).

64. The obvious approach to the problem of drag dependence on bead-surface distance is to simply move the bead far from the surface during calibration. However, refraction at the glass–water interface (Fig. 16) induces spherical aberration in the beam focus that increases progressively as the beam is focused deeper into the sample, thereby weakening the trap strength. Thus, the spring constant measured with the bead far above the surface is significantly different from that very close to the surface. The calibration must therefore be done at the same axial position as the experiment.

65. The method presented here moves the stage intermittently, but the same principle can be applied for sinusoidal or triangle-wave stage motion; see, e.g., Neuman and Block [26]. Another related method involves observing the response to "instantaneously" stepping the trap [26]. The bead will return to the trap center in an exponentially damped manner with time constant $\tau = \gamma / k$ (i.e., the trap pulls the bead toward the center with force proportional to k and is resisted by a force proportional to γ, which slows it down).

66. In principle, by measuring the bead-surface separation at the extrema of the stage motion, any tilt in the cover slip can be calculated and accounted for in the stage movement (i.e., moving not just in x or y but also in z), thereby maintaining a constant value during the calibration.

67. Determining the precise location of the inflection point can be somewhat arbitrary and can influence the final result of calculations. For the purpose of this protocol, the most important issue is reproducibility of the measurement (i.e., precision rather than accuracy). Since the width of the "peak" in the response signal is very consistent, we use this as our "landmark" when fitting the data. We first find the maximum in the response and fit the polynomial in a fixed region surrounding

it (chosen initially by trial and error in order to get the best fit for the region near the inflection point and then set as a fixed value for future fits). We then fit the line from the beginning of the data set until ~100 nm before the beginning of the region fit by the polynomial.

68. Although the range of position detection can be extended considerably beyond the region in which the QPD response signals are linear (*see* also **Note 59**), it is very important to realize that displacement may not be linear with applied force in regions far from the trap center. On the one hand, the optical trapping "potential energy well" is not infinitely wide, so that at large displacements/high forces, it must "flatten out" and allow the bead to escape. In this region, the spring constant is different from that at the trap center (and will eventually become negative [146–148]). Assuming the same spring constant at these positions will therefore yield a significant overestimate of the applied force. On the other hand, trap stiffness may initially *increase* as the bead is displaced [146–148], so that forces would be underestimated. These effects are generally more significant for beads of large diameter (>1 μm) [147, 148], but should be considered for any size microspheres when working far from the center of the trap. The range in which force is linear with displacement can be determined by mapping the potential of the optical trap, e.g., via the viscous drag force calibration method [146, 148] (Fig. 20d), or by using one strong trap (operating exclusively in its linear range) to calibrate a weaker one by displacing the two traps by known distances [147]. A related method employs the AOD to rapidly switch between two positions, effectively creating two separate traps [149], without the need for two separate beams. If the potential has not been fully mapped at large distances, then the spring constant should be adjusted such that the behavior of interest occurs only within the region in which the restoring force is linear with displacement. For the experiments described here, we avoid working at bead–trap separations >~160 nm, and even smaller separations are generally recommended.

Interestingly, recent work [147, 150] (see in particular Farré et al. [150] for a detailed theoretical discussion, including extensive background on BFP detection) has demonstrated that while the QPD response signals are linear with position over only a relatively narrow range, they are linear with force over a much broader one. This is essentially a manifestation of the fact that the direction signals measured by the QPD amount to changes in photon momentum and therefore the force exerted on the bead [150, 151]. In principle, this relationship can be used to accurately measure force even at large displacements and, combined with the aforementioned methods for extending the range of positional detection, could allow the trap to be used over a much wider range.

69. In principle, data collection for equipartition analysis can be done at the same time as that for power spectrum analysis (i.e., the same data set can be use for both methods). We use 65,536 Hz (2^{16} Hz) for the "hybrid" calibration method (because using a sampling rate that is a power of 2 ensures that the spike at $f_{drive} = 32$ Hz falls exactly on a single data point in the resulting spectrum [69]) and therefore use the same sample frequency for all calibration methods. However, if using data that have been low-pass filtered for equipartition analysis (as required for power spectrum analysis), it is very important that the data be sampled at a rate as high as possible (i.e., low-pass filters are set at high frequency). Otherwise, the filters remove a significant amount of power from the signal, thereby reducing the variance and artificially increasing the calculated spring constant. Wong and Halvorsen [152] provide a very detailed analysis of the effects of low-pass filtering/finite detection bandwidth on optical tweezers calibration.

70. $\tau = \gamma / k$ is the characteristic "autocorrelation time" for the trap [153]. At times equal to or shorter than this time scale, bead positions cannot be considered statistically independent (i.e., if the position at one time is known, the position a very short time afterward depends on the initial position and is not random). To obtain an accurate equilibrium measurement of $<x^2>$, data points must therefore be separated by a few τ. The number of data points to discard depends on the trap stiffness (the weaker the trap, the longer the time required between points) and can be predicted based on the input laser power and previous characterization of the expected spring constant (Fig. 21). In practice, discarding correlated values does not have a substantial effect on the computed variance.

71. Low-frequency fluctuations (e.g., due to thermomechanical fluctuations in the optical pathway) can increase the variance of the position data and thereby artificially decrease the calculated spring constant. If these fluctuations can be filtered out, the calculated spring constant will therefore be more accurate. The easiest method for doing this is to take a moving average with a fairly wide window (e.g., 0.2–0.5 s). For a signal without drift, this moving average will be essentially zero, whereas for a signal with slow drift, the resulting average will essentially plot this drift. By subtracting this signal from the original data, the drift can thus be removed. However, this method inevitably leads to inaccuracies (though this may be acceptable if the artifacts introduced are smaller than the original drift). Another method (with essentially the same underlying principle) is to

high-pass filter the data (e.g., at ~2 Hz), although this will likewise remove power from the signal, and some of this power may be due to true thermal motion of the bead. Wong and Halvorsen offer a more robust method (*see* Fig. 4 of ref. [152]), in which the data are high-pass filtered at several different cutoff frequencies, and the results are used to extrapolate the variance at 0 Hz cutoff frequency (i.e., no filter).

72. By the nature of the fast Fourier transform used to compute the power spectral density, the frequency resolution δf is inversely proportional to the measurement time, $\delta f = 1/t_{msr}$. Therefore, dividing the original data set into more subsets will diminish the frequency resolution, but will simultaneously decrease the amplitude of the noise via averaging. This can also be done by block-averaging the final spectrum [26].

73. For water near room temperature, the viscosity η can be calculated to very good approximation [154], in units of pN s/nm^2, as $\eta = (AT_r^a + BT_r^b + CT_r^c + DT_r^d) \times 10^{-12}$, where $T_r = (T_c + 273.15)/300$ (T_c being the magnitude of the temperature in °C), and $A = 280.68$, $B = 511.45$, $C = 61.13$, $D = 45.90 \times 10^{-2}$, $a = {}^-1.9$, $b = {}^-7.7$, $c = {}^-19.6$, and $d = {}^-40.0$. For example, at 25 °C, $\eta = 8.9 \times 10^{-10}$ pN s/nm^2. Since the presence of 1 mg/mL casein contributes only minutely to the solution viscosity [155], we disregard this when making initial estimates.

74. Several experimental issues can lead to a non-Lorentzian power spectral density function in this calibration procedure [156], including aliasing and unintended frequency filtering by the QPD for near-infrared illumination [115, 157] (*see* also **Note 14**) and crosstalk between the QPD response signals. We have taken the approach of experimentally minimizing these artifacts. However, they are well understood theoretically and may be accounted for in the fitting procedure if necessary. Reference [46] provides a free software package to accomplish this. *See* also ref. [158] regarding rigorous parameter estimation from least-squares fitting of power spectra.

75. The motion must be as perfectly sinusoidal as possible (otherwise harmonics—i.e., additional spikes—will appear in the spectrum, and the theory will not be applicable). This precludes, for example, "stepping" the stage by repeated commands from the computer control software. Instead, as smooth a sinusoidal waveform as possible should be sent to the stage with a command to follow the pattern until a stop signal is sent. Wait for a few periods of oscillation before collecting data, so that any initial transients have subsided. The amplitude and frequency of oscillation must be confirmed by recording the stage movement and analyzing it (e.g., via power spectral density analysis of recorded position data) after the

oscillation has completed (rather than just assuming the commanded values). Typically, the actual amplitude is somewhat smaller than the commanded amplitude, and the stage motion lags the commanded movements (i.e., there is a phase lag). The latter is unimportant given the nature of the spectral analysis used to find the spring constant.

76. $[S_{VxVx}(f_{drive}) - S^{fit}_{VxVx}(f_{drive})]$ can be interpreted as the height of the spike, while $1/t_{msr}$, which is related to the separation between neighboring data points in the spectrum, is the width.

77. Many in vitro protocols use a higher paclitaxel concentration. We have found that MTs are stable using the concentrations given here. Paclitaxel is very poorly soluble in water [159] and can form asters and bundles similar in appearance to MTs [160, 161], which we observe often at higher concentrations. These structures can interfere with the optical trapping assay by diffusing into the trap and presumably also deplete the soluble fraction of paclitaxel available for MT binding. After more than ~3 h, paclitaxel-containing solutions should be made fresh from paclitaxel dissolved in DMSO.

78. POC/CS can be stored in the refrigerator at 4 °C overnight if it is not used during one set of experiments. For critical experiments, however, fresh POC should be used.

79. For processive motors such as K560, if 30 % of beads tested in the optical trapping assay exhibit motility, then there is a 95 % chance that any given bead has no more than one motor attached (this probability is derived under the assumption that the number of motors bound to each bead obeys Poissonian statistics) [71]. In fact, given the small likelihood of two motors bound to a microsphere being close enough to engage the MT simultaneously [31, 71, 162], the probability of motion arising from two or more motors under these conditions is well below 1 %. Thus, the condition of 30 % or fewer moving beads is generally regarded as the requirement to attribute observed motility to single molecules. The motor dilution required to achieve this condition varies from one purification to the next and is found by trial and error, starting with an educated guess based on past experience. It should be noted that, especially for a previously uncharacterized motor, it is not valid to assume processivity, and for a nonprocessive motor (by definition), *any* extended motility events must be due to two or more motors simultaneously engaging the MT. To distinguish between these two cases, the motor concentration is varied over several experiments in which the microsphere concentration is kept constant, and the motile fraction of beads is recorded. Plotting the motile fraction P vs. the relative motor-to-bead ratio C, one then attempts to fit the data with one of two Poissonian probability models, $P = 1 - \exp(-\lambda C)$

or $P = 1 - \exp(-\lambda C) - (-)\exp(-\lambda C)$, respectively, where λ is a fitting parameter that can be interpreted as the fraction of motors competent to bind microspheres and contribute to motility. If the data are more consistent with the first model, this suggests that the motor is processive, while if the latter model fits the data more closely, the motor is likely nonprocessive. See, for example, the supplement of ref. [31].

80. The MT has a diameter of 25 nm, whereas the antibody GFP-K560 complex extends a distance of ~20–40 nm from the bead surface [163]. Thus, a ~50–70-nm separation between the cover slip and the bead surface allows the bead to diffuse freely, while allowing interaction between the motor and the MT.

81. During the observation period, the stage z position can be adjusted in small (~10 nm) increments in order to promote binding to the microtubule. Once binding and movement are observed, the stage should be moved slowly back near the original position (the one at which the spring constant was measured).

Acknowledgments

We thank Joshua Shaevitz, Michael Diehl, Erik Schäffer, and Kenneth Jamison for helpful discussions on optical trapping, Marko Swoboda for helpful communications regarding the use of pyranose oxidase in the oxygen scavenging system, and Laura E.K. Nicholas for assistance with photography and figure illustration. The authors are supported by the National Institutes of Health grant R01GM098469. M.P.N. received support from the NIH-funded Medical Scientist Training and Molecular Biophysics Training programs (NIH grants T32GM007288 and T32GM008572, respectively) at the Albert Einstein College of Medicine.

References

1. Verhey KJ, Hammond JW (2009) Traffic control: regulation of kinesin motors. Nat Rev Mol Cell Biol 10:765–777

2. Vallee RB, Williams JC, Varma D et al (2004) Dynein: an ancient motor protein involved in multiple modes of transport. J Neurobiol 58:189–200

3. Johansen KM, Johansen J (2007) Cell and molecular biology of the spindle matrix. Int Rev Cytol 263:155–206

4. Civelekoglu-Scholey G, Scholey JM (2010) Mitotic force generators and chromosome segregation. Cell Mol Life Sci 67: 2231–2250

5. Gaglio T, Saredi A, Bingham JB et al (1996) Opposing motor activities are required for the organization of the mammalian mitotic spindle pole. J Cell Biol 135:399–414

6. Grill SW, Howard J, Schäffer E et al (2003) The distribution of active force generators controls mitotic spindle position. Science 301:518–521

7. Hoffman DB, Pearson CG, Yen TJ et al (2001) Microtubule-dependent changes in assembly of microtubule motor proteins and mitotic spindle checkpoint proteins at PtK1 kinetochores. Mol Biol Cell 12: 1995–2009

8. Winey M, Bloom K (2012) Mitotic spindle form and function. Genetics 190:1197–1224

9. Kiyomitsu T, Cheeseman IM (2012) Chromosome- and spindle-pole-derived signals generate an intrinsic code for spindle position and orientation. Nat Cell Biol 14:311–317

10. Kapitein LC, Peterman EJG, Kwok BH et al (2005) The bipolar mitotic kinesin Eg5 moves on both microtubules that it cross-links. Nature 435:114–118

11. Ferenz NP, Paul R, Fagerstrom C et al (2009) Dynein antagonizes Eg5 by crosslinking and sliding antiparallel microtubules. Curr Biol 19:1833–1838

12. Florian S, Mayer TU (2012) The functional antagonism between Eg5 and dynein in spindle bipolarization is not compatible with a simple push-pull model. Cell Rep 1:408–416

13. Kapitein LC, Kwok BH, Weinger JS et al (2008) Microtubule cross-linking triggers the directional motility of kinesin-5. J Cell Biol 182:421–428

14. Oladipo A, Cowan A, Rodionov V (2007) Microtubule motor Ncd induces sliding of microtubules in vivo. Mol Biol Cell 18:3601–3606

15. Furuta K, Toyoshima YY (2008) Minus-end-directed motor Ncd exhibits processive movement that is enhanced by microtubule bundling in vitro. Curr Biol 18:152–157

16. Fink G, Hajdo L, Skowronek K et al (2009) The mitotic kinesin-14 Ncd drives directional microtubule-microtubule sliding. Nat Cell Biol 11:717–723

17. Mountain V, Simerly C, Howard L et al (1999) The kinesin-related protein, HSET, opposes the activity of Eg5 and cross-links microtubules in the mammalian mitotic spindle. J Cell Biol 147:351–366

18. Szczęsna E, Kasprzak AA (2012) The C-terminus of kinesin-14 Ncd is a crucial component of the force generating mechanism. FEBS Lett 586:854–858

19. Wade RH, Kozielski F (2000) Structural links to kinesin directionality and movement. Nat Struct Mol Biol 7:456–460

20. Rogers GC, Rogers SL, Schwimmer TA et al (2004) Two mitotic kinesins cooperate to drive sister chromatid separation during anaphase. Nature 427:364–370

21. Mennella V, Rogers GC, Rogers SL et al (2005) Functionally distinct kinesin-13 family members cooperate to regulate microtubule dynamics during interphase. Nat Cell Biol 7:235–245

22. Moores CA, Milligan RA (2006) Lucky 13-microtubule depolymerisation by kinesin-13 motors. J Cell Sci 119:3905–3913

23. Ross JL, Wallace K, Shuman H et al (2006) Processive bidirectional motion of dynein-dynactin complexes in vitro. Nat Cell Biol 8:562–570

24. Dixit R, Ross JL, Goldman YE et al (2008) Differential regulation of dynein and kinesin motor proteins by tau. Science 319:1086–1089

25. Hentrich C, Surrey T (2010) Microtubule organization by the antagonistic mitotic motors kinesin-5 and kinesin-14. J Cell Biol 189:465–480

26. Neuman KC, Block SM (2006) Optical trapping. Rev Sci Instrum 75:2787–2809

27. Verdeny I, Farré A, Mas SJ et al (2011) Optical trapping: a review of essential concepts. Opt Pur Apl 44:527–551

28. Svoboda K, Schmidt CF, Schnapp BJ et al (1993) Direct observation of kinesin stepping by optical trapping interferometry. Nature 365:721–727

29. Finer JT, Simmons RM, Spudich JA (1994) Single myosin molecule mechanics: piconewton forces and nanometre steps. Nature 368:113–119

30. Yin H, Wang MD, Svoboda K et al (1995) Transcription against an applied force. Science 270:1653–1657

31. Gennerich A, Carter AP, Reck-Peterson SL et al (2007) Force-induced bidirectional stepping of cytoplasmic dynein. Cell 131:952–965

32. Mallik R, Carter BC, Lex SA et al (2004) Cytoplasmic dynein functions as a gear in response to load. Nature 427:649–652

33. Simmons RM, Finer JT, Chu S et al (1996) Quantitative measurements of force and displacement using an optical trap. Biophys J 70:1813–1822

34. Rohrbach A (2005) Stiffness of optical traps: quantitative agreement between experiment and electromagnetic theory. Phys Rev Lett 95:168102–168104

35. Mahamdeh M, Pérez Campos C, Schäffer E (2011) Under-filling trapping objectives optimizes the use of the available laser power in optical tweezers. Opt Express 19:11759–11768

36. Ashkin A, Schütze K, Dziedzic JM et al (1990) Force generation of organelle transport measured in vivo by an infrared laser trap. Nature 348:346–348

37. Soppina V, Rai A, Ramaiya A et al (2009) Tug-of-war between dissimilar teams of

microtubule motors regulates transport and fission of endosomes. Proc Natl Acad Sci U S A 106:19381–19386

38. Gross SP (2003) Application of optical traps in vivo. Methods Enzymol 361:162–174

39. Block SM, Goldstein LS, Schnapp BJ (1990) Bead movement by single kinesin molecules studied with optical tweezers. Nature 348:348–352

40. Piggee C (2009) Optical tweezers: not just for physicists anymore. Anal Chem 81:16–19

41. Vale RD, Reese TS, Sheetz MP (1985) Identification of a novel force-generating protein, kinesin, involved in microtubule-based motility. Cell 42:39–50

42. Case RB, Pierce DW, Hom-Booher N et al (1997) The directional preference of kinesin motors is specified by an element outside of the motor catalytic domain. Cell 90:959–966

43. Swoboda M, Henig J, Cheng H-M et al (2012) Enzymatic oxygen scavenging for photostability without pH Drop in single-molecule experiments. ACS Nano 6:6364–6369

44. Tolić-Nørrelykke IM, Berg-Sørensen K, Flyvbjerg H (2004) MatLab program for precision calibration of optical tweezers. Comput Phys Commun 159:225–240

45. Hansen P, Tolicnorrelykke I, Flyvbjerg H et al (2006) tweezercalib 2.0: faster version of MatLab package for precise calibration of optical tweezers. Comput Phys Commun 174:518–520

46. Hansen P, Tolicnorrelykke I, Flyvbjerg H et al (2006) tweezercalib 2.1: Faster version of MatLab package for precise calibration of optical tweezers. Comput Phys Commun 175:572–573

47. Osterman N (2010) TweezPal – optical tweezers analysis and calibration software. Comput Phys Commun 181:1911–1916

48. Lee WM, Reece PJ, Marchington RF et al (2007) Construction and calibration of an optical trap on a fluorescence optical microscope. Nat Protoc 2:3226–3238

49. Selvin PR, Lougheed T, Hoffman MT et al. (2007) Equipment setup for fluorescence imaging with one-nanometer accuracy (FIONA) Cold Spring Harb Protoc 2007, pdb.ip45

50. Van Mameren J, Wuite GJL, Heller I (2011) Introduction to optical tweezers: background, system designs, and commercial solutions. Methods Mol Biol 783:1–20

51. Bornhorst JA, Falke JJ (2000) Purification of proteins using polyhistidine affinity tags. Methods Enzymol 326:245–254

52. Stock MF, Hackney DD (2001) Expression of kinesin in *Escherichia coli*. Methods Mol Biol 164:43–48

53. Gennerich A, Reck-Peterson SL (2011) Probing the force generation and stepping behavior of cytoplasmic Dynein. Methods Mol Biol 783:63–80

54. Hermanson GT (2008) Bioconjugate techniques, 2nd edn. Academic, New York

55. Abramoff MD, Magalhães PJ, Ram SJ (2004) Image processing with ImageJ. Biophotonics International 11:36–42

56. Schneider CA, Rasband WS, Eliceiri KW (2012) NIH Image to ImageJ: 25 years of image analysis. Nat Methods 9:671–675

57. Visscher K, Gross SP, Block SM (1996) Construction of multiple-beam optical traps with nanometer-resolution position sensing. IEEE J Sel Top Quantum Electron 2: 1066–1076

58. Allersma MW, Gittes F, deCastro MJ et al (1998) Two-dimensional tracking of ncd motility by back focal plane interferometry. Biophys J 74:1074–1085

59. Gittes F, Schmidt CF (1998) Interference model for back-focal-plane displacement detection in optical tweezers. Opt Lett 23:7–9

60. Pralle A, Prummer M, Florin EL et al (1999) Three-dimensional high-resolution particle tracking for optical tweezers by forward scattered light. Microsc Res Tech 44:378–386

61. Rohrbach A, Kress H, Stelzer EHK (2003) Three-dimensional tracking of small spheres in focused laser beams: influence of the detection angular aperture. Opt Lett 28:411–413

62. Neuman KC, Abbondanzieri EA, Block SM (2005) Measurement of the effective focal shift in an optical trap. Opt Lett 30:1318–1320

63. Schäffer E, Nørrelykke SF, Howard J (2007) Surface forces and drag coefficients of microspheres near a plane surface measured with optical tweezers. Langmuir 23: 3654–3665

64. Vermeulen KC, van Mameren J, Stienen GJ et al (2006) Calibrating bead displacements in optical tweezers using acousto-optic deflectors. Rev Sci Instrum 77:013704–013706

65. Florin EL, Pralle A, Stelzer EHK et al (1998) Photonic force microscope calibration by thermal noise analysis. Appl Phys A 66:75–78

66. Gittes F, Schmidt CF (1997) Signals and noise in micromechanical measurements. Methods Cell Biol 55:129–156

67. Svoboda K, Block SM (1994) Biological applications of optical forces. Annu Rev Biophys Biomol Struct 23:247–285

68. Appleyard DC, Vandermeulen KY, Lee H et al (2007) Optical trapping for undergraduates. Am J Phys 75:5

69. Tolić-Nørrelykke SF, Schäffer E, Howard J et al (2006) Calibration of optical tweezers with positional detection in the back focal plane. Rev Sci Instrum 77:103101–103111

70. Guzmán C, Flyvbjerg H, Köszali R et al (2008) In situ viscometry by optical trapping interferometry. Appl Phys Lett 93:184102–184103

71. Svoboda K, Block SM (1994) Force and velocity measured for single kinesin molecules. Cell 77:773–784

72. Wright WH, Sonek GJ, Berns MW (1993) Radiation trapping forces on microspheres with optical tweezers. Appl Phys Lett 63:715–717

73. Nieminen TA, Knöner G, Heckenberg NR et al (2007) Physics of optical tweezers. Methods Cell Biol 82:207–236

74. Roichman Y, Waldron A, Gardel E et al (2006) Optical traps with geometric aberrations. Appl Opt 45:3425–3429

75. Dutra RS, Viana NB, Maia Neto PA et al (2012) Absolute calibration of optical tweezers including aberrations. Appl Phys Lett 100:131115

76. Viana NB, Rocha MS, Mesquita ON et al (2007) Towards absolute calibration of optical tweezers. Phys Rev E Stat Nonlin Soft Matter Phys 75:021914

77. Böhm K, Steinmetzer P, Daniel A (1997) Kinesin-driven microtubule motility in the presence of alkaline–earth metal ions: indication for a calcium ion-dependent motility. Cell Motil Cytoskeleton 37:226–231

78. Swan IDA (1972) The inhibition of hen egg-white lysozyme by imidazole and indole derivatives. J Mol Biol 65:59–62

79. Grabski AC (2009) Advances in preparation of biological extracts for protein purification. Methods Enzymol 463:285–303

80. Schweitzer C, Schmidt R (2003) Physical mechanisms of generation and deactivation of singlet oxygen. Chem Rev 103:1685–1758

81. Davies MJ (2004) Reactive species formed on proteins exposed to singlet oxygen. Photochem Photobiol Sci 3:17–25

82. Nilsson R, Merkel PB, Kearns DR (1972) Unambiguous evidence for the participation of singlet oxygen (1δ_) in photodynamic oxidation of amino acids. Photochem Photobiol 16:117–124

83. Vigers GP, Coue M, McIntosh JR (1988) Fluorescent microtubules break up under illumination. J Cell Biol 107:1011–1024

84. Guo H, Xu C, Liu C et al (2006) Mechanism and dynamics of breakage of fluorescent microtubules. Biophys J 90:2093–2098

85. Redmond RW, Gamlin JN (1999) A compilation of singlet oxygen yields from biologically relevant molecules. Photochem Photobiol 70:391–475

86. Landry MP, McCall PM, Qi Z et al (2009) Characterization of photoactivated singlet oxygen damage in single-molecule optical trap experiments. Biophys J 97:2128–2136

87. Yildiz A, Forkey JN, McKinney SA et al (2003) Myosin V walks hand-over-hand: single fluorophore imaging with 1.5-nm localization. Science 300:2061–2065

88. Joo C, McKinney SA, Nakamura M et al (2006) Real-time observation of RecA filament dynamics with single monomer resolution. Cell 126:515–527

89. Blanchard SC, Kim HD, Gonzalez RL et al (2004) tRNA dynamics on the ribosome during translation. Proc Natl Acad Sci U S A 101:12893–12898

90. Dempsey GT, Wang W, Zhuang X (2009) Fluorescence imaging at sub-diffraction-limit resolution with stochastic optical reconstruction microscopy. In: Hinterdorfer P, Oijen A (eds) Handbook of single-molecule biophysics, pp. 95–127. Springer, US

91. Rasnik I, McKinney SA, Ha T (2006) Nonblinking and long-lasting single-molecule fluorescence imaging. Nat Methods 3:891–893

92. Sambongi Y, Iko Y, Tanabe M et al (1999) Mechanical rotation of the c subunit oligomer in ATP synthase (F0F1): direct observation. Science 286:1722–1724

93. Adachi K, Yasuda R, Noji H et al (2000) Stepping rotation of F1-ATPase visualized through angle-resolved single-fluorophore imaging. Proc Natl Acad Sci U S A 97:7243–7247

94. Harada Y, Sakurada K, Aoki T et al (1990) Mechanochemical coupling in actomyosin energy transduction studied by in vitro movement assay. J Mol Biol 216:49–68

95. Guydosh NR, Block SM (2009) Direct observation of the binding state of the kinesin head to the microtubule. Nature 461:125–128

96. Spudich JA, Rice SE, Rock RS et al (2011) The optical trapping dumbbell assay for non-processive motors or motors that turn around filaments. Cold Spring Harb Protoc 2011:1372–1374

97. Arai Y, Yasuda R, Akashi K et al (1999) Tying a molecular knot with optical tweezers. Nature 399:446–448

98. Batters C, Veigel C (2011) Using optical tweezers to study the fine details of myosin ATPase mechanochemical cycle. In: Mashanov

GI, Batters C (eds) Single molecule enzymology, pp. 97–109. Humana Press, New York

99. Neuman KC, Chadd EH, Liou GF et al (1999) Characterization of photodamage to *Escherichia coli* in optical traps. Biophys J 77:2856–2863

100. Perkins TT, Li H-W, Dalal RV et al (2004) Forward and reverse motion of single RecBCD molecules on DNA. Biophys J 86: 1640–1648

101. Keilin D, Hartree EF (1948) Properties of glucose oxidase (notatin). Biochem J 42: 221–229

102. Wong CM, Wong KH, Chen XD (2008) Glucose oxidase: natural occurrence, function, properties and industrial applications. Appl Microbiol Biotechnol 78:927–938

103. Bankar SB, Bule MV, Singhal RS et al (2009) Glucose oxidase—an overview. Biotechnol Adv 27:489–501

104. Aitken C, Marshall R, Puglisi J (2007) An oxygen scavenging system for improvement of dye stability in single-molecule fluorescence experiments. Biophys J 94:1826–1835

105. Patil PV, Ballou DP (2000) The use of protocatechuate dioxygenase for maintaining anaerobic conditions in biochemical experiments. Anal Biochem 286:187–192

106. Shi X, Lim J, Ha T (2010) Acidification of the oxygen scavenging system in single-molecule fluorescence studies: in situ sensing with a ratiometric dual-emission probe. Anal Chem 82:6132–6138

107. Eigel WN, Butler JE, Ernstrom CA et al (1984) Nomenclature of proteins of cow's milk: fifth revision. J Dairy Sci 67:1599–1631

108. Hofland GW, van Es M, van der Wielen LAM et al (1999) Isoelectric precipitation of casein using high-pressure CO_2. Ind Eng Chem Res 38:4919–4927

109. Plumeré N, Henig J, Campbell WH (2012) Enzyme-catalyzed O2 removal system for electrochemical analysis under ambient air: application in an amperometric nitrate biosensor. Anal Chem 84:2141–2146

110. Giffhorn F (2000) Fungal pyranose oxidases: occurrence, properties and biotechnical applications in carbohydrate chemistry. Appl Microbiol Biotechnol 54:727–740

111. Pazarlioglu NK, Akkaya A, Tahsinsoy D (2009) Pyranose 2-oxidase (P2O): production from trametes versicolor in stirred tank reactor and its partial characterization. Prep Biochem Biotechnol 39:32–45

112. Artolozaga MJ, Kubátová E, Volc J et al (1997) Pyranose 2-oxidase from *Phanerochaete chrysosporium*– further biochemical characterisation. Appl Microbiol Biotechnol 47:508–514

113. Leitner C, Volc J, Haltrich D (2001) Purification and characterization of pyranose oxidase from the white rot fungus trametes multicolor. Appl Environ Microbiol 67:3636–3644

114. van Dijk MA, Kapitein LC, van Mameren J et al (2004) Combining optical trapping and single-molecule fluorescence spectroscopy: enhanced photobleaching of fluorophores. J Phys Chem B 108:6479–6484

115. Berg-Sørensen K, Oddershede L, Florin E-L et al (2003) Unintended filtering in a typical photodiode detection system for optical tweezers. J Appl Phys 93:3167

116. Peterman EJG, van Dijk MA, Kapitein LC et al (2003) Extending the bandwidth of optical-tweezers interferometry. Rev Sci Instrum 74:3246–3249

117. Madadi E, Samadi A, Cheraghian M et al (2012) Polarization-induced stiffness asymmetry of optical tweezers. Opt Lett 37: 3519–3521

118. Dutra RS, Viana NB, Neto PAM et al (2007) Polarization effects in optical tweezers. J Opt A Pure Appl Opt 9:S221–S227

119. Ashkin A (1992) Forces of a single-beam gradient laser trap on a dielectric sphere in the ray optics regime. Biophys J 61:569–582

120. Hecht E (2001) Optics, 4th edn. Addison-Wesley, Boston

121. Brasselet S, Aït-Belkacem D, Gasecka A et al (2010) Influence of birefringence on polarization resolved nonlinear microscopy and collagen SHG structural imaging. Opt Express 18:14859–14870

122. Schön P, Munhoz F, Gasecka A et al (2008) Polarization distortion effects in polarimetric two-photon microscopy. Opt Express 16:20891–20901

123. Brideau C, Stys PK (2012) Automated control of optical polarization for nonlinear microscopy. Proc SPIE. doi:10.1117/12.908995

124. Axelrod D (1989) Fluorescence polarization microscopy. Methods Cell Biol 30:333–352

125. Axelrod D (1979) Carbocyanine dye orientation in red cell membrane studied by microscopic fluorescence polarization. Biophys J 26:557–573

126. Chou CK, Chen WL, Fwu PT et al (2008) Polarization ellipticity compensation in polarization second-harmonic generation microscopy without specimen rotation. J Biomed Opt 13:014005–014005

127. Abbondanzieri EA, Greenleaf WJ, Shaevitz JW et al (2005) Direct observation of base-pair stepping by RNA polymerase. Nature 438:460–465

128. Carter AR, King GM, Ulrich TA et al (2007) Stabilization of an optical microscope to 0.1 nm in three dimensions. Appl Opt 46:421–427

129. Nugent-Glandorf L, Perkins TT (2004) Measuring 0.1-nm motion in 1 ms in an optical microscope with differential back-focal-plane detection. Opt Lett 29:2611–2613

130. Mahamdeh M, Schäffer E (2009) Optical tweezers with millikelvin precision of temperature-controlled objectives and base-pair resolution. Opt Express 17:17190–17199

131. Valentine MT, Guydosh NR, Gutiérrez-Medina B et al (2008) Precision steering of an optical trap by electro-optic deflection. Opt Lett 33:599–601

132. Tsien RY (1998) The green fluorescent protein. Annu Rev Biochem 67:509–544

133. Johnson BH, Hecht MH (1994) Recombinant proteins can be isolated from E. coli cells by repeated cycles of freezing and thawing. Biotech 12:1357–1360

134. James GT (1978) Inactivation of the protease inhibitor phenylmethylsulfonyl fluoride in buffers. Anal Biochem 86:574–579

135. Walker JM (ed) (1996) The protein protocols handbook. Humana Press, Totowa, NJ

136. Robinson J, Engelborghs Y (1982) Tubulin polymerization in dimethyl sulfoxide. J Biol Chem 257:5367–5371

137. Brouhard GJ (2010) Quality control in single-molecule studies of kinesins and microtubule-associated proteins. Methods Cell Biol 97:497–506

138. Novotny L, Grober RD, Karrai K (2001) Reflected image of a strongly focused spot. Opt Lett 26:789–791

139. Perrone S, Volpe G, Petrov D (2008) 10-fold detection range increase in quadrant-photodiode position sensing for photonic force microscope. Rev Sci Instrum 79:106101–106101

140. Martínez IA, Petrov D (2012) Back-focal-plane position detection with extended linear range for photonic force microscopy. Appl Opt 51:5973–5977

141. Lang MJ, Asbury CL, Shaevitz JW et al (2002) An automated two-dimensional optical force clamp for single molecule studies. Biophys J 83:491–501

142. Dreyer JK, Dreyer JK, Berg-Sorensen K, Oddershede L (2004) Improved axial position detection in optical tweezers measurements. Appl Opt 43:1991–1995

143. Deufel C, Wang MD (2006) Detection of forces and displacements along the axial direction in an optical trap. Biophys J 90:657–667

144. Reif F (2008) Fundamentals of statistical and thermal physics. Waveland Press Inc, Long Grove, IL

145. Turner LE (1976) Generalized classical equipartition theorem. Am J Phys 44:104

146. Richardson AC, Reihani SNS, Oddershede LB (2008) Non-harmonic potential of a single beam optical trap. Opt Express 16:15709–15717

147. Jahnel M, Behrndt M, Jannasch A et al (2011) Measuring the complete force field of an optical trap. Opt Lett 36:1260–1262

148. Godazgar T, Shokri R, Reihani SNS (2011) Potential mapping of optical tweezers. Opt Lett 36:3284–3286

149. Martínez IA, Petrov D (2012) Force mapping of an optical trap using an acousto-optical deflector in a time-sharing regime. Appl Opt 51:5522–5526

150. Farré A, Marsà F, Montes-Usategui M (2012) Optimized back-focal-plane interferometry directly measures forces of optically trapped particles. Opt Express 20:12270–12291

151. Smith SB, Cui Y, Bustamante C (2003) Optical-trap force transducer that operates by direct measurement of light momentum. Methods Enzymol 361:134–162

152. Wong WP, Halvorsen K (2006) The effect of integration time on fluctuation measurements: calibrating an optical trap in the presence of motion blur. Opt Express 14:12517–12531

153. Lukić B, Jeney S, Tischer C et al (2005) Direct observation of nondiffusive motion of a Brownian particle. Phys Rev Lett 95:160601–160604

154. Pátek J, Hrubý J, Klomfar J et al (2009) Reference correlations for thermophysical properties of liquid water at 0.1 MPa. J Phys Chem Ref Data 38:21–29

155. Colas B, Gobin C, Lorient D (1988) Viscosity and voluminosity of caseins chemically modified by reductive alkylation with reducing sugars. J Dairy Res 55:539–546

156. Berg-Sørensen K, Flyvbjerg H (2004) Power spectrum analysis for optical tweezers. Rev Sci Instrum 75:594

157. Berg-Sørensen K, Peterman EJG, Weber T et al (2006) Power spectrum analysis for optical tweezers. II: laser wavelength dependence of parasitic filtering, and how to achieve high bandwidth. Rev Sci Instrum 77:063106

158. Nørrelykke SF, Flyvbjerg H (2010) Power spectrum analysis with least-squares fitting: amplitude bias and its elimination, with application to optical tweezers and atomic force microscope cantilevers. Rev Sci Instrum 81:075103

159. Mathew AE, Mejillano MR, Nath JP et al (1992) Synthesis and evaluation of some water-soluble prodrugs and derivatives of taxol with antitumor activity. J Med Chem 35:145–151

160. Foss M, Wilcox BWL, Alsop GB et al (2008) Taxol crystals can masquerade as stabilized microtubules. PLoS ONE 3:e1476

161. Castro JS, Deymier PA, Trzaskowski B et al (2010) Heterogeneous and homogeneous nucleation of Taxol crystals in aqueous solutions and gels: effect of tubulin proteins. Colloids Surf B Biointerfaces 76:199–206

162. Vershinin M, Carter BC, Razafsky DS et al (2006) Multiple-motor based transport and its regulation by Tau. Proc Natl Acad Sci U S A 104:87–92

163. Kerssemakers J, Howard J, Hess H et al (2006) The distance that kinesin-1 holds its cargo from the microtubule surface measured by fluorescence interference contrast microscopy. Proc Natl Acad Sci U S A 103:15812–15817

164. Humphrey W, Dalke A, Schulten K (1996) VMD: visual molecular dynamics. J Mol Graph 14(33–38):27–28

165. Shaevitz JW (2004) The biophysics of molecular motors: optical trapping studies of kinesin and RNA polymerase. Stanford University, Stanford

Chapter 11

Seeded Microtubule Growth for Cryoelectron Microscopy of End-Binding Proteins

Sebastian P. Maurer, Franck J. Fourniol, Andreas Hoenger, and Thomas Surrey

Abstract

End-binding proteins (EBs) have the ability to autonomously track the ends of growing microtubules, where they recruit several proteins that control various aspects of microtubule cytoskeleton organization and function. The structural nature of the binding site recognized by EBs at growing microtubule ends has been a subject of debate. Recently, a fluorescence microscopy assay used for the study of dynamic end tracking in vitro was adapted for cryoelectron microscopy (cryo-EM). In combination with single-particle reconstruction methods, this modified assay was used to produce the first subnanometer-resolution model of how the microtubule-binding domain of EBs binds to microtubules grown in the presence of GTPγS. A GTPγS microtubule can be considered a static mimic of the transiently existing binding region of EBs at a microtubule end growing in the presence of GTP. Here we describe in detail the procedure used to generate these samples. It relies on the polymerization of microtubules from preformed stabilized and quantum dot-labeled microtubule seeds. This allows the cryo-EM analysis of proteins bound to paclitaxel-free microtubules. It provides freedom for using different GTP analogues during microtubule elongation independent of their nucleation properties. This assay could also be useful for the cryo-EM analysis of other microtubule-associated proteins.

Key words *cryo-EM*, Microtubule structure, GMP-CPP seeds, GTPγS, EB1, Mal3, Kinesin

1 Introduction

The complex properties of the microtubule cytoskeleton depend on the activities of multiple microtubule-associated proteins (MAPs). Understanding these activities requires combining high-resolution structural information of how MAPs interact with the microtubule surface and kinetic data characterizing the dynamic properties of these interactions. Because microtubules are too large for NMR spectroscopy and too heterogeneous in length for crystallography, cryoelectron microscopy (cryo-EM) is currently the method of choice for high-resolution structural analysis of microtubule-MAP interactions (for a review, *see*, e.g., [1–3]).

David J. Sharp (ed.), *Mitosis: Methods and Protocols*, Methods in Molecular Biology, vol. 1136,
DOI 10.1007/978-1-4939-0329-0_11, © Springer Science+Business Media New York 2014

Rapid freezing and vitrification preserves structural detail to atomic resolution. Where possible, extensive averaging of identical particles in typically low-contrast 2D cryo-EM images can yield 3D reconstructions with nanometer or subnanometer resolution, in which secondary structure elements of tubulin and MAPs can be resolved [4–18]. Dynamic information, for example, about microtubule polymerization properties, binding/unbinding kinetics of MAPs, or the motility of molecular motors, is typically obtained by recording movies using optical microscopy, in most cases fluorescence microscopy (for a review, *see*, e.g., [19, 20]). The spatial resolution of standard fluorescence microscopy is in the range of several 100 nm. Modern super-resolution techniques can achieve higher resolution [21] but require sacrifices with respect to temporal resolution, observed sample size, and/or recording duration. In general, static high-resolution cryo-EM data and dynamic, typically lower-resolution data from fluorescence microscopy provide highly complementary information.

Samples that are prepared for cryo-EM of MAPs and microtubules are often quite different from the ones prepared for fluorescence microscopy studies. A standard in vitro assay for the study of dynamic microtubules and MAPs by fluorescence microscopy relies on the immobilization on a glass surface of short and stable "microtubule seeds" that are pre-grown with the non-hydrolysable GTP analogue GMP-CPP. The extension of these seeds in the presence of fluorescently labeled tubulin, GTP, and fluorescently labeled MAPs can then be monitored by total internal reflection fluorescence (TIRF) microscopy [19, 20]. The local excitation of fluorophores close to the glass surface (~100 nm) by this type of microscopy offers excellent suppression of background fluorescence (optical sectioning). Therefore, this assay lends itself to both the visualization of dynamic microtubules, even in the presence of high tubulin concentrations, and to single molecule imaging of MAPs on such dynamic microtubules.

For high-resolution cryo-EM reconstructions of microtubules, data averaging requires samples that are as homogenous as possible and contain microtubules that are as completely decorated by MAPs as possible. Most structural studies have used microtubules grown in the presence of paclitaxel to produce stabilized microtubules [1–3]. However, paclitaxel-stabilized microtubules do not have exactly the same structural properties as dynamic microtubules [16, 22, 23], which in principle can affect the interactions with MAPs [22, 24, 25]. To facilitate the nucleation of microtubules in the absence of paclitaxel, purified cellular nucleators such as centrosomes and axonemes have been used, but this often led to the generation of dense microtubule bundles [26], which are not suitable for high-resolution image reconstruction. In the special case of the microtubule-stabilizing protein doublecortin, microtubules could be nucleated from solution at moderate tubulin concentrations in the absence of paclitaxel, which produced one of the rare examples of a subnanometer-resolution structure of a MAP bound to a GDP-microtubule lattice that was free of paclitaxel [11].

A class of MAPs that has recently attracted increasing interest is the conserved protein family of microtubule end-binding (EB) proteins (reviewed in [27–30]). Microtubule end tracking is a conserved phenomenon found in many organisms from plants to mammals. EBs are critical, because they autonomously bind to growing microtubule end regions and recruit the majority of other end-tracking proteins [31–33]. Together they fulfill a variety of functions from regulation of microtubule dynamic instability to linking microtubules to subcellular structures as actin, membranes, or the kinetochore [30]. Recently, the combination of fluorescence microscopy of in vitro reconstituted microtubule end tracking [31, 32] and cryo-EM of decorated microtubules [16, 24] significantly advanced our understanding of the molecular mechanism of microtubule end tracking by EBs.

Because EBs are sensitive to the conformational state of tubulin in the microtubule [24, 25], it was important to develop an assay that allowed the cryo-EM analysis of EBs bound to paclitaxel-free microtubules grown in the presence of different guanine nucleotides. To this end, microtubules were elongated from GMP-CPP seeds, analogous to the TIRF microscopy experiments (*see* **Note 1**). Nucleation from seeds allows the use of only moderate tubulin concentrations despite the absence of paclitaxel (in comparison to nucleation directly from solution). Furthermore, microtubule elongation can take place in the presence of a range of different nucleotides. Because it is critical to be able to distinguish between GMP-CPP segments and microtubule segments that elongate from the seeds, the seeds were labeled with quantum dots that are easily identifiable in the electron micrographs.

This assay allowed the study of how EBs interact with microtubules grown either in the presence of GTP or GTPγS, a GTP analogue that was found by TIRF microscopy to induce the high-affinity state of microtubule-incorporated tubulin for EB binding [24]. A subnanometer reconstruction from cryoelectron micrographs of the microtubule-binding domain (calponin homology domain) of the fission yeast EB Mal3 bound to GTPγS microtubules can therefore be considered to show how EBs bind to their typical binding site at growing microtubule ends. The EB calponin homology domain binds between protofilaments, except at the microtubule seam, at the corner of four neighboring tubulin dimers, suggesting an ideal position to sense events taking place during the GTP hydrolysis cycle.

The procedure described in the following is a three-step protocol (Fig. 1). In the first step (Subheading 3.1), short, biotinylated GMP-CPP seeds are polymerized. They are then (Subheading 3.2) labeled with streptavidin-linked quantum dots. These quantum dot-labeled seeds (Fig. 2) are then purified to remove excess of quantum dots, unpolymerized tubulin, and GMP-CPP.

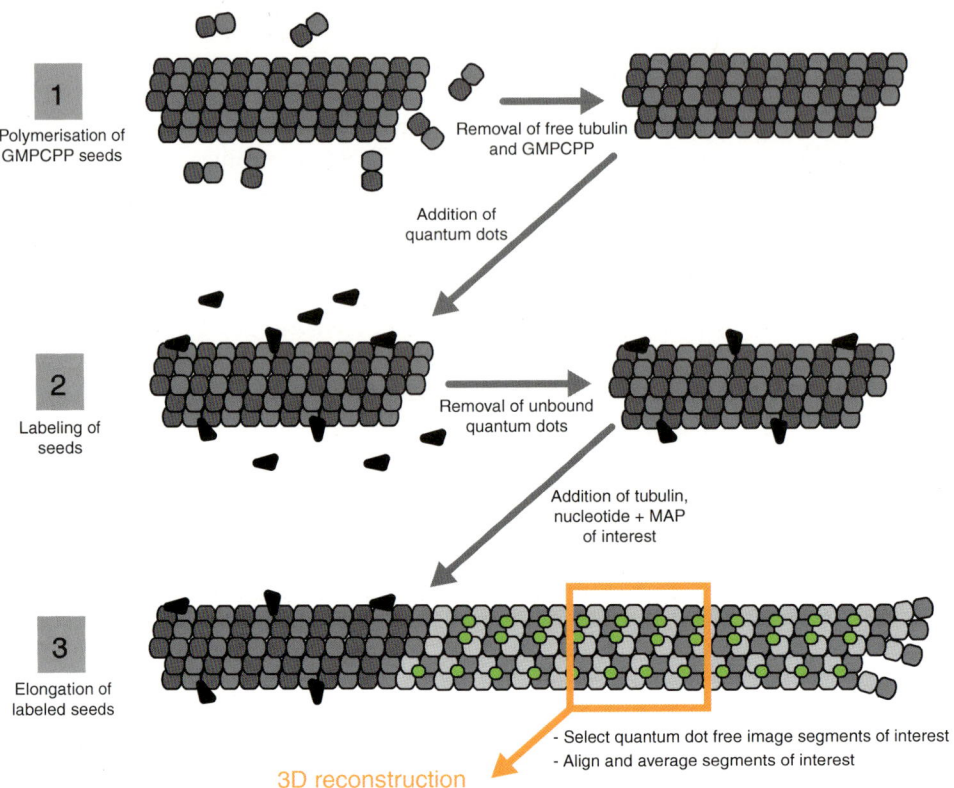

Fig. 1 Overview of the preparation process of microtubules from labeled seeds for cryo-EM. *Step 1*: GMP-CPP seeds are grown from a mixture of biotinylated tubulin and unlabeled tubulin, and unpolymerized proteins are removed by centrifugation. *Step 2*: The biotinylated seeds are incubated with streptavidin-conjugated quantum dots. Unbound quantum dots are removed by centrifugation. *Step 3*: Directly prior to freezing samples, microtubules are grown from a mixture of quantum dot-labeled seeds, unlabeled tubulin, and the MAP(s) of interest. In micrographs, the electron-dense quantum dots distinguish the GMP-CPP seeds from the elongated microtubule section that is subject to analysis through averaging and 3D reconstruction

In the following step (Subheading 3.3), the labeled GMP-CPP seeds are elongated in the presence of microtubule-binding proteins and either GTPγS or GTP. To illustrate how proteins can respond differently to microtubules grown in the presence of different guanine nucleotides, we show examples with the Mal3

Fig. 2 (continued) EB1-GFP signal is shown in *green*, Cy5-microtubule signal in *red*. The GMP-CPP seeds have been polymerized in the presence of a higher concentration of Cy5-labeled tubulin than the microtubule segments extending from the seeds, allowing them to be distinguished. (**b**) Kymographs (time-space plots) of the microtubules shown in (**a**) demonstrating how these microtubules grow (and shrink). (**c**) Electron micrographs of negatively stained samples of biotinylated GMP-CPP seeds. Overview (*left*) and close up view (*right*). (**d**) Same as (**c**) after labeling the seeds with quantum dots. Two examples are shown in the *left* and *right panel*. (**e**) Comparison of GMP-CPP seeded GTPγS microtubules visualized by fluorescence microscopy and by negative-stain electron microscopy, *Upper panel*: TIRF microscopy image of a Cy5-labeled GTPγS microtubule grown from a brightly labeled GMP-CPP seed. *Bottom panel*: EM image of a negative-stain sample showing a quantum dot-labeled GMP-CPP seed which has been elongated with a non-labeled microtubule grown in the presence of GTPγS

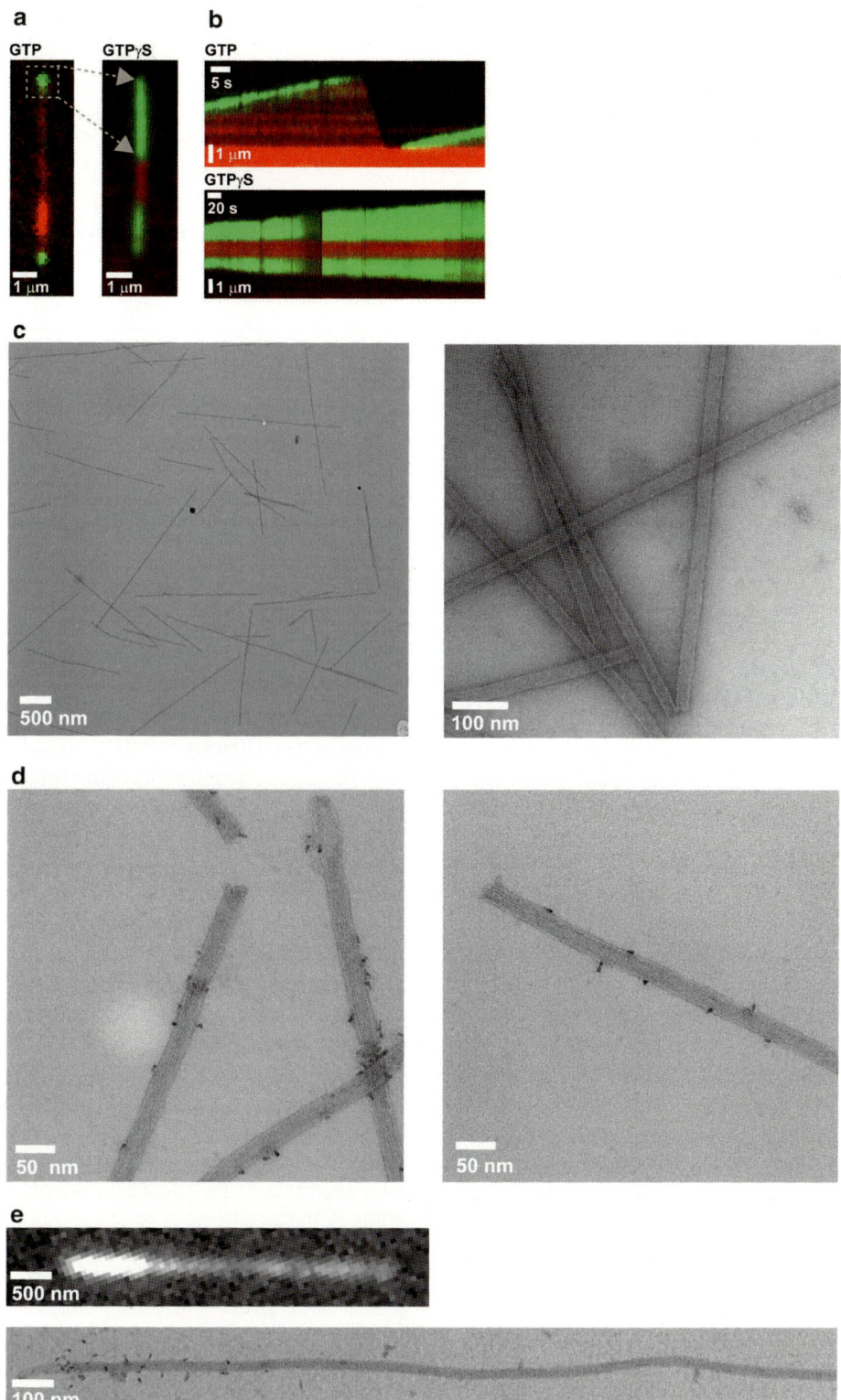

Fig. 2 Adaption of a TIRF microscopy assay for electron microscopy. (**a**) TIRF microscopy image of microtubules grown in the presence of 50 nM EB1-GFP and 1 mM GTP (*left*) or 1 mM GTPγS (*right*). The *dashed box* highlights the position of the high-affinity EB-binding site present only at the very end of microtubules growing in the presence of GTP. The *arrows* indicate that high-affinity binding sites exist all along GTPγS-microtubule segments.

calponin homology domain and a kinesin-1 motor domain in this cryo-EM assay (Fig. 3).

This assay is based on microtubule elongation from quantum dot-labeled GMP-CPP seeds. It offers the general possibility to investigate how MAPs bind to paclitaxel-free microtubules by cryo-EM in a configuration that is very similar to established TIRF microscopy assays.

2　Materials

2.1　Polymerization of GMP-CPP Seeds

1. BRB80 buffer: 80 mM K-PIPES pH 6.8 at room temperature (RT), 2 mM $MgCl_2$, 1 mM EGTA. BRB80 should be filtered through a 0.22 μm bottletop filter (e.g., Steritop™, Millipore) and stored at 4 °C for no longer than 6 weeks.

2. Pig brain tubulin in BRB80, purified as described [34], flash-frozen, and stored in single-use aliquots at a concentration of 15–20 mg/ml at liquid nitrogen temperature.

3. Biotinylated tubulin in BRB80, labeled as described [35], flash-frozen, and stored in single-use aliquots at a concentration of 10–15 mg/ml at liquid nitrogen temperature.

4. Guanosine-50-[(α,β)-methylene]triphosphate (GMP-CPP): 10 mM solution from Jena Bioscience stored in single-use aliquots at –80 °C.

5. Heat block for 1.5 ml Eppendorf Tubes®.

6. Tabletop centrifuge (e.g., Heraeus Pico 17, Thermo Scientific).

Fig. 3 Cryo-EM images of GMP-CPP-seeded GTPγS microtubules decorated with different microtubule-associated proteins. Cryo-EM image of GTPγS microtubules grown from quantum dot-labeled GMP-CPP seeds and decorated with Mal3₁₄₃ (**a**) or the motor domain of rat kinesin (rigor mutant T93N) (**d**). The *black boxes* mark quantum dot-labeled (*left*) and unlabeled (*right*) microtubule parts. The quantum dot-labeled GMP-CPP-microtubule segments shown here contain 14 protofilaments, while the unlabeled GTPγS-microtubule segments are formed of 13 protofilaments [16]. *Black boxes* and *arrows* point to the diffraction patterns (**b**, **e**) and Fourier-filtered images (**c**, **f**) of the corresponding microtubule segments. In the diffraction patterns, layer lines corresponding to a 4 nm repeat reflect the axial repeat of tubulin monomers, while layer lines corresponding to an 8 nm repeat are characteristic of MAPs binding every dimer. Interestingly, while the rigor mutant motor domain of rat kinesin binds both GMP-CPP and GTPγS-microtubule parts (**d–f**), the microtubule-binding domain of Mal3 binds selectively GTPγS-microtubule segments (**a–c**). (**g**) TIRF microscopy images of an EB1-GFP decorated GTPγS microtubule grown from a brightly labeled GMP-CPP seed. The *left panel* shows the fluorescence signal of the Cy5-labeled microtubule and the *right panel* the corresponding GFP signal of EB1-GFP. (**h**) 3D-reconstruction of a GTPγS microtubule decorated with the calponin homology domain of Mal3 [16]. This reconstruction was generated from 2D cryo-EM images as shown in (**d**) using a single-particle method developed by C.V. Sindelar [7, 14]. The microtubule is shown in *gray*, while the microtubule-binding domain of EBs is highlighted in *green*. Note the 8 nm repeat of tubulin dimers and calponin homology domains along the microtubule axis

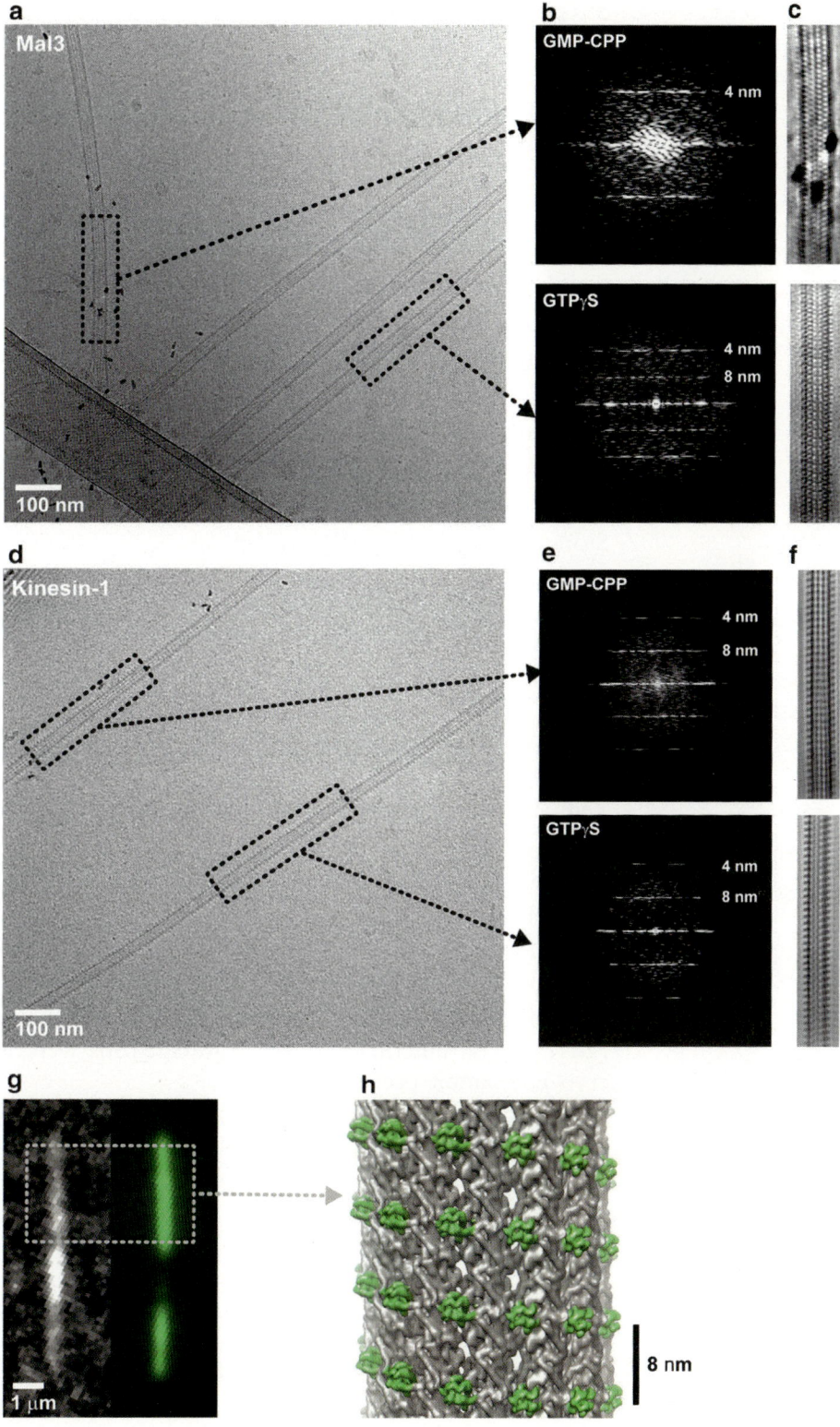

2.2 Labeling of Biotinylated Seeds with Quantum Dots

1. Streptavidin-conjugated quantum dots 1 μM stock solution (Qdot® 655 Streptavidin Conjugate, Invitrogen). Stored at 4 °C.

2. BRB80 buffer warmed to room temperature.

3. Tabletop centrifuge (*see* 2.1).

2.3 Sample Preparation for Cryo-EM

1. Holey carbon TEM grids (Lacey 300 mesh Cu, Agar http://www.agarscientific.com).

2. Stainless steel tweezers.

3. Slot screw driver.

4. Glow-discharge unit (e.g., EMITECH K100X).

5. Vitrobot (FEI Company). For this work, a Vitrobot Mark III and Mark IV have been used.

6. Blotting paper for Vitrobot (FEI Company) (alternatively use Wathman Nr3).

7. Storage boxes for cryo-EM grids.

8. Storage dewar for cryo-EM grids.

9. Ethane gas cylinder.

10. Liquid nitrogen.

11. Heat block for 1.5 ml Eppendorf Tubes®.

12. Zebaspin desalting columns (Zeba Micro Spin Desalting Columns, 7 K MWCO, 75 μl).

13. Temperature-controllable tabletop centrifuge (e.g., Heraeus Fresco 17, Thermo Scientific).

14. GTPγS solution (GTPγS lithium salt, Roche). Stored in single-use aliquots at –80 °C.

15. 100 mM GTP solution in 10 mM Tris–HCl pH 7.0. Stored in single-use aliquots at –80 °C.

16. The MAP of interest. In this article, Mal3$_{143}$ (described in [24]) and rat kinesinT93N rigor mutant [36] were used for cryo-EM sample preparation.

17. BRB40 buffer: 40 mM K-PIPES pH 6.8 at RT, 2 mM MgCl$_2$, 1 mM EGTA; *see* **Note 2**.

3 Methods

3.1 Polymerization of GMP-CPP Seeds

1. Start warming a 1.5 ml aliquot of BRB80 buffer in a heat block set to 37 °C 5 min before starting.

2. Pipet seed mix on ice: total volume 40 μl in BRB80 buffer, 18 μM tubulin, 11 μM biotinylated tubulin, 0.5 mM GMP-CPP. Mix by pipetting up and down ten times.

3. Incubate seed mix for 3 min at 37 °C.

4. Add 400 µl pre-warmed BRB80, mix by slowly pipetting up and down ten times.

5. Spin for 8 min at RT in tabletop centrifuge at $15,800 \times g$.

6. Remove supernatant and resuspend in 80 µl pre-warmed BRB80 (*see* **Note 3**).

3.2 Labeling of Biotinylated GMP-CPP Seeds with Quantum Dots

1. Add 10 µl quantum dots to 40 µl resuspended seeds from Subheading 3.1 to a final concentration of 200 nM quantum dots. Mix by pipetting up and down ten times.

2. Incubate at RT for 8 min.

3. Add 400 µl pre-warmed BRB80, mix by pipetting up and down ten times.

4. Centrifuge for 8 min at RT in tabletop centrifuge at $15,800 \times g$ (*see* **Note 4**). Afterwards there should be a sharp orange dot visible as pellet.

5. Remove supernatant, resuspend in 50 µl pre-warmed BRB80.

6. Centrifuge for 8 min at RT in tabletop centrifuge at $15,800 \times g$. Afterwards a smooth, slightly orange pellet should be visible.

7. Remove supernatant, resuspend in 50 µl pre-warmed BRB80.

8. Spin for 8 min at RT in tabletop centrifuge at $15,800 \times g$. Afterwards the pellet should be barely visible.

9. Resuspend in 50 µl pre-warmed BRB80 and use for sample preparation within the next 3 h. Longer time periods might be possible but have not been tested. Keep seeds and labeled seeds at RT.

3.3 Cryo-EM Sample Preparation of GMP-CPP Seeded Microtubules

The following protocol describes the preparation of $Mal3_{143}$ decorated GTPγS microtubules. This method, however, can also be used to produce MAP-decorated GDP microtubules, which requires different sample preparation parameters. A list of different sample preparation conditions which have been successfully used in our lab is given in Table 1.

1. Start the Vitrobot, set temperature to 38 °C and humidity to 85 % (*see* **Note 5**). Install fresh blotting paper. Allow Vitrobot chamber to equilibrate for 30 min before starting to plunge freeze samples.

2. Depending on the storage buffer composition of the MAP of interest, the protein might have to be desalted (*see* **Notes 6** and **7**):

    (a) Cool tabletop centrifuge to 4 °C.

    (b) Equilibrate Zebaspin desalting column by washing it 3× with cold (4 °C) BRB40 (or any other chosen sampling

Table 1
Polymerization conditions

Nucleotide	Nucleotide concentration (mM)	Tubulin concentration (μM)	MAP	MAP concentration (μM)	Incubation time (min)	References
GTPγS	2	15	Mal3$_{143}$	45	3	Figure 3a–c
GTPγS	2	15	rKinT93N	20	5	Figure 3d–f
GTPγS	2	25	–	–	25	[16]
GTP	2	10	Mal3$_{143}$	67	1	[16]
GTP	2	20	–	–	2	[16]

buffer) according to the manufacturer's protocol. Make sure column orientation in the cooled tabletop centrifuge is always the same.

(c) Run the protein(s) through the desalting column(s). The required protein volume depends on the desalting column used. In this case, 10 μl Mal3$_{143}$ was desalted using a single column.

3. Glow-discharge grids. Use a batch of glow-discharged grids maximally for 30 min after glow-discharging; if necessary glow-discharge again.

4. Pipette the premix together: total volume 40 μl in BRB40 buffer, 2 mM GTPγS, 15 μM tubulin, 45 μM Mal3$_{143}$ (*see* **Note 8**).

5. Transfer the premix to a heat block set to 30 °C and warm it up for 1 min.

6. Generate the final mix by adding 1 μl of quantum dot-labeled seeds. Mix by pipetting.

7. Incubate the final mix for 3 min at 30 °C, allowing for microtubule elongation from the seeds. Incubation time and tubulin concentration varies depending on nucleotides and MAPs used for sample preparation (*see* Table 1).

8. Mount a glow-discharged grid into the Vitrobot 1 min before applying the sample to allow the grid to warm up.

9. Pipet the final mix up and down once before applying 4 μl to the grid already mounted in the Vitrobot (*see* **Note 9**).

10. Incubate the final mix for 30 s on the grid.

11. Blot 3–3.5 s depending on blot force used. The optimal blot force has to be established for each Vitrobot and pair of tweezers, by trial and error. Samples shown in this article have been plunge-frozen without delay time between blotting and plunging.

12. Store the frozen grid in a cryo-sample box (e.g., Cryo Grid Boxes, Ted Pella Inc.) in liquid nitrogen.

13. To freeze additional grids, repeat **steps 8–12**. Keep the final mix with seeds at 30 °C (*see* **Note 10**).

14. Grid boxes can be stored in a liquid nitrogen storage container and/or transferred into a cryo-TEM for imaging (described elsewhere [2]). In brief, atomic resolution data can be collected on a TEM equipped with a field-emission gun and operating at 200 kV or higher voltage. Depending on the microtubule architectures present in the sample (*see* **Note 11**), specific computational approaches can be employed to process images and yield subnanometer-resolution 3D reconstructions [2, 7, 14].

4 Notes

1. Unlike GMP-CPP, GTPγS does not promote nucleation of microtubules at low tubulin concentration. Therefore, the use of GMP-CPP microtubule seeds as nucleators is critical to polymerize GTPγS microtubules.

2. Always use buffers that were filtered through a 0.22 μm filter, and we recommend storing them at 4 °C for no longer than approximately 6 weeks.

3. At 4 °C, GMP-CPP microtubules depolymerize. Make sure quantum dots (and other reagents) were warmed up to RT for 1 min before adding them to the seeds. Labeled seeds keep at least 3 h at RT.

4. Do not spin seeds faster than indicated in this protocol as unbound quantum dots might then pellet as well. The quantum dots are orange in color; therefore, the extent to which they pellet can be estimated by visual inspection: as unbound quantum dots are gradually removed with successive washes, the microtubule pellet should get lighter in color.

5. It is important to plunge grids using a device controlling environmental parameters (e.g., a Vitrobot). The microtubule sample should be kept at 37 °C. Also, samples should be plunge-frozen in a high humidity environment—this ensures that the sample is subject to as little evaporation as possible, so that the buffer composition stays constant.

6. Microtubule-MAP associations depend on electrostatic interactions. Keep the ionic strength as low as possible in the sampling buffer. Desalt proteins if necessary.

7. Cryoprotectants such as glycerol should be kept at a minimum, because they raise the density of the solution which lowers the

contrast dramatically and because they increase the sensitivity of the sample to electron beam damage. Desalt proteins if necessary.

8. Cryo-EM requires a much more concentrated sample compared to TIRF-M. A considerable amount of proteins is blotted away during the preparation of cryo-EM samples or can be trapped on the carbon or at the water–air interface. In TIRF microscopy experiments, GTPγS microtubules are typically fully decorated in the presence of 50 nM full-length Mal3, while in cryo-EM experiments, complete decoration with the Mal3 monomer is typically observed at ~50 μM (at comparable tubulin concentrations of ~10–20 μM).

9. Pipetting up and down can help to homogenize the sample and break microtubule bundles, before applying the sample to the EM grid. How much mixing of sample is needed might depend on the nucleotide and MAP used and has to be tested for each new case individually.

10. It is not recommended to use a final sample mix incubated at 30 °C for longer than 10 min, because microtubules growing too long tend to bundle, which results in cryo-samples crowded with filaments and formation of thick ice, causing poor image contrast. Optimal incubation times vary, because microtubules grow with different velocities depending on conditions—*see* Table 1.

11. Microtubule polymerization in the presence of microtubule-associated proteins might affect microtubule architecture, reflecting potentially important regulatory aspects (e.g., Mal3 and DCX [11, 16]). Most GTPγS microtubules grown with the Mal3 calponin homology domain have 13-protofilaments and one seam, i.e., the microtubule architecture found in most cells [37]. These microtubules were processed using single-particle reconstruction scripts developed by C.V. Sindelar [7, 14].

Acknowledgements

We thank Julia Cope, Rachel A. Santarella, and Cindi L. Schwartz for training in electron microscope operation and cryoelectron microscope sample preparation. We thank the electron microscopy facility at EMBL Heidelberg for support. S.P.M. was supported by Marie Curie (PIEF-GA-2009-253043) and EMBO (ALTF 1032-2009) fellowships. F.J.F. was supported by an EMBO Long-Term Fellowship (ALTF 219-2011). A.H. and the Boulder 3-D lab are supported by grant P41-GM103431 (NIH-NIGMS).

References

1. Amos LA, Hirose K (2007) Studying the structure of microtubules by electron microscopy. Methods Mol Med 137:65–91

2. Fourniol FJ, Moores CA (2011) Snapshots of kinesin motors on microtubule tracks. Methods Mol Biol 778:57–70

3. Hoenger A, Gross H (2008) Structural investigations into microtubule-MAP complexes. Methods Cell Biol 84:425–444

4. Li H, DeRosier DJ, Nicholson WV, Nogales E, Downing KH (2002) Microtubule structure at 8 A resolution. Structure 10:1317–1328

5. Kikkawa M, Hirokawa N (2006) High-resolution cryo-EM maps show the nucleotide binding pocket of KIF1A in open and closed conformations. EMBO J 25:4187–4197

6. Hirose K, Akimura E, Akiba T, Endow SA, Amos LA (2006) Large conformational changes in a kinesin motor catalyzed by interaction with microtubules. Mol Cell 23:913–923

7. Sindelar CV, Downing KH (2007) The beginning of kinesin's force-generating cycle visualized at 9-Å resolution. J Cell Biol 177:377–385

8. Dietrich KA, Sindelar CV, Brewer PD, Downing KH, Cremo CR, Rice SE (2008) The kinesin-1 motor protein is regulated by a direct interaction of its head and tail. Proc Natl Acad Sci U S A 105:8938–8943

9. Bodey AJ, Kikkawa M, Moores CA (2009) 9-Angstrom structure of a microtubule-bound mitotic motor. J Mol Biol 388:218–224

10. Peters C, Brejc K, Belmont L, Bodey AJ, Lee Y, Yu M, Guo J, Sakowicz R, Hartman J, Moores CA (2010) Insight into the molecular mechanism of the multitasking kinesin-8 motor. EMBO J 29:3437–3447

11. Fourniol FJ, Sindelar CV, Amigues B, Clare DK, Thomas G, Perderiset M, Francis F, Houdusse A, Moores CA (2010) Template-free 13-protofilament microtubule-MAP assembly visualized at 8 A resolution. J Cell Biol 191:463–470

12. Alushin GM, Ramey VH, Pasqualato S, Ball DA, Grigorieff N, Musacchio A, Nogales E (2010) The Ndc80 kinetochore complex forms oligomeric arrays along microtubules. Nature 467:805–810

13. Sui H, Downing KH (2010) Structural basis of interprotofilament interaction and lateral deformation of microtubules. Structure 18:1022–1031

14. Sindelar CV, Downing KH (2010) An atomic-level mechanism for activation of the kinesin molecular motors. Proc Natl Acad Sci U S A 107:4111–4116

15. Goulet A, Behnke-Parks WM, Sindelar CV, Major J, Rosenfeld SS, Moores CA (2012) The structural basis of force generation by the mitotic motor kinesin-5. J Biol Chem 287(53):44654–44666

16. Maurer SP, Fourniol FJ, Bohner G, Moores CA, Surrey T (2012) EBs recognize a nucleotide-dependent structural cap at growing microtubule ends. Cell 149:371–382

17. Redwine WB, Hernández-López R, Zou S, Huang J, Reck-Peterson SL, Leschziner AE (2012) Structural basis for microtubule binding and release by dynein. Science 337:1532–1536

18. Yajima H, Ogura T, Nitta R, Okada Y, Sato C, Hirokawa N (2012) Conformational changes in tubulin in GMPCPP and GDP-taxol microtubules observed by cryoelectron microscopy. J Cell Biol 198:315–322

19. Bieling P, Telley IA, Hentrich C, Piehler J, Surrey T (2010) Fluorescence microscopy assays on chemically functionalized surfaces for quantitative imaging of microtubule, motor, and +TIP dynamics. Methods Cell Biol 95:555–580

20. Gell C, Bormuth V, Brouhard GJ, Cohen DN, Diez S, Friel CT, Helenius J, Nitzsche B, Petzold H, Ribbe J et al (2010) Microtubule dynamics reconstituted in vitro and imaged by single-molecule fluorescence microscopy. Methods Cell Biol 95:221–245

21. Toomre D, Bewersdorf J (2010) A new wave of cellular imaging. Annu Rev Cell Dev Biol 26:285–314

22. Vale RD, Coppin CM, Malik F, Kull FJ, Milligan RA (1994) Tubulin GTP hydrolysis influences the structure, mechanical properties, and kinesin-driven transport of microtubules. J Biol Chem 269:23769–23775

23. Elie-Caille C, Severin F, Helenius J, Howard J, Muller DJ, Hyman AA (2007) Straight GDP-tubulin protofilaments form in the presence of taxol. Curr Biol 17:1765–1770

24. Maurer SP, Bieling P, Cope J, Hoenger A, Surrey T (2011) GTPgammaS microtubules mimic the growing microtubule end structure recognized by end-binding proteins (EBs). Proc Natl Acad Sci U S A 108:3988–3993

25. Zanic M, Stear JH, Hyman AA, Howard J (2009) EB1 recognizes the nucleotide state of tubulin in the microtubule lattice. PLoS One 4:e7585

26. Chretien D, Fuller SD, Karsenti E (1995) Structure of growing microtubule ends: two-dimensional sheets close into tubes at variable rates. J Cell Biol 129:1311–1328

27. Akhmanova A, Steinmetz MO (2010) Microtubule + TIPs at a glance. J Cell Sci 123:3415–3419

28. Galjart N (2010) Plus-end-tracking proteins and their interactions at microtubule ends. Curr Biol 20:R528–R537

29. Kumar P, Wittmann T (2012) +TIPs: SxIPping along microtubule ends. Trends Cell Biol 22:418–428

30. Duellberg C, Fourniol FJ, Maurer SP, Roostalu J, Surrey T (2012) End-binding proteins and Ase1/PRC1 define local functionality of structurally distinct parts of the microtubule cytoskeleton. Trends Cell Biol 23(2):54–63

31. Bieling P, Laan L, Schek H, Munteanu EL, Sandblad L, Dogterom M, Brunner D, Surrey T (2007) Reconstitution of a microtubule plus-end tracking system in vitro. Nature 450:1100–1105

32. Bieling P, Kandels-Lewis S, Telley IA, van Dijk J, Janke C, Surrey T (2008) CLIP-170 tracks growing microtubule ends by dynamically recognizing composite EB1/tubulin-binding sites. J Cell Biol 183:1223–1233

33. Honnappa S, Gouveia SM, Weisbrich A, Damberger FF, Bhavesh NS, Jawhari H, Grigoriev I, van Rijssel FJ, Buey RM, Lawera A et al (2009) An EB1-binding motif acts as a microtubule tip localization signal. Cell 138:366–376

34. Castoldi M, Popov AV (2003) Purification of brain tubulin through two cycles of polymerization-depolymerization in a high-molarity buffer. Protein Expr Purif 32:83–88

35. Hyman A, Drechsel D, Kellogg D, Salser S, Sawin K, Steffen P, Wordeman L, Mitchison T (1991) Preparation of modified tubulins. Methods Enzymol 196:478–485

36. Crevel IM, Nyitrai M, Alonso MC, Weiss S, Geeves MA, Cross RA (2004) What kinesin does at roadblocks: the coordination mechanism for molecular walking. EMBO J 23: 23–32

37. McIntosh JR, Morphew MK, Grissom PM, Gilbert SP, Hoenger A (2009) Lattice structure of cytoplasmic microtubules in a cultured mammalian cell. J Mol Biol 394:177–182

Chapter 12

The Segmentation of Microtubules in Electron Tomograms Using Amira

Stefanie Redemann, Britta Weber, Marit Möller, Jean-Marc Verbavatz, Anthony A. Hyman, Daniel Baum, Steffen Prohaska, and Thomas Müller-Reichert

Abstract

The development of automatic tools for the three-dimensional reconstruction of the microtubule cytoskeleton is crucial for large-scale analysis of mitotic spindles. Recently, we have published a method for the semiautomatic tracing of microtubules based on 3D template matching (Weber et al., J Struct Biol 178:129–138, 2012). Here, we give step-by-step instructions for the automatic tracing of microtubules emanating from centrosomes in the early mitotic *Caenorhabditis elegans* embryo. This approach, integrated in the visualization and data analysis software Amira, is applicable to tomographic data sets from other model systems.

Key words Microtubules, Segmentation, Automatic tracing, *C. elegans*, Electron tomography, Three-dimensional reconstruction

1   Introduction

During mitosis, cells form bipolar spindles to align the chromosomes on the metaphase plate and segregate them to the daughter cells during anaphase. Mitotic spindles are bipolar, self-organizing structures, composed of microtubules and numerous microtubule-interacting proteins, including microtubule polymerases, depolymerases, and stabilizers, and additional motor proteins.

Within the bipolar spindle, microtubules show a distinct organization. The minus ends are organized at microtubule organizing centers (MTOCs), such as spindle poles or centrosomes. The plus ends of microtubules radiate out from the MTOCs towards the cell cortex (astral microtubules) or the aligned chromosomes (kinetochore microtubules). In addition, some microtubules emanating from one of the MTOCs interact with microtubules from the opposite pole (interdigitating or interpolar microtubules).

David J. Sharp (ed.), *Mitosis: Methods and Protocols*, Methods in Molecular Biology, vol. 1136,
DOI 10.1007/978-1-4939-0329-0_12, © Springer Science+Business Media New York 2014

The function of mitotic spindles is highly conserved across species and so are the proteins that form the spindle. However, the size, volume, shape, and dynamics of spindles vary dramatically. For example, the mitotic spindle in a one cell-stage embryo of the amphibian *Xenopus laevis* [1] has a volume approximately 50,000 times larger than that of the spindle in the budding yeast cell *Saccharomyces cerevisiae* [2]; and this diversity is mirrored in the different structural organization of these spindles [3]. Surprisingly, the dynamic properties of mitotic spindle assembly have been studied intensively by light microscopy, but the detailed three-dimensional architecture of kinetochore, interpolar, and astral microtubules has been described in only a very small number of model systems, namely, in budding yeast [4, 5], in the early *C. elegans* embryo [6], and in mammalian tissue culture cells [7–9].

Undoubtedly, such a 3D analysis of the microtubule cytoskeleton is essential for a deeper understanding of the mechanisms underlying spindle formation.

Here, we briefly describe the preparation of *C. elegans* early embryos for a reconstruction of the first mitotic spindle, followed by a detailed step-by-step description of microtubule tracing and modeling in tomograms using the Amira software.

Currently we use the EM PACT2 + RTS high-pressure freezer (Leica Microsystems, Vienna, Austria) for correlative light and electron microscopy of single early *C. elegans* embryos [10–12]. The rapid transfer system (RTS) is used for a quick loading of embryos contained in cellulose capillary tubes, thus allowing a precise staging of embryos prior to cryo-immobilization and reducing the time window between light microscopic observation and preservation by high-pressure freezing. Electron tomography is then applied to study the three-dimensional organization of spindles and their components with a resolution of approximately 5–8 nm in 3D [13, 14].

We use the IMOD software package [15] to calculate tomograms and the Amira software for automatic tracing of microtubules [16, 17]. In Amira the obtained 3D models of traced microtubules can be edited. For example, microtubules can be segregated into different classes, for example depending on their origin or end morphology.

2 Materials

2.1 High-Pressure Freezing and Freeze Substitution

1. EM PACT2 + RTS (Leica Microsystems, Vienna, Austria) (*see* **Note 1**).

2. Stereomicroscope with light source.

3. Inverted light microscope (phase contrast, DIC, epifluorescence).

4. Cellulose capillary tubes with an inner diameter of 200 μm (Spectrum, 23022 La Cadena Dr., Suite 100, Laguna Hills, CA 92653, USA, and Leica Microsystems, Vienna, Austria).

5. Micropipettor (0.5–1 μl size) and gel loader tips.

6. Nail polish.

7. *C. elegans* strain of interest.

8. M9 buffer: 22 mM potassium phosphate monobasic (KH_2PO_4), 19 mM NH_4Cl, 48 mM sodium phosphate dibasic (Na_2HPO_4), 9 mM NaCl.

9. 20 % BSA (w/v) in M9 buffer (*see* **Note 2**).

10. Tools, for example, two syringe needles for cutting worms open to release the embryos, scalpels for cutting dialysis tubing, good forceps.

11. Microscope slides and high-precision cover slips.

12. Specimen holders for high-pressure freezer: 100 μm deep membrane carriers [10].

13. Filter paper wedges for removing fluids.

14. EM-grade acetone.

15. Freeze substitution cocktail: 1 % OsO_4 + 0.1 % uranyl acetate in acetone [18].

16. Automated freeze-substitution device (Leica EM AFS2).

17. Epon/Araldite resin: 6.2 g Epon 812 substitute, 4.4 g Araldite, 12.2 g DDSA, and 0.55 ml DMP-30.

18. Teflon-coated glass slides for thin-layer embedding: Slides are coated with a Teflon® solution (MS-143V, Miller-Stephenson Chemical Co., Inc., Danbury, CT, USA).

19. "Dummy" blocks for remounting and fast glue.

2.2 Electron Tomography

1. For electron tomography of semi-thick (200–400 nm) sections, we use a microscope operating at intermediate voltage (300 kV; TECNAI F30; FEI Company, Eindhoven, The Netherlands). Automated image acquisition programs are available from commercial vendors (FEI Company, Eindhoven, The Netherlands; Tietz Video and Image Processing systems (TVIPS), Gauting, Germany). In addition, several software packages are freely available (TOM, UCSF Tomography and SerialEM, Boulder Laboratory for 3D electron microscopy of cells).

2. Copper slot grids.

3. 0.7 % (w/v) Formvar in ethylene dichloride.

4. 2 % uranyl acetate in 70 % methanol.

5. Rinse solution of 70, 50, and 20 % methanol.

6. Reynold's lead citrate.

7. 10 or 15 nm colloidal gold.

2.3 Three-Dimensional Reconstruction

1. We routinely use the IMOD software package (http://bio3d. colorado.edu/imod), which contains all of the programs needed for calculating tomograms [15]. The IMOD software package runs on multiple platforms, including Linux, Mac OSX, and Windows. The programs used for tomographic reconstruction are managed by a graphical user interface, eTomo. The eTomo interface guides users through the various steps of the process. This program can be run by command line; it contains windows for image display and slicer tool for rotating slices of image data and for modeling features of interest in the reconstruction.

2.4 Modeling

We currently use an extension to the filament editor of the visualization and data analysis software Amira [16] to automatically model and trace microtubules in our tomograms. This extension is freely available (*see* **Note 3**).

---

3 Methods

3.1 High-Pressure Freezing and Freeze Substitution

The combination of high-pressure freezing with freeze substitution has evolved into a routine technique to prepare *C. elegans* for electron tomography. In principle, either whole worms or isolated embryos can be cryo-immobilized within milliseconds [19, 20]. In general, not only *C. elegans* embryos are a suitable system for this method of high-pressure freezing but also basically any organism of a size below 200 μm.

3.1.1 Cryo-Immobilization of Isolated Embryos

1. Prepare a loading device for collecting isolated early embryos into cellulose capillary tubes by mounting a piece of tubing (approx. 2 cm) into a pipette tip and using nail polish to seal. Dry before using [11].

2. Cut worms open in a drop of M9 buffer on a microscope slide. Select an early embryo under the dissecting scope, and suck the selected embryo into the capillary tube.

3. Submerge the capillary into the buffer, and use the scalpel to cut off the region of the cellulose capillary containing the embryo. Shorten the tube to a size that will fit the specimen carrier. The ends of the tube should be crimped so that the tube is sealed and the embryo will not leak out. This can be achieved by using the blunt side of a scalpel.

4. Transfer the capillary from the microscope slide to a drop of M9 on a high-precision cover slip, and observe the early development of the embryo by DIC or fluorescence microscopy.

5. At an appropriate mitotic stage, transfer the cellulose capillary containing the embryo from the cover slip to a specimen carrier, prefilled with 20 % BSA in M9 buffer, freeze, and store in LN_2.

1. Transfer samples to precooled (–90 °C) cryovials containing the freeze substitution cocktail composed of 1 % OsO_4 plus 0.1 % uranyl acetate in acetone. Maintain temperature for 8–24 h at –90 °C.

2. Warm samples to room temperature at a rate of 5 °C/h.

3. Rinse samples in pure acetone and infiltrate with Epon/Araldite.

4. Process samples for thin-layer embedding, and remount selected specimens on dummy blocks for ultramicrotomy [19].

5. Cut ribbons of serial semi-thick sections (300–400 nm) through the embryos, and collect the ribbons of sections (serial sections) on Formvar-coated copper slot grids.

6. Stain sections with uranyl acetate, followed by Reynold's lead citrate.

7. Apply 10–15 nm gold fiducials to the samples by placing the slot grids on top of a drop of gold solution for 2 min, blotting excess fluid, turning the grid over, and repeating for the other side. Wash the slot grids briefly in a drop of water (*see* **Note 4**).

8. Examine semi-thick sections in a standard TEM operating at 100 or 120 kV to identify the sections containing the regions of interest. Map the features of interest through the serial sections by imaging at a low magnification.

3.2 Acquisition of Tomographic Data

1. Image a tilt series of semi-thick sections in an intermediate-voltage (200–300 kV) EM equipped with a eucentric tilting stage. Collect serial tilted views of the section every degree over a ±60° or 70° range. If a full spindle should be acquired, montages have to be acquired instead of single frames. This can be done using the SerialEM software. To acquire a montage, go to "*file*" and "*New montage.*" A new window, the "*Montage Control*" box will appear, and the size of the montage, 1×1, 2×2, 3×3, etc. has to be chosen to cover the region of interest. For a large-scale 3D reconstruction it is essential to cut serial sections. As an example, for a reconstruction of the first mitotic spindle in *C. elegans*, each section needs 12 single-tilt tomograms to cover the pole-to-pole area of the mitotic spindle in *x*- and *y*- and 20 consecutive sections to acquire the *z*-dimension of the spindle (*see* **Note 5**).

2. After the first tilt series has been acquired, rotate the grid 90° to image a second tilt series over a ±60° or 70° range.

3. Calculate a double-tilt tomogram using IMOD [21]. The calculated tomogram will be stored as the so-called .rec file. In order to use the automatic microtubule tracing in Amira it is essential to flatten the tomogram at the end of the reconstruction process and to trim off grey areas (*see* **Note 6**).

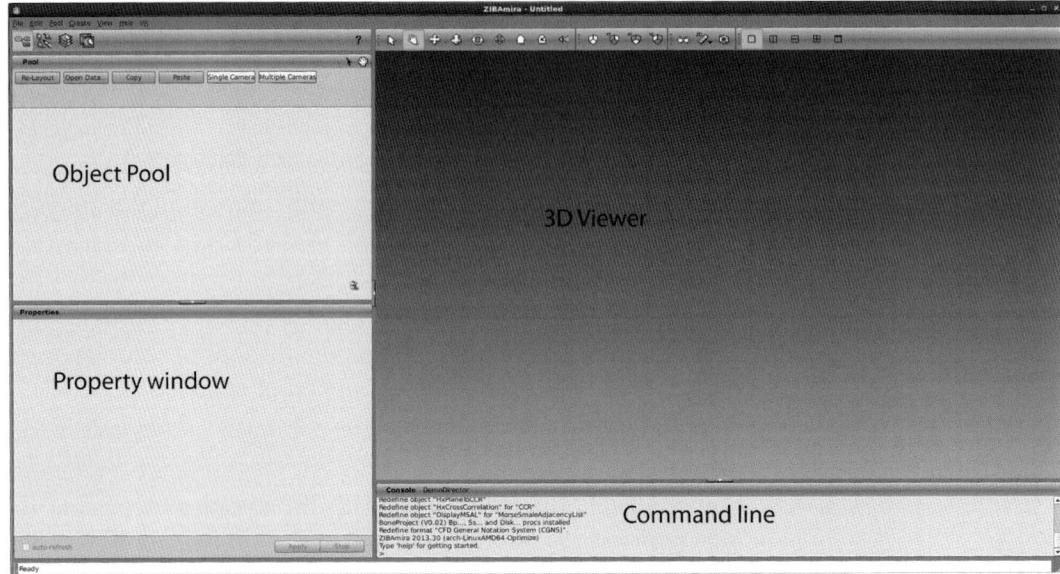

Fig. 1 Screenshot of the Amira interface. The interface of Amira can be divided into four major windows: the object pool, in which all the working files will be located; the property window, where values and parameters for different tasks can be adjusted; the 3D-viewer, which can display either the tomogram or any 3D model; and the command line

3.3 Automatic Tracing and Modeling of Spindle Microtubules

1. Open Amira, and open the reconstructed tomogram. This can be done by either choosing "*File*" and "*open*" from the taskbar or dragging and dropping the *.rec* file, which needs to be flattened and trimmed in IMOD, into the *object pool* (Fig. 1). Depending on the data size, a pop-up menu will appear. Always choose "*Read complete volume into memory*" from the buttons.

2. The first step for automatically tracing microtubules is a computation of normalized cross correlation of the tomogram with an idealized cylinder, resulting in a correlation and vector field. Clicking the right mouse inside of the *object pool* will open up a window with a menu. Choose *Create*, then *Microtubule Tracing*, and then *Cylinder Template Matching CUDA* from the dragdown menu (Fig. 2). Adjust the settings in the properties windows by changing the *blocksize* to X:256 Y:256 and Z:256. Click apply. Computation is performed on a graphics card using the CUDA API [22]. You need an NVIDIA graphics card. For a $2k \times 2k \times 200$ tomogram, computation takes approximately 6 h on a GeForce 400 series.

3. The successful computation will generate a *correlation field* and a *vector field*, which will show up in the *object pool*. Note: Always save those files as Amira has no autosave function. For automatic microtubule tracing right click into the *object pool*. Choose *Create*, then *Microtubule Tracing*, and finally

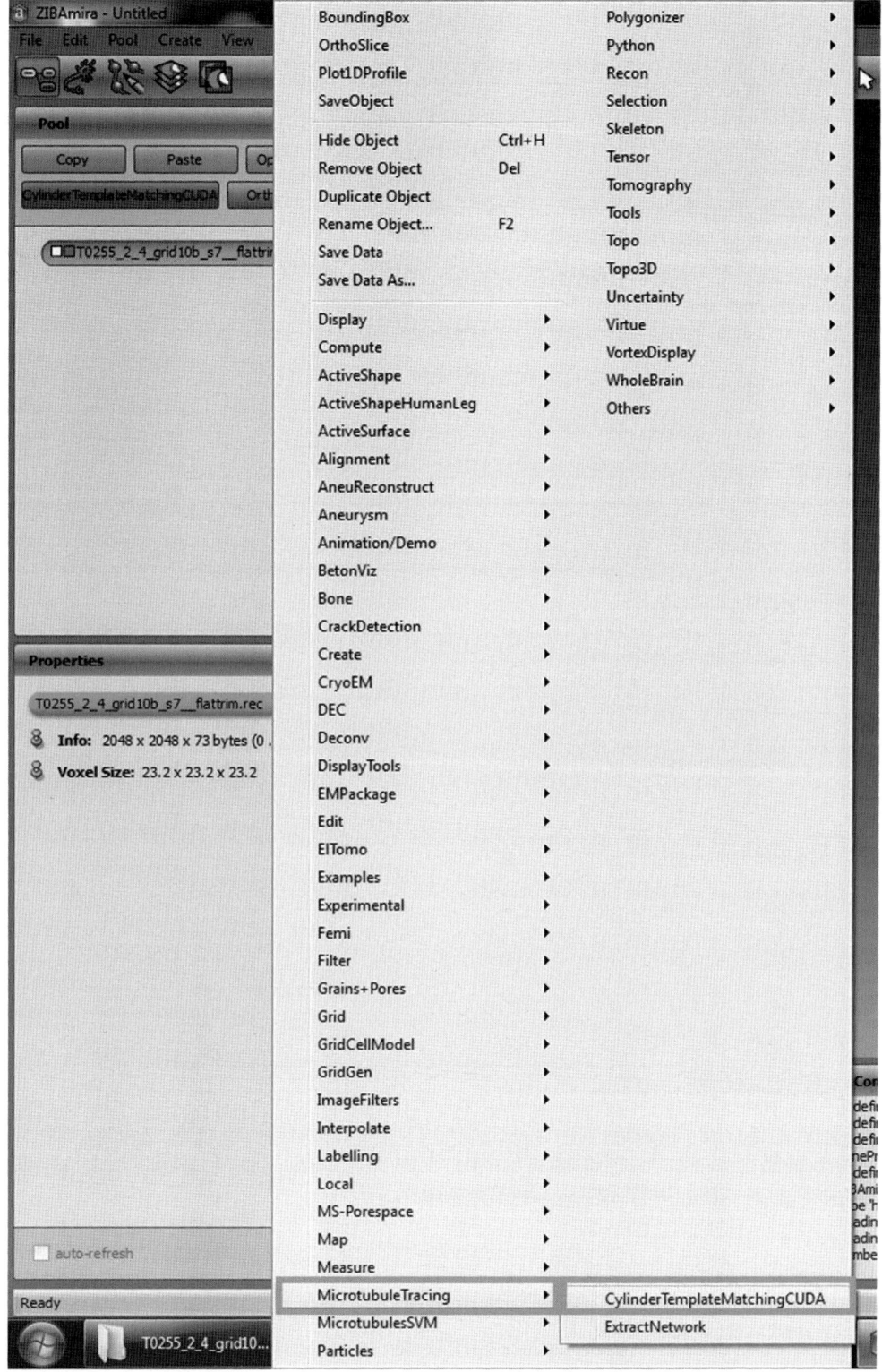

Fig. 2 CylinderTemplateMatchingCUDA. After the tomogram is loaded into the object pool, a right click inside the object pool opens up a drag-down menu. There the user first has to choose "*MicrotubuleTracing*" followed by "*CylinderTemplateMatchingCUDA*" to start the segmentation. Settings like the blocksize can be adjusted in the properties window

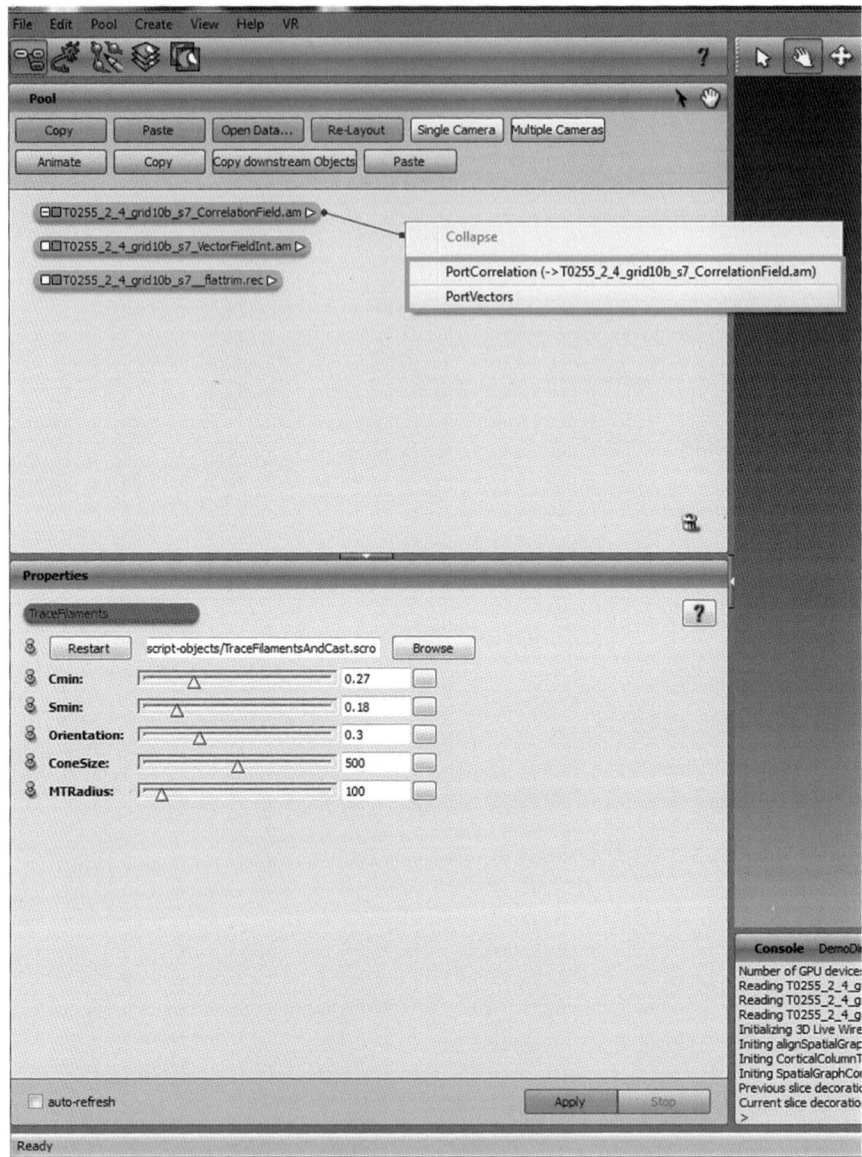

Fig. 3 *PortCorrelation* and *PortVector.* After a successful *CylinderTemplateMatchingCUDA* Amira will generate a vector field and a correlation field. For the automatic microtubule tracing those fields have to be linked to the corresponding ports, called *PortVectors* and *PortCorrelations*

TraceFilaments from the menu. A new box named *TraceFilaments* will appear in the *object pool*. Click on the little square on the left side of the box, and connect the port named *PortCorrelation* with the correlation field and the port named *PortVectors* with the vector field (Fig. 3). Click apply. This step is usually fast and will create another file in the *object*

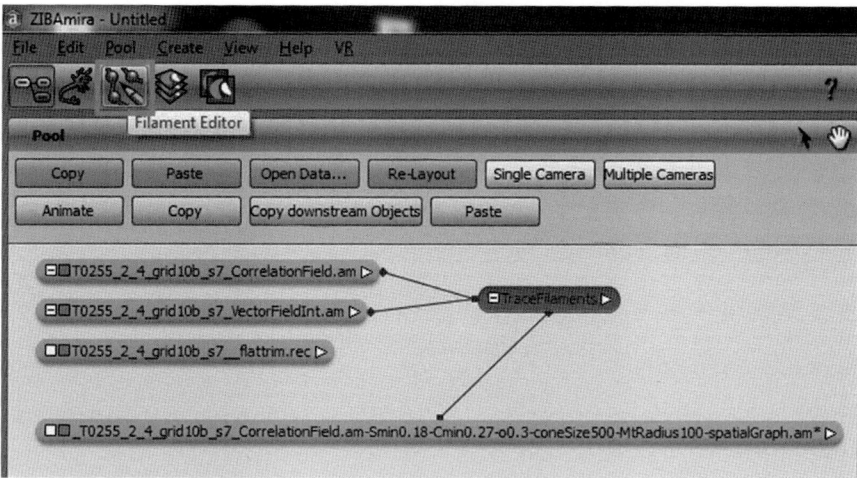

Fig. 4 Microtubule tracing. Successful microtubule tracing will generate the so-called *SpatialGraph* in the object pool, which subsequently can be opened and edited with the *filament editor*

pool, containing the traced centerlines. The lines are stored in a data structure called *SpatialGraph*.

For microtubules this *SpatialGraph* will only contain Edges, which represent microtubule centerlines and pairs of nodes that mark the endpoints of an edge.

3.4 Editing of the Traced Microtubules

1. Load the .rec file, the vector, and the correlation field and the traced microtubules (*SpatialGraph*). After loading, start the *filament editor* by clicking on the icon marked in Fig. 4.

2. The *filament editor* is organized in several panels (Fig. 5), which are briefly described here. *MPR-viewer*: Displays an oblique slice through the tomogram. Additionally, any crossing of elements of the currently processed *SpatialGraph* with the current slice is rendered. Usually, this window is used for drawing and editing lines.

   3D-viewer: The 3D-viewer displays the *SpatialGraph* containing the modeled microtubules in three-dimensional space. Along with the model, you can also display the orientation of the slice currently rendered in the MPR-viewer. You can configure the size of the MPR- and 3D-viewer by clicking on the separating bar and moving it.

   Data management: This lets you choose, which tomogram and *SpatialGraph* is be displayed in the viewers. In the first row, you can choose the tomogram; in the second row, the corresponding model of microtubules can be chosen; and in the third row the contrast for displaying the tomogram can be adjusted. The fourth row allows the configuration of the maximum intensity projection of the current slice.

Data management   Navigation tools   Editing/Selection tools

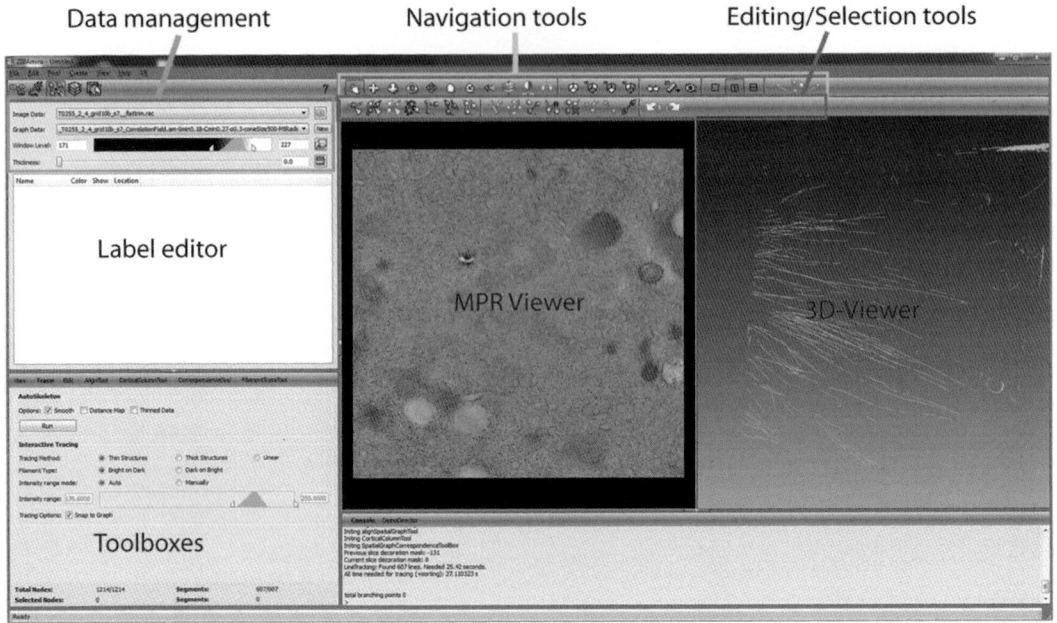

Fig. 5 Screenshot of the *filament editor*. The *filament editor* is organized into different panels, which are depicted here. It consists of a data management panel, where the working tomogram and corresponding *SpatialGraph* can be chosen; the *Label editor*, which is needed if microtubules will be annotated according to origin or end morphology; the Toolboxes, in which parameters can be adjusted; the MPR-viewer, which shows the tomogram; the 3D-viewer, in which the model is shown; and the navigation and editing/selection tools to edit the microtubule model

        Label editor: Modeled edges and nodes can be labeled according to properties (for further description *see* below).

        Tools: Tools offer the functionality to navigate and to edit a traced graph.

        Toolbox: Toolboxes are used to configure tools and run algorithms.

3. You can navigate in your tomogram by changing the slice currently displayed in the MPR-viewer or in the 3D-viewer. The navigation tools are located in the upper right bar, and their availability and functionality depend on the currently selected viewer. Table 1 explains the most common tools used for navigation and their function in the two viewers. To use the tools, first select a viewer by clicking in the viewer. A red border marks the active viewer. Activate a tool by clicking on it.

4. Because the automatic tracing of microtubules can make mistakes, verifying the tracing is essential; one can manually edit nodes and edges. Occasionally, one can find microtubules in the MPR-viewer, which have not been detected by the tracing program. In this case, it is possible to manually trace the microtubules.

Table 1
Description of the navigation tools in the MPR- and 3D-viewer. This table shows a description of the most common navigation tools of the MPR-viewer and the 3D-viewer

Icon	Function in MPR-viewer	Function in 3D-viewer
	Rotate slice around the center of image (marked by a yellow cross)	Rotate scene around the center of volume
	Rotate slice around the center of image (marked by a yellow cross)	Rotate scene around the center of volume
	Translate image	Translate scene
	Zoom in and out	Zoom in and out
	Small rotation	Small rotation
	Local zoom	
	Return to initial view of image	Return to initial view of volume
		Switch to orthographic view
	Change the plane of view	
	Tools for measurements and snapshots	
	Arrangement of windows	

For this, open the *semiautomatic tracing toolbox* (*see* Fig. 6) and choose the proper fields in the menus correlation and vector field. The parameters need not be configured for tracing microtubules, except for the adjustment radius. Fill in the radius of a microtubule (10 nm). Note that the voxel size of the volume might be measured in Angstroms: in this case, the adjustment radius needs to be set to 100. The semiautomatic tracing is activated by pressing the button indicated in Table 2. The *filament editor* will try to

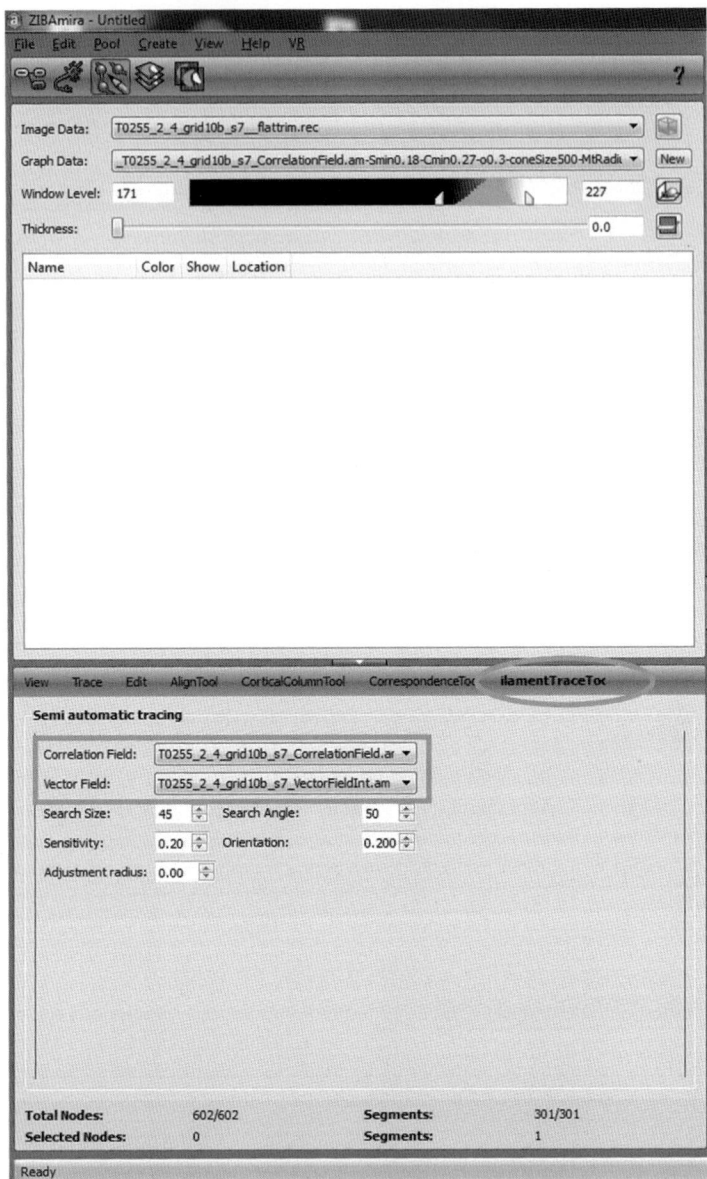

Fig. 6 Semiautomatic tracing toolbox. For a functional semiautomatic tracing of microtubules, which in the course of editing of the model is needed, the appropriate correlation and vector fields have to be inserted into the *Filament Trace Tool* in the toolbox

find a microtubule centerline whenever you click at the slice in the MPR-viewer (it will also do so even if you clicked somewhere where there is no microtubule).

5. To systematically check each line in a model for correctness, the *edge stepper tool* (Fig. 7) is helpful. This tool allows you to systematically step through all lines and show their embedding

Table 2
Description of the editing tools in the *filament editor*. This table shows a description of the most important editing tools of the *filament editor*, which can be used in the MPR-viewer and the 3D-viewer. Hotkeys for the different functions are also shown in this table

Icon	Function	Key
	Selected tool for picking and selecting a node or a segment (shift selects both)	e
	Draw a line around an area, and select everything within the area	q
	Selects all items in the spatial graph	a
	Deselects all items	c
	Semiautomatic tracing tool, start tracing or drawing a new segment	
	Connect tool, select two nodes/edges, and connect them	s
	Extract line, click on any point in the MPR viewer, and line is shortened or prolonged (+shift)	n
	Select a segment, go to the points that should be separated, then split	

in the volume. Stepping through the edges is triggered by increasing/decreasing the edge number in the *Segment* input field. The edge stepper will automatically focus the beginning of the edge in the MPR-viewer. To step through the edge points, you can increase/decrease the point number in the *Points* field. The slice in the MPR-viewer will automatically be adjusted to an optimal view of the currently selected edge point. When stepping through edges, the chosen edge will automatically be selected. When stepping through points, only the current point will be selected.

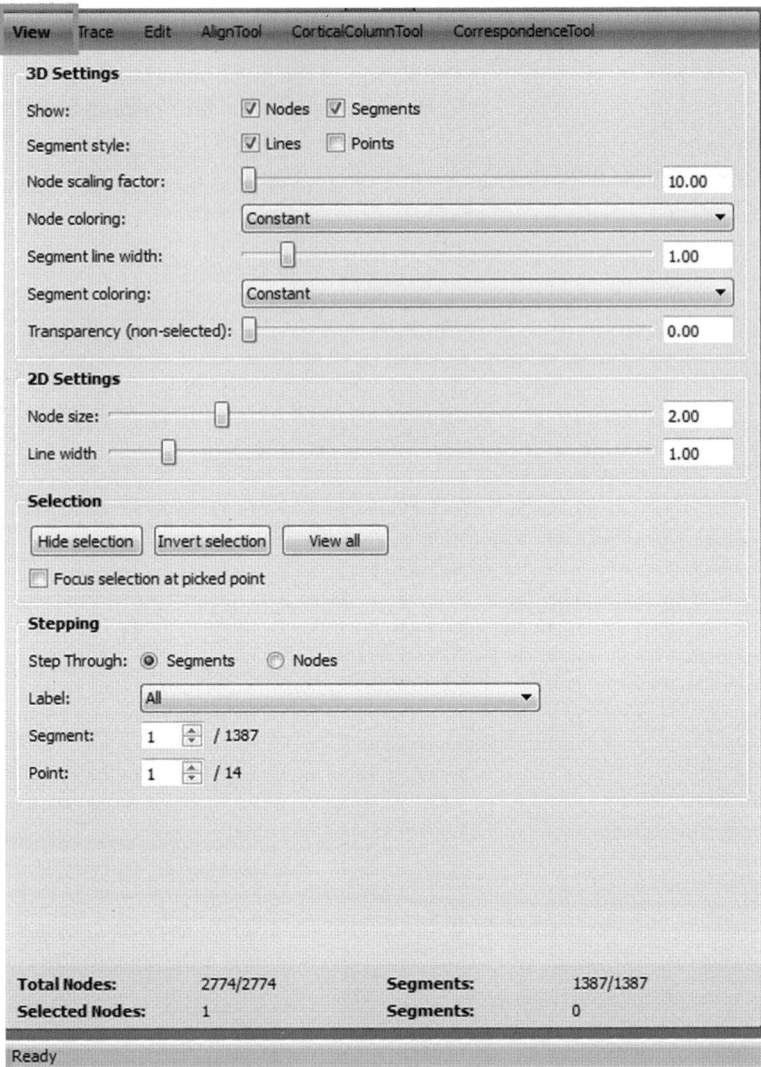

Fig. 7 The *edge stepper* tool. This tool allows the user to systematically step through all lines and show their embedding in the volume. Stepping through the edges is triggered by increasing/decreasing the edge number in the segment input field. To step through the edge points, the point number in the points field can be increased/decreased

6. Another important step is to check the correctness of traced lines and correct errors. To correct a traced microtubule, it needs to be selected first. To do so, push the select button (Table 2) and then click on the edge or the node of interest in the MPR-viewer. Clicking and pressing *shift* simultaneously selects nodes and edges of a microtubule. In the case that the traced line overshoots a microtubule end, you can jump from one end of the microtubule to the other with the keys *k* and *h*. Pressing *p* will remove the last endpoint. If the semiautomatic

tracing tool is active, you can click at a position on the edge where the microtubule ends and the microtubule are shortened up to this point. A falsely detected microtubule can be selected (*shift*+select to select edges and nodes) and then deleted by pressing *d*. If the traced line stops before the true microtubule end, you can select the end that needs to be lengthened. If the semiautomatic tracing tool is active, you can then click on a point where the microtubule continues and the microtubule will be continued. If you prefer not to use the automatic tracing, press *shift* and click on the point where the microtubule ends. The current endpoint will then be connected to the clicked point by a straight line. Occasionally, you will need to connect two edges to form a single microtubule. This can be done by pressing *ctrl* and selecting both microtubules. Then press the connect icon (Table 2).

7. You can assign attributes to edges and nodes of the *SpatialGraph*. For example, a microtubule end can be closed or open. You might also want to distinguish between different kinds of edges, since some might, for example, represent a filament and others the centrosome border. In the filament editor the attributes are called labels. A *SpatialGraph* can have three kinds of labels, NodeLabels, SegmentLabels, and GraphLabels, that can be assigned to nodes and segments. The labels are organized in label groups. A label group represents a property of an item, such as end type of microtubule or spindle part, and contains different labels such as plus, minus, or for the second case actin, microtubule, and centriole border. To create a label, right click in the *label editor* and select the kind of label group you want to create. Add labels by right clicking again on the label groups, and add labels as you wish. You can choose a color for each label by clicking on the color icon next to the name. The created labels will be saved together with the *SpatialGraph*. To assign a label, select the items you want to assign a label to. Right click in the label you want to assign in the label editor, and select assign selection. Leaves in the label tree can be assigned to selected items by the keys 1–9 (Fig. 8).

4 Summary and Outlook

In this chapter, we briefly described the cryo-immobilization of *C. elegans* embryos and further preparation for electron microscopy and tomography, followed by a more extensive introduction on how to use the automatic microtubule tracing tool in Amira. We have used Amira to successfully trace and model microtubules in tomograms of semi-thick sections of the *C. elegans* one-cell embryo (Fig. 9). A tool to stitch tomograms and models from several consecutive serial sections is currently developed to enable a

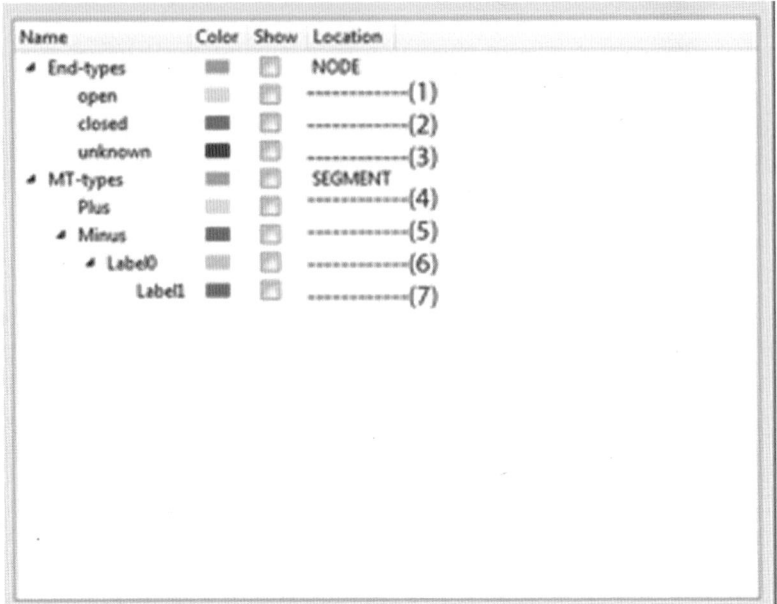

Fig. 8 *Label editor.* Labels (annotations) of microtubules can be added with the *label editor* by a right click into the *label editor*. Labels can be named individually. A color for each label can be chosen by clicking on the color icon next to the name. The created labels will be saved together with the *SpatialGraph*. Leaves in the label tree can be assigned to selected items by the keys 1–7

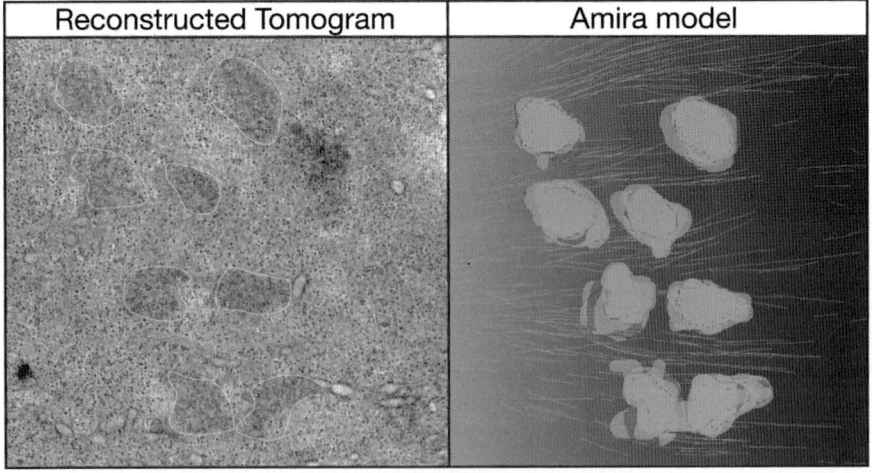

Fig. 9 Model of microtubules within a tomogram created with Amira. In the *left panel* one section of the reconstructed tomogram is shown; the *right panel* shows a snapshot of the microtubules (*green*) and chromosomes (*yellow*) created using the automatic microtubule tracing in Amira (Color figure online)

full reconstruction of larger volumes. This chapter can only be considered as an introduction on how to use the Amira program for microtubule tracing. Depending on the specific project and question asked, other operations and tools within the software package, including the filament editor, might be of more importance to other users. The Amira program offers a help file with several tutorials, which the users might find helpful.

5 Notes

1. The EM PACT2 + RTS (Leica) high-pressure freezer is a portable machine that can be easily moved to the microscope, where staging of the isolated embryo is performed. The RTS allows fast loading of the specimen into a preloaded high-pressure freezer under standardized conditions [10, 12].

2. The use of 20 % BSA gives reproducibly good freezing results, also for other samples such as Drosophila embryos and tissue culture cells [10]. However, we have also made good experiences with BSA concentrations as low as 10 %.

3. Information and help for SerialEM can be found here: http://bio3d.colorado.edu/SerialEM/. Information on how to use IMOD, as well as test data, can be obtained from http://bio3d.colorado.edu/imod/. IMOD as well as SerialEM are freely available. The packages needed for the described microtubule tracing and editing functionality can be obtained from http://www.zib.de/en/visual/software/MicrotubuleTracing.html. A free test license for Amira can be requested at the same site. In addition, test data described in Weber et al. [16] are provided at http://publications.mpi-cbg.de/4629-data.

4. It is advisable to put the gold on the sections and grids before they have been observed with an electron microscope. Exposure to the electron beam seems to severely affect the affinity of gold to the sections.

5. The Navigator window of SerialEM can be used to acquire several positions or sections automatically in one go. This can save a lot of time as the acquisition can be run overnight.

6. Prior to segmentation and modeling in Amira the reconstructed tomograms have to be flattened and trimmed to avoid problems with the microtubule tracing. The best way to trim a flattened tomogram is by command line using the command

   newstack −sec zstart-zend inputfile outputfile
   zstart, zend are a list of integer ranges

References

1. Wuhr M, Chen Y, Dumont S, Groen AC, Needleman DJ, Salic A, Mitchison TJ (2008) Evidence for an upper limit to mitotic spindle length. Curr Biol 18(16):1256–1261. doi:10.1016/j.cub.2008.07.092

2. Maddox PS, Bloom KS, Salmon ED (2000) The polarity and dynamics of microtubule assembly in the budding yeast *Saccharomyces cerevisiae*. Nat Cell Biol 2(1):36–41. doi:10.1038/71357

3. Heath IB (1980) Variant mitoses in lower eukaryotes: indicators of the evolution of mitosis. Int Rev Cytol 64:1–80

4. O'Toole ET, Winey M, McIntosh JR (1999) High-voltage electron tomography of spindle pole bodies and early mitotic spindles in the yeast *Saccharomyces cerevisiae*. Mol Biol Cell 10(6):2017–2031

5. Winey M, Mamay CL, O'Toole ET, Mastronarde DN, Giddings TH Jr, McDonald KL, McIntosh JR (1995) Three-dimensional ultrastructural analysis of the *Saccharomyces cerevisiae* mitotic spindle. J Cell Biol 129(6): 1601–1615

6. O'Toole ET, McDonald KL, Mantler J, McIntosh JR, Hyman AA, Muller-Reichert T (2003) Morphologically distinct microtubule ends in the mitotic centrosome of *Caenorhabditis elegans*. J Cell Biol 163(3):451–456. doi:10.1083/jcb.200304035

7. Dong Y, Vanden Beldt KJ, Meng X, Khodjakov A, McEwen BF (2007) The outer plate in vertebrate kinetochores is a flexible network with multiple microtubule interactions. Nat Cell Biol 9(5):516–522. doi:10.1038/ncb1576

8. Mastronarde DN, McDonald KL, Ding R, McIntosh JR (1993) Interpolar spindle microtubules in PTK cells. J Cell Biol 123(6 Pt 1): 1475–1489

9. McDonald KL, O'Toole ET, Mastronarde DN, McIntosh JR (1992) Kinetochore microtubules in PTK cells. J Cell Biol 118(2):369–383

10. McDonald KL, Morphew M, Verkade P, Muller-Reichert T (2007) Recent advances in high-pressure freezing: equipment- and specimen-loading methods. Methods Mol Biol 369:143–173

11. Muller-Reichert T, Srayko M, Hyman A, O'Toole ET, McDonald K (2007) Correlative light and electron microscopy of early *Caenorhabditis elegans* embryos in mitosis. Methods Cell Biol 79:101–119. doi:10.1016/S0091-679X(06)79004-5

12. Verkade P (2008) Moving EM: the rapid transfer system as a new tool for correlative light and electron microscopy and high throughput for high-pressure freezing. J Microsc 230(Pt 2):317–328. doi:10.1111/j.1365-2818.2008.01989.x

13. McEwen BF, Marko M (2001) The emergence of electron tomography as an important tool for investigating cellular ultrastructure. J Histochem Cytochem 49(5):553–564

14. McIntosh R, Nicastro D, Mastronarde D (2005) New views of cells in 3D: an introduction to electron tomography. Trends Cell Biol 15(1):43–51. doi:10.1016/j.tcb.2004.11.009

15. Kremer JR, Mastronarde DN, McIntosh JR (1996) Computer visualization of three-dimensional image data using IMOD. J Struct Biol 116(1):71–76. doi:10.1006/jsbi.1996.0013

16. Weber B, Greenan G, Prohaska S, Baum D, Hege HC, Muller-Reichert T, Hyman AA, Verbavatz JM (2012) Automated tracing of microtubules in electron tomograms of plastic embedded samples of *Caenorhabditis elegans* embryos. J Struct Biol 178(2):129–138. doi:10.1016/j.jsb.2011.12.004

17. Stalling D, Westerhoff M, Hege H-C (2005) Amira: a highly interactive system for visual data analysis. In: Charles DH, Chris RJ (eds) Visualization handbook. Butterworth-Heinemann, Burlington, pp 749–767

18. McDonald K, Muller-Reichert T (2002) Cryomethods for thin section electron microscopy. Methods Enzymol 351:96–123

19. Muller-Reichert T, Hohenberg H, O'Toole ET, McDonald K (2003) Cryoimmobilization and three-dimensional visualization of *C. elegans* ultrastructure. J Microsc 212(Pt 1): 71–80

20. Muller-Reichert T, Mantler J, Srayko M, O'Toole E (2008) Electron microscopy of the early *Caenorhabditis elegans* embryo. J Microsc 230(Pt 2):297–307. doi:10.1111/j.1365-2818.2008.01985.x

21. Mastronarde DN (1997) Dual-axis tomography: an approach with alignment methods that preserve resolution. J Struct Biol 120(3):343–352. doi:10.1006/jsbi.1997.3919

22. NVIDIA (2010) NVIDIA CUDA Programming Guide 3.1. http://developer.download.nvidia.com/compute/cuda/3_1/toolkit/docs/NVIDIA_CUDA_C_ProgrammingGuide_31pdf. Accessed 04 Aug 2011

Part IV

Extras

Chapter 13

Manipulating Cell Shape by Placing Cells into Micro-fabricated Chambers

Fred Chang, Erdinc Atilgan, David Burgess, and Nicolas Minc

Abstract

Cell shape is an important cellular parameter that influences the spatial organization and function of cells. However, it has often been challenging to study the effects of cell shape because of difficulties in experimentally controlling cell shape in a defined way. We describe here a method of physically manipulating sea urchin cells into specified shapes by inserting them into micro-fabricated chambers of different shapes. This method allows for generation of large systematic and quantitative data sets and may be adaptable for different cell types and contexts.

Key words Cytokinesis, Sea urchin, Mitosis, Cell shape, Micro-fabrication, Microtubules

1 Introduction

Cell shape is a critical parameter that influences cellular function and organization. The shape of a cell, for instance, provides geometric cues that dictate the future plane of cell division in many cell types. Although geometric considerations lay at the heart of spatial control in cells, the effects of cell shape on cellular processes remain understudied. One reason why cell shape has not been studied intensely could be that there are methodological hurdles that have made studying cell shape effects challenging. In some cell types, the shape and size of cells can be highly nonuniform. It has also been difficult to reproducibly manipulate the shape of cells. Genetic perturbations, for instance, can cause changes in cell shape, but these genetic changes may also affect the very cellular processes that are under study. There is a history of classic studies in marine eggs that involve physically manipulating the shape of cells one at a time using needles and other devices [1, 2]; however, these methods are heavily dependent on the skill of the investigator and can only generate small sample numbers. In recent years, micro-fabricated devices for physically manipulating the shape of individual cells have been developed. Thery, Bornens, and colleagues have manipulated

David J. Sharp (ed.), *Mitosis: Methods and Protocols*, Methods in Molecular Biology, vol. 1136,
DOI 10.1007/978-1-4939-0329-0_13, © Springer Science+Business Media New York 2014

cell shape by applying cell adhesive molecules in specific patterns onto the substrate [3]. These studies reveal that in vertebrate tissue culture cells, the pattern of adhesive contacts influences spindle orientation and ultimately the cell division plane [4].

Here we describe the use of micro-fabricated chambers to manipulate cell shape in a systematic way. Arrays of cell-sized wells of specified shapes are generated in PDMS using photolithography. Cells are simply placed into these wells to mold them into specific and predictable shapes. Using these arrays, the shapes and dimensions of the cells can be precisely and reproducibly controlled in large numbers of cells. Advantages of this micro-chamber approach are the following: (1) it is possible to quickly obtain large data sets necessary for quantitative analysis; (2) the approach is simple and inexpensive; and (3) this method should be adaptable for many different contexts and cell types.

In our initial studies, we use sea urchin eggs to study the effects of cell shape on cell division [5, 6]. The eggs of the sea urchin are large, spherical single cells of about 120 μm in diameter. At an hour after fertilization, all the zygotes enter mitosis and then undergo cytokinesis in a synchronous manner. These eggs are ideal for these cell shape experiments as they are malleable, are sturdy, of uniform size, and can be obtained easily in large quantities as single cells in seawater (in contrast to dissected mouse or worm oocytes, for instance). These cells are highly responsive to their cell shape; they are non-adherent and non-polarized and thus lack the adhesion and polarity cues that may override the cues from cell shape. The chambers do not appear to significantly affect cell physiology, as the zygotes continue to divide and form embryos in the chambers. We have used this micro-chamber approach to study how cells use microtubules to sense cell shape to orient and position their spindle and to signal to the cell cortex to specify the plane of cell division. Similar chambers have been used to manipulate the shape of microbial cells such as fission yeast and bacteria [7–10].

2 Materials

2.1 Making SU8 Masters

1. Silicon wafer, 2 in. diameter.
2. Petri dish, 6 cm diameter.
3. SU8 50–100 resist and developer (MicroChem, Newton, MA).

2.2 Making PDMS Wells

1. Sylgard 184 PDMS kit with base and curing agent (Dow Corning Corp.)
2. Plastic syringes, cup, stirring rod.
3. Surgical blade, tweezers.
4. Silane (tridecafluoro-(1,1,2,2-tetrahydrooctyl)-1-trichlorosilane, United Chemical Technologies, T2492).

2.3 Preparing Sea Urchin Zygotes

1. Live sea urchins *Lytechinus pictus* (Marinus Scientific, CA).

2. Filtered seawater.

3. Calcium-free seawater: 26.24 g NaCl, 0.671 KCl, 4.687 Na_2SO_4, 4.3 ml $NaHCO_3$ (0.5 M), 10 ml Tris–Cl (1 M, pH 8.0), 10 ml EGTA (0.25 M, pH 8.0), H_2O to 1 L.

4. Calcium-free seawater + 5 mM 4-aminobenzoic acid (PABA).

5. 0.5 M potassium chloride (KCl).

6. 118 μm Nitex mesh.

7. Beakers, 15 ml Falcon tubes, aspiration pipettes.

8. Microscope glass slides and cover slips (22×22 mm and 50×50 mm).

9. Kim wipes.

10. Hoechst 33342 (dye for DNA, Molecular Probes, Life Technologies).

2.4 Staining Sea Urchins

1. Fixation buffer: 100 mM Hepes, pH 6.9 with potassium hydroxide (KOH), 50 mM ethylene glycol tetraacetic acid (EGTA), 10 mM magnesium sulfate ($MgSO_4$), 2 % formaldehyde, 0.2 % glutaraldehyde, 0.2 % Triton X-100, and 400 mM dextrose.

2. Phosphate-buffered saline (PBS).

3. PBS + 0.1 % Triton X-100 (PBT).

4. PBT + 0.1 % bovine serum albumin (BSA).

5. Anti-alpha tubulin antibody (clone DMA-1), mouse monoclonal (Santa Cruz Biotechnology, CA).

6. AlexaFluor 488 Phalloidin (Molecular Probes, Life Technologies).

7. Labeled anti-mouse secondary antibody.

3 Methods

3.1 Designing the Wells

We designed wells of different shapes in X–Y and of uniform height in Z. The wells were approximately the volume of the *L. pictus* sea urchin eggs, which are normally spheres of around 120 μm in diameter. We typically make wells 50–90 μm in depth, so that the cells are slightly flattened. For example, for a rectangular shape with a 1.5 length-to-width ratio ($H = 80$ μm), the width is $W = 85$ μm and the length is $L = 125$ μm. We also use different scales of the same shapes (for instance 95–110 %). In each array, 100 chambers can be drawn in a 10-by-10 grid, with a 50 μm separation between each well. Each slide contains many repeating arrays of these patterns (Fig. 1a, c). The design of patterns is done on a computer-assisted drawing (CAD) program and can be converted to a format (generally *.dxf) that can be used as input for a laser writer (*see* **Note 1**).

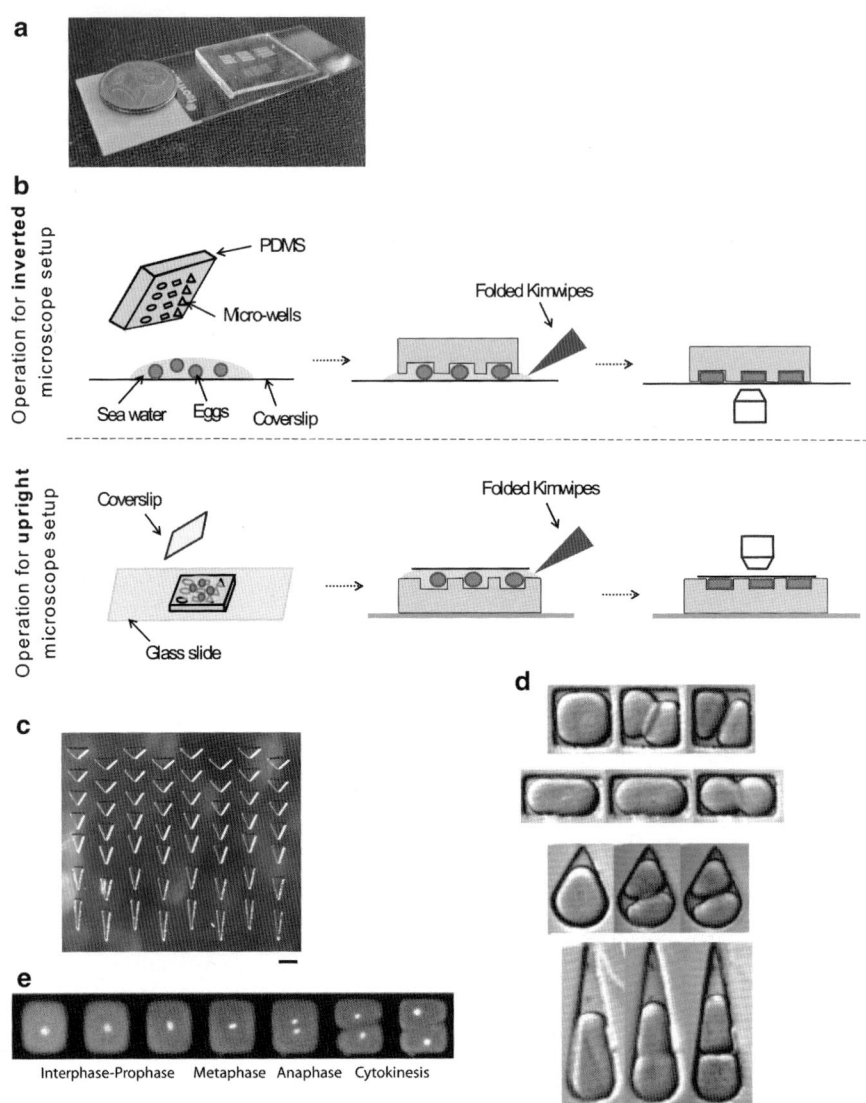

Fig. 1 Shaping sea urchin eggs using chambers. (**a**) Image of PDMS slab containing arrays of chambers on a glass side. (**b**) Schematic showing two ways of inserting sea urchin zygotes into the PDMS chambers. The *top* shows the operation to image with an inverted microscope (which can also be adapted for immunofluorescence and perfusion of eggs in chambers). The *bottom* represents the operation, for an upright microscope. (**c**) Image of an array of chambers of different shapes. (**d**). DIC images of sea urchin zygotes dividing over time in the chambers. (**e**) Time-lapse images of a dividing square-shaped cell stained with Hoechst 33342 to visualize the DNA. Images in (**d**) and (**e**) were reprinted from Minc et al., 2011, with permission from the journal

3.2 Formation of the Master Using Photolithography

Chambers for shaping cells are made from PDMS using a positive master composed of SU8. A positive master is a hardened structure of SU8 such that the desired geometric shapes extend as posts from a relatively smooth surface (generally silicon wafer or cover slip glass). SU8 is an epoxy-based negative photoresist: when exposed to UV light it becomes insoluble to the photoresist developer.

Masters are prepared using standard lithography methods for SU-8 micro-fabrication [7, 11] (*see* **Note 2**).

1. Clean the silicon wafer by rinsing with acetone, isopropyl alcohol and water. Dry it with nitrogen gas and bake at 200 °C for 5 min on hot plate.

2. Spin coat SU8 onto the wafer. Place SU8 on wafer using a syringe (approximately 1 ml of SU8 for each square inch). Spin parameters depend on the type of SU8 used and desired thickness of the thin coating and is likely to be laboratory dependent (*see* **Note 3**).

3. Bake the resist on a level hot plate for about 6 min at 65 °C and 20 min at 95 °C. This creates a thin layer of SU8 coating the wafer with desired thickness.

4. Create patterns on the SU8 by exposing it to patterns of UV light. This is done by microscope-like instruments used for photolithography or "laser writing" (Heidelberg μPG 101 Laser Writer).

5. The sample is then baked for 5–10 min at 95 °C, washed with developer, rinsed with isopropyl alcohol, and dried with nitrogen or air.

6. The master is then silanized by exposing it to silane vapors. The masters are incubated with an open flask containing 0.2 μl silane liquid in a vacuum desiccator overnight. This treatment prevents the PDMS from sticking to the SU-8, and extends the lifetime of the master.

7. The master can be stored in a petri dish.

3.3 Creating PDMS Chambers from a Master

Polydimethylsiloxane (PDMS) chambers can be fabricated multiple times from a positive SU8 master. PDMS is a clear elastomere used for fabrication of a wide range of microanalytical systems. PDMS is highly suitable for biological applications as it is optically clear, gas permeable, and generally nontoxic to cells. These procedures can be done easily in the standard molecular biology laboratory (*see* **Note 4**).

1. Mix Sylgard-84 PDMS base and curing agent in a 10:1 ratio. Stir for 2 min with a plastic stirrer bar.

2. Degas mixture in a vacuum chamber. Apply vacuum until bubbles disappear.

3. Clean wafer with nitrogen or air. Apply PDMS slowly onto the center of the mold. Once wafer is coated completely by a few millimeters of PDMS, let it settle onto wafer for 10 min. Bake at 60 °C for at least 4 h. Cool for at least 10 min. The PDMS can be stored for long periods of time.

4. Cut the PDMS (typically a 2×2 cm piece that fits on a standard glass slide) with a blade, and carefully peel it off the master

using a tweezer. Place the PDMS slab with the wells facing up on a glass slide (Fig. 1a).

5. Just before use, activate the PDMS with a plasma cleaner (Harrick Plasma, Ithaca, NY) (*see* **Note 5**). Insert the PDMS into the chamber of the plasma cleaner for 45–60 s under vacuum. Immediately cover the chamber with two transfer pipette drops of seawater to prevent the chamber to become hydrophobic again. Dry immediately prior to placing egg suspension on chambers.

3.4 Shaping Sea Urchin Eggs

Samples are kept at 17–19 °C throughout the experiment.

1. Sea urchins are kept in a seawater aquarium at 17 °C. Gametes are collected by injecting 0.5 M potassium chloride (KCl) into the coelom of the adult urchin. Females shed eggs, which can be collected in a small beaker of seawater. Eggs are used on the day of collection. Males shed sperm, which are collected by pipette and can be stored in an Eppendorf tube at 4 °C for several days without seawater (as the sperm are activated by seawater).

2. For fertilization, dilute the sperm 1:1,000 in 10 ml seawater. Check that the sperm are motile. Add two drops of activated diluted sperm onto an egg suspension in 5 ml seawater. Monitor the formation of fertilization envelopes in 2–5 min using a dissection microscope and estimating the percentage of cells forming the envelope. At an optimal sperm concentration, approximately 90 % of the cells form envelopes (*see* **Note 6**).

3. Remove fertilization envelopes within 2 min after fertilization by pouring the eggs through 118 μm Nitex mesh in PABA CaFree seawater twice. Check the cells under microscope if the envelopes have been removed in most eggs. The eggs are then placed back in CaFree seawater.

4. DNA can be labeled by adding the fluorescent dye Hoechst 33342 at a dilution of 1 μg/ml to the fertilized egg suspension. This labeling of the chromosomes allows for monitoring the health of the cells and events in cell division and identifying abnormal polyspermic embryos (*see* **Note 7**). At this concentration, this dye has no major effect on developmental timing or physiology.

5. Depending on the microscope used, two setups are possible for introducing the fertilized eggs into the chambers (Fig. 1b):

(a) If the microscope is an upright microscope, the PDMS slab is placed on a glass slide or chamber, with the wells facing up. One transfer pipette drop (about 100 μl) of concentrated fertilized eggs in seawater are placed on the top of the PDMS chambers and allowed to settle onto the PDMS for about a minute. The number of eggs should be similar to the number of wells; a vast excess of eggs can cause problems with cell debris from broken eggs that do not fall into the wells. Gently place a glass cover slip (22×22 mm) on top.

(b) If an inverted microscope is used, the suspension can be first placed on a large cover slip (50×50 mm), and the PDMS slab is applied with the wells facing down. We have also used custom-built chambers in which the bottom is made of thin glass.

(c) Imaging the eggs through the cover slip provides better images.

6. In both cases, the water between the cover slip and PDMS slab is then wicked from the sides by capillary action using Kim wipes. This wicking step is usually repeated several times until the eggs have filled completely the well and have adopted their shape. With practice, this step can be done directly under the microscope, in about 2–3 min. Thus same cell can be imaged before and after shape manipulation can be each filled.

7. Typically up to 50 % of the wells can be filled with a single cell.

8. Cells are imaged with 10× to 40× objectives. The nucleus, spindle poles, mitotic spindle, and cytokinesis furrow can be monitored using DIC imaging (Fig. 1d) or with Hoechst 33342 staining and imaging in the UV channel (Fig. 1e) (*see* **Notes 8** and **9**).

9. In general the eggs in the chambers divide with the same timing and fidelity as controls kept in a beaker. The embryos will develop until late stages of cleavages (up to 64–128 cells).

3.5 Perfusion Chambers for Staining Cells in Wells

PDMS chambers can be modified into simple perfusion chambers to fix and stain or to apply drugs to the cells while still in wells.

1. Make holes all the way through the PDMS slab containing the wells using a punch (like the type used in leather work). Typically we make two 0.5 mm holes approximately 5 mm apart adjacent to the arrays of chambers. These holes allow for solutions to traverse the top and bottom of the PDMS slab and perfuse the cells in the wells. This allows for introduction of liquid from the chamber in one hole and removal of liquid from the other.

2. For experiments using an inverted microscope, place the eggs on a large cover slip and cover with the PDMS slab. Some of the extra seawater usually fills part of the large hole and can be removed by gently pipetting through the holes, using thin gel-loading pipette tips. Water is wicked using Kim wipes from the hole and also from the side of the chamber. This helps to introduce the eggs into the wells.

3. Liquid (containing drugs, fixatives, buffers, etc.) can then be gently added in the hole using gel-loading tips. It is particularly important to add fluid slowly to avoid the fluid from lifting the PDMS slab from the cover slip and causing the eggs to escape the well. The eggs can be directly observed with the microscope while performing such operations. Adding liquid into the hole will create a gentle flow, which can be visualized by the movements of residual dust. Note that the volume that needs to be added to exchange most of the media surrounding the eggs in the chamber is on the order of few microliters, and thus fluid exchange is expected to be fast and efficient.

4. Using this method, the cells can be bathed with inhibitors such as nocodazole or latrunculin to inhibit the microtubule and actin cytoskeletons, respectively.

5. For immunostaining, the same chamber setup is used, and all rinsing steps are made under the microscope to ensure that the eggs do not change shape. Cells are typically fixed for 1 h in fixative buffer with formaldehyde and glutaraldehyde. Fixed eggs should drastically change color/transparency when examined by microscope. The eggs are then rinsed three times in PBT (PBS + 0.1 % Triton X-100) 10 min each and then perfused with PBS containing 0.1 % $NaBH_4$ in PBS (made fresh 5 min before) to remove autofluorescence and to quench excess glutaraldehyde. Wash once with PBS and then PBT and block in 5 % normal goat serum in PBT with 0.1 % BSA 30′ (blocking is probably unnecessary for actin staining).

6. For actin visualization, the eggs are incubated in Alexa- phalloidin 3–10 U/ml in PBT for 1 h and then washed in PBT for 10 min and then in PBS for 10 min.

7. For staining microtubules, the eggs in chambers antibody are rinsed with 1:8,000 anti α-tubulin in PBT + 0.1%BSA. The chambers are then placed in a large petri dish containing wet Kim wipes. The petri dish is then sealed with parafilm and incubated overnight at 4 °C. The chambers are perfused twice with PBT and then with Alexa Fluor-labeled anti-mouse secondary antibody at 1:750 in PBT + BSA 0.1 % for 3–4 h. A final rinse in PBS is then made prior to imaging.

4 Notes

1. Available drawing programs include AutoCAD and DraftSight (freeware).

2. Details of SU8 properties and preparation can be found in data sheets available online on the MicroChem website.

3. For spin coating SU8, we suggest as a starting point the following procedure for 50 μm structure thickness: (a) Spin with an acceleration of 100 rpm/s up to 500 rpm. (b) Increase to 2,000 rpm with an acceleration of 300 rpm/s and maintain for 30 s.

4. We use a clean room facility for preparation of the SU8 master by photolithography, but then do the subsequent steps such as preparation of the PDMS chambers in a typical molecular biology lab setting. The cleanliness of these preparations is not as critical for these biological applications as it might be for some engineering applications. The clean room is used primarily because of the availability of the photolithography equipment there at our institution.

5. Plasma cleaning makes the wells hydrophilic. The treatment is critical, because without it, the wells are hydrophobic and will fill with air bubbles. The treatment is best performed just before use.

6. Avoid too high sperm concentrations that yield >90 % of fertilization, as this can lead to polyspermy (fertilization with multiple sperm), which leads to high incidence of abnormal mitoses and divisions.

7. Cell division abnormalities (multiple furrows or cell cycle arrests) may arise from an abnormal batch of eggs, general experimental conditions (polyspermy, temperature of the room, UV damage), or effects of more extreme cell shapes.

8. The dose of UV in imaging should be kept at a minimum to avoid damage and stress on the cell. Therefore, it is important to avoid long periods of UV exposure during focusing or searching for appropriate fields.

9. An automated X–Y microscope stage allows for imaging of cells in many fields in each experiment, which is very useful for acquiring large data sets.

Acknowledgments

We thank the members of the Burgess and Chang laboratories for discussion, the Columbia U. Center for Integrated Science and Engineering clean room for instruction and use of their micro-fabrication facility, and the Marine Biological Laboratory Whitman Summer Investigators program. This work was supported by the National Institutes of Health Grants GM 069670 to F.C. and G.M, 093978 to D.B., an M.B.L. Erik B. Fries and the Colwin Endowed Summer Research Fellowship to F.C., and M.B.L. E.B. Wilson Summer Research Fellowship to D.B.

References

1. Rappaport R (1986) Establishment of the mechanism of cytokinesis in animal cells. Int Rev Cytol 101:245–281

2. Rappaport R (1996) Cytokinesis in animal cells. Cambridge University Press, Cambridge

3. Thery M et al (2005) The extracellular matrix guides the orientation of the cell division axis. Nat Cell Biol 7(10):947–953

4. Thery M et al (2006) Anisotropy of cell adhesive microenvironment governs cell internal organization and orientation of polarity. Proc Natl Acad Sci U S A 103(52): 19771–19776

5. Minc N, Burgess D, Chang F (2011) Influence of cell geometry on division-plane positioning. Cell 144(3):414–426

6. Atilgan E, Burgess D, Chang F (2012) Localization of cytokinesis factors to the future cell division site by microtubule-dependent transport. Cytoskeleton 69(11):973–982

7. Weibel DB, Diluzio WR, Whitesides GM (2007) Microfabrication meets microbiology. Nat Rev 5(3):209–218

8. Minc N, Boudaoud A, Chang F (2009) Mechanical forces of fission yeast growth. Curr Biol 19(13):1096–1101

9. Minc N, Bratman SV, Basu R, Chang F (2009) Establishing new sites of polarization by microtubules. Curr Biol 19(2):83–94

10. Takeuchi S, DiLuzio WR, Weibel DB, Whitesides GM (2005) Controlling the shape of filamentous cells of Escherichia coli. Nano Lett 5(9):1819–1823

11. Qin D, Xia Y, Whitesides GM (2010) Soft lithography for micro- and nanoscale patterning. Nat Protoc 5(3):491–502

Chapter 14

Four-Color FISH for the Detection of Low-Level Aneuploidy in Interphase Cells

Francesca Faggioli, Jan Vijg, and Cristina Montagna

Abstract

FISH (fluorescent in situ hybridization) is a molecular cytogenetic technique established in the early 1980s that allows for the detection of DNA copy number changes (gains and losses) mapping to genomic regions of interest (Langer-Safer et al. Proc Natl Acad Sci USA 79:4381–4385, 1982). This technology has been extensively applied to research-based investigations and is routinely used in prenatal diagnosis and oncology. Here we describe a modification of the standard FISH protocol adapted for the detection of low-frequency mosaic aneuploidy in interphase cells. This approach represents a straightforward method for the measurement of aneuploidy levels in mammalian cells. This system combines four probes mapping to two different chromosomes. The choice of probes is essential for the successful performance of this approach. It greatly reduces the enumeration of false-positive signals that are challenging in the enumeration of ploidy changes (particularly if these are complex and/or involve a significant increase of chromosome number).

Key words FISH (fluorescent in situ hybridization), Fluorochromes, Aneuploidy, Interphase FISH, Chromosomes

1 Introduction

FISH is based on the use of fluorescently labeled probes (plasmid, BACs, PCR fragments) that bind to homologous regions on target chromosomes and allow for the visualization of copy number changes in metaphase chromosomes and interphase cells. Aneuploidy has been commonly linked to pathological states. It is a hallmark of spontaneous abortions and birth defects and is also observed virtually in every human tumor. Therefore, FISH probes are routinely used in clinical diagnostic laboratories, in prenatal diagnosis for the screening of chromosomal abnormalities, and in clinical oncology for the detection of both numerical and structural chromosomal rearrangements involving oncogenes and tumor suppressor genes relevant to therapeutic decision-making (i.e., HER2/neu) [2, 3]. Despite the well-known association of aneuploidy with pathological states, recent studies from our group

David J. Sharp (ed.), *Mitosis: Methods and Protocols*, Methods in Molecular Biology, vol. 1136, DOI 10.1007/978-1-4939-0329-0_14, © Springer Science+Business Media New York 2014

and others have revealed that mosaic aneuploidy is also found in disease-free tissues at a frequency much higher than previously believed to be compatible with appropriate cell function under physiological conditions [4–8]. The consequences of mosaic aneuploidy in adult tissues are largely unknown but one may hypothesize as underlying causes of tissue dysfunction and degeneration. For instance, our group, using the techniques described in this book chapter, has recently shown that aneuploidy accumulates in a chromosome-specific manner during aging, reaching a frequency as high as 50 % of all nuclei in 28-month-old murine cortices [4]. Because the frequency of mosaic aneuploidy in normal tissues is lower than commonly observed in cancer and because it is predicted not to be clonal, several modifications to routine FISH protocols are required to allow sensitivity suitable for the quantification of low-level aneuploidies. In this chapter we will describe specific modifications to standard FISH protocols that allow for maximum sensitivity and specificity for the measurement of mosaic aneuploidy in disease-free tissues.

2   Materials

2.1   Reagents

2.1.1   Slide Preparation

1. Hypotonic solution: 0.075 M KCl in H_2O. Autoclave before use, the solution can be stored at room temperature for at least 6 months.

2. Fixative: Combine methanol and acetic acid at a 3:1 ratio (vol/vol) (*see* **Notes 1** and **2**).

2.1.2   Probe Labeling by Nick Translation

1. *1 kb* DNA *Ladder*.

2. dNTPs: 0.5 mM each of dATP, dCTP, and dGTP, and 0.05 mM dTTP.

3. DNase I from bovine pancreas: 1 mg/ml dissolved in 0.15 M NaCl, 50 % glycerol.

4. DNA Polymerase I: 10 U/μl, Thermo Scientific EP0042.

5. Modified dUTPs directly labeled: We routinely use dyes from Dyomics, DY-590-dUTP (#590-34), DY-505-dUTP (#505-34), DY-415-dUTP (#415-34), and DY-647-dUTP1 (#647-34) (*see* **Notes 3** and **4**).

6. Water, sterile.

7. β-Mercaptoethanol: 0.1 M in sterile water.

2.1.3   Probe Precipitation

1. Master Mix: 50 % Dextran sulfate in 2× SSC, pH 7.0.

2. Ethanol, absolute.

3. Deionized formamide: 70 % in 2× SSC.

4. Cot-1 DNA: 1 mg/ml (*see* **Note 5**).

5. 20× SSC.

6. Sodium acetate (NaOAc): 3 M NaOAc pH 5.2.

7. Water, sterile.

2.1.4 Slide Pretreatment

1. PBS: 1×, pH 7.4, without calcium and without MgCl$_2$ (cat. no. 10010, GIBCO, Invitrogen).

2. Solution 1: RNase A 20 mg ml in sterile water (store aliquots at –20 °C).

3. Solution 2: pepsin 100 mg ml in sterile water (store aliquots at –20 °C).

   Working solution: Add 10 μl pepsin stock solution inside an empty, clean 100 ml glass beaker, then add 100 ml pre-warmed 0.01 M HCl; mix well and pour into a Coplin jar (*see* **Note 6**).

4. Solution 3: 1× PBS/0.05 M MgCl$_2$ (add 50 ml of 1 M MgCl$_2$ to 950 ml of 1× PBS).

5. Solution 4: 1 % (vol/vol) formaldehyde in 1× PBS/MgCl$_2$ (add 2.7 ml of 37 % formaldehyde to 97.3 ml 1× PBS/MgCl$_2$) (*see* **Note 7**).

2.1.5 Slide Denaturation and Hybridization

1. Ethanol: 70, 90, and 100 % (*see* **Note 8**).

2. Deionized Formamide: 70 % vol/vol deionized formamide/2×SSC, or FA/SSC (20 ml dH$_2$O, 10 ml of 20× SSC and 70 ml deionized formamide). Adjust to pH 7.25 with 1.0 N HCl. Mix well; store as 1-ml aliquots at –20 °C (*see* **Note 9**).

3. Rubber cement.

4. 20× SSC.

2.1.6 Detection

1. Ethyl alcohol (ethanol), anhydrous 200 proof: 70, 90, and 100 %.

2. Wash Solution 1: 50 % vol/vol formamide/2×SSC (combine 100 ml formamide with 20 ml 20×SSC and 80 ml ddH$_2$O. Adjust pH to 7.0 using 1.0 N HCl. Mix well and use fresh).

3. Wash Solution 2: 1×SSC (add 25 ml of 20×SSC to 475 ml dH$_2$O).

4. Wash Solution 3: 4×SSC/Tween 20 (add 200 ml of 20×SSC to 799 ml of dH$_2$O and add 1 ml of polyoxyethylene sorbitan monolaurate-Tween 20) (*see* **Note 10**).

5. Mounting media: Antifade (e.g., Vectashield Mounting Medium with DAPI, Vector Laboratories #H-1200).

2.2 Equipment

1. Thermotron humidity chamber (model CDS-5, Thermotron).

2. Slide warmer (e.g., model 77, Fisher Scientific).

3. Thermomixer (e.g., Eppendorf, Thermomixer comfort 1.5 ml).

4. Temperature-controlled hybridization chamber (e.g., Boekel Slide Moat model 240000).

5. Benchtop centrifuge (e.g., Allegra 6R).

6. Biological safety cabinet.

7. Reciprocal shaking water baths (e.g., model 2870, Thermo Scientific).

8. Vortex mixer.

9. Microcentrifuge (e.g., model 5418, VWR Scientific).

10. Microscope slides, glass (e.g., cat no. 12-544-7, Fisher Scientific).

11. Microscope cover glass, 18 mm × 18 mm (e.g., cat. no. 2865-22, Corning) and microscope cover glass, 24 mm × 60 mm (e.g., cat. no. 3322, Thermo Fisher Scientific).

12. Fluorescent microscope equipped with filters corresponding to the wavelength excitation/emission of dUTP indicated in Subheading 2.1.2, **item 5** (e.g., Zeiss Axiovert 200 equipped with Chroma Technology specific filters).

13. Inverted tissue culture microscope.

3 Methods

Figure 1: Flowchart of experimental steps and timeline

3.1 Preparation of Slides

1. Single-cell suspensions of tissues of interest can be obtained virtually from every tissue. It is beyond the scope of this book chapter to provide protocols for tissue dissection and digestion to obtain single-cell suspensions. We recommend consulting specific literature and books dedicated to this process [9]. Single-cell aneuploidy studies can also be performed using cultured cell lines; in this case, we suggest growing cell lines under recommended conditions and use exponentially growing cells.

2. We suggest starting from a minimum of 1×10^{-4} cells resuspended in 1 ml of fresh tissue culture medium.

3. Spin cells in a clinical centrifuge at $217 \times g$ for 10 min at room temperature.

4. Remove the supernatant carefully by vacuum aspiration, leaving 0.2 ml in the tube, and flick the tube with finger to loosen the pellet.

5. Add 1 ml hypotonic solution to the Eppendorf tube containing the cells in suspension and incubate at 37 °C temperature for 15–25 min (*see* **Notes 11** and **12**).

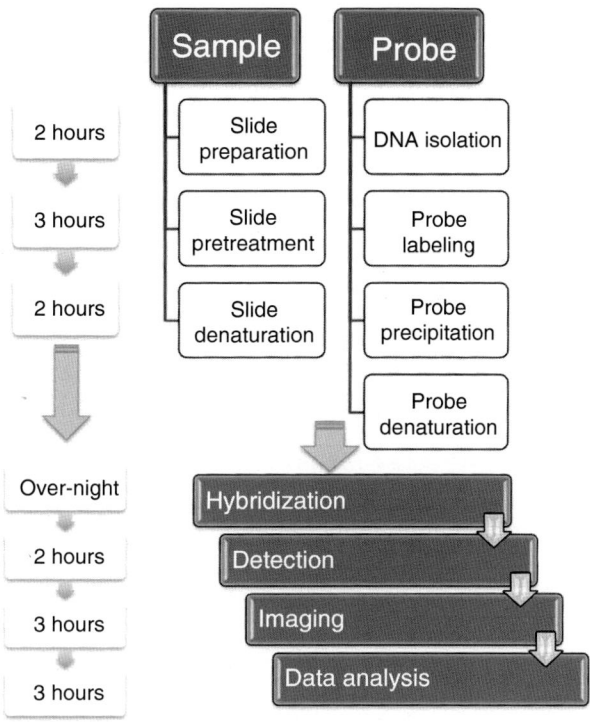

Fig. 1 Flowchart indicating the steps required for sample processing and for the generation of the locus-specific probes. The *boxes* on the left indicate the time required to carry on each step in the hybridization procedure

6. Add 0.01 ml of fresh methanol/acetic acid fixative (3:1 vol/vol) to the tube and invert to mix (this step will stop the hypotonic process and prefix the cells).

7. Spin cell suspension using a clinical centrifuge at $217 \times g$ for 10 min at room temperature.

8. Remove the supernatant carefully, leaving 0.1 ml liquid, flick the tube to loosen the pellet and fully resuspend all cells.

9. Add 1 ml fresh fixative very slowly, gently flicking pellet.

10. Pellet the cells at $217 \times g$ for 10 min at room temperature.

11. Remove the supernatant and flick the tube to loosen the pellet and fully resuspend the cells.

12. Add 1 ml fresh fixative to wash the cell preparations.

13. Pellet the cells at $217 \times g$ for 10 min at room temperature.

14. Remove the supernatant and flick the tube to loosen the pellet and fully resuspend the cells.

15. Add 0.5 ml fresh ice-cold fixative along the wall of the tube.

16. Set the Thermotron at 48 % humidity and 24 °C temperature, drop ~20 μl of the fixed cell suspension onto a clean microscope slide, and allow the slide to fully dry by evaporation (approximately 5 min).

17. View the dried slide with a ×20 high-dry phase objective on a light microscope to determine final cell density (*see* **Note 13**).

18. Sample slides can be heated for 1 h at 60 °C for immediate use (*see* **Note 14**).

3.2 Slide Pretreatment and Denaturation

1. Equilibrate slides in a Coplin jar containing 2× SSC for 5 min at RT.

2. Dilute the RNase stock solution (1:200) in 2× SSC.

3. Apply 120 μl of RNase working dilution to 24×60 mm coverslip (*see* **Note 15**).

4. Incubate slides in a moist temperature-controlled hybridization chamber at 37 °C for 45 min.

5. Carefully remove coverslips.

6. Wash slides three times for 5 min in a Coplin jar containing 2× SSC at room temperature shaking.

7. Incubate slides in Coplin jar containing pepsin at working dilution at 37 °C for 5 min.

8. Wash twice for 5 min each in 1× PBS at room temperature, shaking.

9. Wash once for 5 min in 1× PBS/MgCl$_2$.

10. Place slide in 50 ml Coplin jar containing 1 % formaldehyde/1× PBS/MgCl$_2$ for 10 min at room temperature (without shaking).

11. Wash slides for 5 min in 1× PBS at room temperature, shaking.

12. Dehydrate slide in ethanol series: 70, 90, 100 % ethanol, 3 min each.

13. Air-dry slide. Slides are now ready for denaturation (*see* **Note 16**).

14. Apply 120 μl of 70 % deionized formamide/2× SSC to a 24×60 mm coverslip. Touch the slide to the coverslip.

15. Denature slide at 80 °C on a hot plate for 1 min and 30 s (*see* **Note 17**).

16. Immediately let coverslip slide off and place slide in 70 % precooled ethanol for 3 min, followed by 90 % ethanol and 100 % ethanol for 3 min each at room temperature.

17. Let slide air-dry.

3.3 Generation and Labeling of Probes for Enumeration of Whole Chromosome Aneuploidy (See Note 18)

1. Genomic regions of interest can be selected using publicly available tools and repositories. We recommend the use of bacterial artificial chromosomes (BAC clones), which can be purchased through the BACPAC Resources Center at Children's Hospital Oakland Research Institute in Oakland, California, in the United States (https://bacpac.chori.org/about.htm). Clones corresponding to the region of interest can be visualized using the UCSC (University of California Santa Cruz) Genome Browser website (http://genome.ucsc.edu/ [10]) (*see* **Note 19**).

2. BACs culture and DNA isolation can be carried on using standard protocols [11].

3. High molecular weight DNA should be visible on a 1.2 % agarose gel (Fig. 3a).

4. For each DNA sample, prepare one Eppendorf tube containing:

   2 μg DNA.

   10 μl 10× DNA PolI buffer.

   10 μl dNTP.

   10 μl 0.1 M ß-mercaptoethanol.

   4 μl modified dUTP (1 mM).

   X μl sterile water (the total volume should be 100 μl) (*see* **Note 20**).

5. Vortex, centrifuge, and place tubes on ice.

6. Add 2 μl PolI first.

7. Add 6 μl of DNase (1 mg/ml, dilute stock 1:1,000 using ice-cold water).

8. Vortex and centrifuge.

9. Incubate at 16 °C for 1.5 hours.

10. Prepare gel electrophoresis (1.2 % agarose in 1× TAE).

11. Run 5 μl of each sample and 5 μl of *1 kb* DNA *Ladder* (*see* **Note 21**).

12. Stop the Nick translation with 1 μl of 0.5 M EDTA and incubate at 65 °C for 10 min.

13. Combine probes for DNA precipitation:

    20 μl (200–500 ng DNA) Nick-translated probes DNA.

    80 μl Cot-1 DNA (1 mg/ml).

    40 μl DNAse-free water.

    20 μl of NaOAc 3 M.

    500 μl of absolute ethanol (*see* **Notes 22** and **23**).

14. Vortex; store at −20 °C overnight or at −80 °C for at least 15–30 min.

15. Centrifuge at 29582 × g at 4 °C for 30 min.

16. Pour off supernatant and speed vac for 5–10 min to dry pellet.

17. Add 6 μl deionized formamide (pH 7.5).

18. Incubate at 37 °C for 30 min, shaking (*see* **Note 24**).

19. Add 6 μl Master Mix, vortex, and centrifuge briefly.

20. Denature probe DNA at 76 °C for 5 min and centrifuge briefly.

21. Pre-anneal at 37 °C for 1–2 h (*see* **Note 25**).

3.4 Slides Hybridization and Detection for Single-Cell Aneuploidy Analyses

1. After pre-annealing (*see* above **step 21** of Subheading 3.3), apply the denatured probe to the denatured slide (*see* above **step 17** of Subheading 3.2).

2. Cover with 18 × 18 mm coverslip.

3. Seal coverslip with rubber cement (*see* **Note 26**).

4. Place slides in the hybridization chamber and hybridize at 37 °C over night.

5. Remove the slide form the hybridization.

6. Carefully remove the rubber cement from the slide, taking care not to drag the cover glass across the slide, thereby scratching the metaphase preparations (*see* **Note 27**).

7. Place the slide into a Coplin jar containing wash Solution 1 and wash slides while shaking at 45 °C for 5 min (*see* **Note 28**).

8. Repeat **step 7** two more times, with shaking, using fresh wash Solution 1 each time.

9. Wash the slides with wash Solution 2 a total of three times, 5 min each, with shaking (*see* **Note 29**).

10. Discard wash Solution 2 and wash once for 5 min at 45 °C with wash Solution 3, with shaking.

11. Wash the slide for 5 min in a Coplin jar containing 2×SSC, with shaking, at room temperature.

12. Dehydrate the slide sequentially in three different Coplin jars containing 70 % (room temperature), 90 %, and 100 % ethanol, respectively, without shaking, keeping the slides out of ambient light, for 3 min in each solution.

13. Air-dry the slide in the dark (15–20 min).

14. Add 35 μl of antifade solution containing DAPI to a 24 mm × 60 mm cover glass, invert the slide metaphase-spread-side

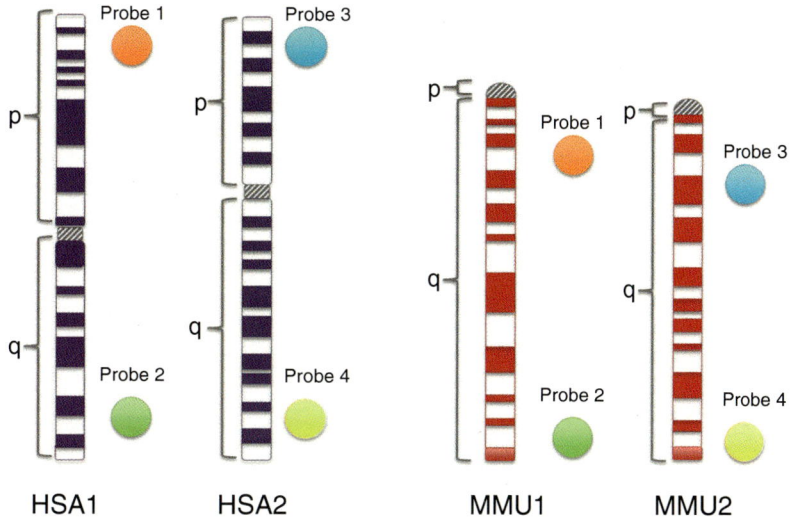

Fig. 2 Ideograms of human (HSA, *purple*) and mouse (MMU, *orange*) chromosomes. Autosomes 1 and 2 have been randomly selected for representative purposes. For both species the p and q chromosome arms are pinpointed. For human chromosomes we suggest to select one probe for each arm of the same chromosome (spectrum *orange* and spectrum *green* for HSA1 and spectrum *aqua* and *far red-yellow* for HSA2). Because murine chromosomes are all acrocentric, locus-specific probes for this specie should be selected at the extremities of the same autosome (spectrum *orange* proximal and spectrum *green* distal on MMU1 and spectrum *aqua* proximal and *far red-yellow* distal on MMU2)

3.5 Image Acquisition and Analysis

down onto the liquid-bearing cover glass, and immediately reinvert the slide. Carefully remove any air bubbles (*see* **Note 30**).

1. The precise procedure for image acquisition will vary depending on the microscope and available software. We have extensive experience using the Zeiss Axiovert 200 inverted fluorescence microscope with a fine focusing oil immersion lens (×40, NA 1.3 oil and ×60, NA 1.35 oil). The microscope is equipped with a high-resolution CCD Camera Hall 100 and the images are acquired using the FISHView application of the Applied Spectral Imaging software [12] (*see* **Note 31**).

2. For each hybridization multiple focal planes are acquired within each channel to ensure that signals on different focal planes are included (*see* **Note 32**).

3. Using the probes shown in Fig. 2, fluorescence emission images are collected using 470 nm (for DAPI), 600 nm (for spectrum orange), 449 nm (for spectrum aqua), 540 nm (for AlexaFluor488), and 700 nm (for far red) filters (*see* **Note 33**).

Table 1
Template of cell scoring

Cell	SO	SA	SG	FR	Ploidy
1	2	2	2	2	$2n$
2	4	4	4	4	$4n$
3	5	5	4	4	Aneuploid
4	1	1	2	2	Aneuploid
5	12	12	10	10	Aneuploid
⋮					
500	8	8	8	8	$>4n$

From left to right each column indicates, respectively, the cell analyzed; the number of signals in the Spectrum Orange (SO), Spectrum Aqua (SA), Spectrum Green (SG), and Far Red (FR), and the ploidy estimated for that cell based on the number of signals

Table 2
Summary of cell scoring

Ploidy	Sample A		Sample B	
	Number of cells	Frequency (%)	Number of cells	Frequency (%)
$2n$	410	82	496	99.2
$4n$	15	3	2	0.4
$>4n$	20	4	1	0.2
Aneuploid	55	11	1	0.2
Total	500	100	500	100

For two samples analyzed (sample A and sample B), the table summarizes the number of cells found with any given ploidy (diploid = $2n$, tetraploid = $4n$, polyploid $\geq 4n$, and aneuploid). Frequencies are also indicated as percentage of cells with any given ploidy over the total number of cells analyzed

4. After acquisition signals for each fluorochrome are visually inspected and manually counted and for each cell. Plotting of raw counts is summarized as shown in Table 1.

5. Diploid nuclei will have two signals for each probe (in our human chromosome example one red and one green signal for autosome 1 and one aqua and one yellow signal for autosome 2). Likewise, polyploidy nuclei can be identified by the presence of red/green and aqua/yellow signals matching in number (i.e., four for tetraploid cells). Aneuploid cells are enumerated when the numbers of red/green and aqua/yellow signals do not match (*see* **Note 34**).

6. Once all nuclei have been inspected, data can be summarized as in Table 2 (*see* **Note 35**).

4   Notes

1. Methanol is highly flammable. Use only in chemical fume hood, wear gloves when handling.

2. Fixative should be freshly prepared before use and can be stored before procedure (1–2 h) at 4 °C.

3. Do not repeatedly freeze and thaw modified dUTPs. Store aliquots according to manufacturer's directions and record expiration dates.

4. All steps involving the use of fluorochromes (hybridization and detection) should be performed in the dark.

5. Cot-1 DNA should match the genome of the species used for hybridization.

6. The time of pepsin treatment and amount of pepsin stock solution to be used is dependent on

   (a) The amount of cytoplasm (i.e., tissue dependent).

   (b) The age of the slide.

   Slides with excess cytoplasm require longer treatment with pepsin (>10 min) and higher concentrations of stock pepsin ranging from 10 to 30 µl. Slides older than 6 months may also require more intensive pretreatment.

   It is very important that the pepsin is added to an empty clean beaker first and not directly into the acid solution. Direct addition of the pepsin to the acid solution will cause the pepsin to precipitate and it will not dissolve properly into the acid solution.

7. Make Solution 4 fresh for each experiment, mix well and store at room temperature until use. Formaldehyde is flammable, carcinogenic, and poisonous. Formaldehyde should be dispensed in a chemical fume hood and only handled when wearing gloves.

8. Pre-cool 70 % ethanol in ice.

9. The use of deionized formamide is important, as it contains fewer impurities than conventional formamide. Formamide is a known mutagen that causes eye and skin irritation. Wear gloves when handling and dispense under a chemical fume hood.

10. Solutions 1, 2, and 3 should be mixed well and prepared fresh; pre-warm in a water bath at 45 °C until use.

11. The volumes indicated in Subheading 3.1, points 3 and 4, are recommended for 1×10^{-4} cells; for higher number of cells, these volumes need to be scaled up accordingly.

12. Detection of mosaic aneuploidy in single cells is performed in interphase cells; thus, hypotonic exposure is not strictly required. However, we suggest performing hypotonic exposure to reduce the amount of cytoplasm and improve hybridization signal.

13. Only interphase cells will be visible at this stage, we do not expect the presence of metaphase chromosomes. The integrity of chromatin structure is dependent on the evaporation rate of the fixative, as determined by the percent humidity, temperature, and success of the hypotonic procedure. Interphase nuclei should appear light gray in color and should not show shiny edges or have a bright halo around them.

14. For better FISH performance we suggest aging slides in a 37 °C drying oven for 3–7 days. This process yields optimum results.

15. Drop the RNAse to a coverslip and touch slide to coverslip.

16. Visually inspect slides under a bright-field microscope to ensure that nuclei are not damaged.

17. The denaturation time and temperature depends on the age of the slide, the species, and cell type. For example, mouse preparations usually require lower temperature and reduced denaturing time.

18. We suggest using four-color FISH using two probes for each chromosome selected. One subcentromeric and one distal probe mapping to the same chromosome should be labeled with two different dyes (*see* Fig. 2 for a schematic representation of probe selection).

19. Select from the pull-down menu the desired organism and the most updated genome assembly. If BAC clones are not visible in the display window, make sure that the BAC End Pairs and the FISH clones tracks are selected (full display).

20. We suggest using DY-590-dUTP (#590-34) diluted 1:5 while all other modified dUTP indicated in Subheading 2.1.2 **item 5** should be used without dilution.

21. Ideally the length of the DNA should be 500–1,000 bp after Nick translation (*see* Fig. 3b). If DNA is too large, add more DNase and incubate at 15 °C for additional 10–30 min.

22. All four probes mapping to the two chromosomes of interest should be combined in one tube and precipitated together (for a total of 80 µl of labeled DNA).

23. Cot-1 DNA should match the species used in the experimental design.

24. Vortex a few times during the 30 min incubation

25. The probe is now ready for hybridization onto denatured slides.

26. Make sure that no air bubbles are present by gently applying pressure to the coverslip (e.g., with forceps) to remove them. It is important that the coverslip is completely sealed with rubber cement to avoid drying of the coverslip that would otherwise damage the probe and result in high background.

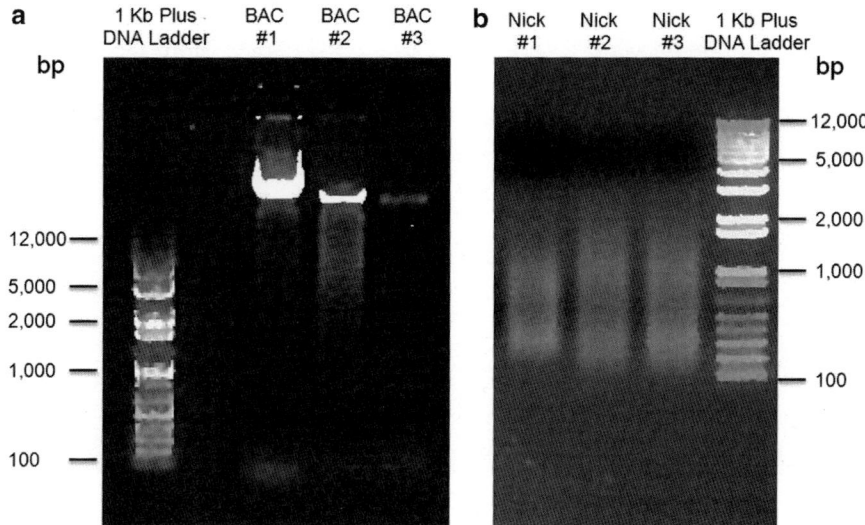

Fig. 3 Gel images representative of: (**a**) genomic DNA extracted from three different BAC clones. For demonstration we choose three BACs with different plasmid copy number (high on the *left*, medium in the *middle*, and low on the *right*); in (**b**) the normalized DNA concentration from (**a**) was labeled using the Nick translation protocol described in this book chapter

27. In case the coverslip does not slide off easily, dip the slide with the coverslip in a pre-warmed (45 °C) Coplin jar containing detection Solution 3. This will loosen the hybridization mix and aid the removal of the coverslip.

28. Make sure, using a thermometer, that the detection solutions are used at the temperature indicated.

29. Because we use probes directly labeled with modified dUTP, the preferred wash steps employs 1×SSC at 45 °C. If high background is observed, we suggest increasing the stringency of slide washing by using 0.1× SSC at 45 °C. The stringency can be further increased by washing the slides at 60 °C. However, this last step should be carried out with caution since this may damage the fluorochromes resulting in poor signal.

30. Let the slides sit at room temperature for at least 30 min before visualization to allow the DAPI to uniformly stain the chromatin. Representative images for one diploid and on aneuploidy nucleus are shown in Fig. 4.

31. Hybridized slides can be stored at 4 °C until visualization using a fluorescence microscope. Because signal intensity diminishes with time, we recommend image acquisition to be carried on within 1 week from detection.

32. For single mammalian cells we suggest starting by acquiring five focal planes spaced 1 μm and merging the resulting images.

33. Images representing a minimum of 500 cells for each hybridization should be analyzed. Images must be randomly acquired.

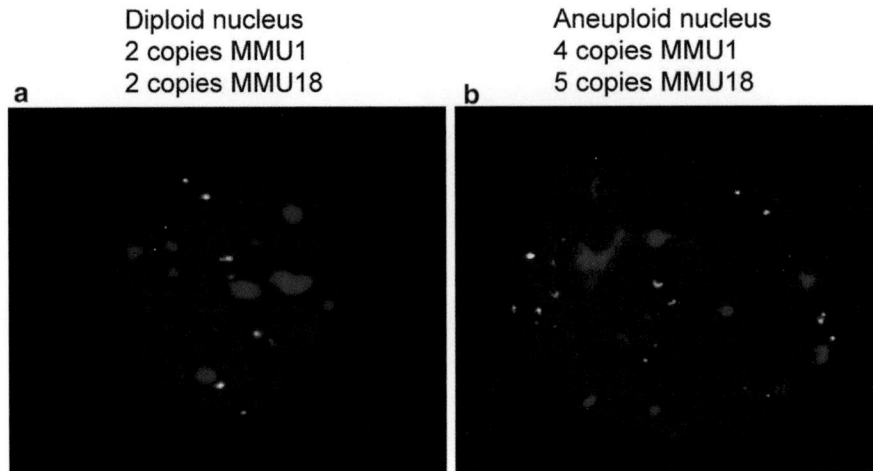

Fig. 4 Representative images of (**a**) a diploid nucleus where two copies for each autosome are visible (*aqua* and *yellow* for MMU1 and *green* and *red* for MMU18); in (**b**) one aneuploidy nucleus is shown where four copies of MMU1 and five copies of MMU18 are visible

34. In the FISH design presented in this chapter, we specifically measure chromosome copies by the presence or absence of two signals for each autosome (red/green and aqua/yellow). We recognize that this practice may lead to an underestimation of aneuploidy levels (due to the fact that structural rearrangement may lead to chromosome breakage between two probes mapping to the same chromosome); however, this approach results in a highly sensitive system that has been proven reproducible for the quantification of low-level aneuploidy in somatic cells [4].

35. Frequencies of aneuploidy can be analyzed using standard statistical approaches [4].

References

1. Langer-Safer PR, Levine M, Ward DC (1982) Immunological method for mapping genes on *Drosophila* polytene chromosomes. Proc Natl Acad Sci U S A 79:4381–4385

2. Hudziak RM, Lewis GD, Winget M, Fendly BM, Shepard HM et al (1989) p185HER2 monoclonal antibody has antiproliferative effects in vitro and sensitizes human breast tumor cells to tumor necrosis factor. Mol Cell Biol 9:1165–1172

3. Slamon DJ, Leyland-Jones B, Shak S, Fuchs H, Paton V et al (2001) Use of chemotherapy plus a monoclonal antibody against HER2 for metastatic breast cancer that overexpresses HER2. N Engl J Med 344:783–792

4. Faggioli F, Wang T, Vijg J, Montagna C (2012) Chromosome-specific accumulation of aneuploidy in the aging mouse brain. Hum Mol Genet 21:5246–5253

5. Faggioli F, Vezzoni P, Montagna C (2011) Single-cell analysis of ploidy and centrosomes underscores the peculiarity of normal hepatocytes. PLoS One 6:e26080

6. Duncan AW, Taylor MH, Hickey RD, Hanlon Newell AE, Lenzi ML et al (2010) The ploidy conveyor of mature hepatocytes as a source of genetic variation. Nature 467:707–710

7. Rehen SK, McConnell MJ, Kaushal D, Kingsbury MA, Yang AH et al (2001) Chromosomal variation in neurons of the developing and adult

mammalian nervous system. Proc Natl Acad Sci U S A 98:13361–13366

8. Westra JW, Peterson SE, Yung YC, Mutoh T, Barral S et al (2008) Aneuploid mosaicism in the developing and adult cerebellar cortex. J Comp Neurol 507:1944–1951

9. Freshney RI (2010) Culture of animal cells: a manual of basic technique and specialized applications. 6th ed. New York: Wiley-Liss; 187–206

10. Kent WJ, Sugnet CW, Furey TS, Roskin KM, Pringle TH et al (2002) The human genome browser at UCSC. Genome Res 12: 996–1006

11. Sambrook J, Fritsch EF, Maniatis T (1989) Molecular Cloning: A Laboratory Manual, 3rd edn. Cold Spring Harbor Laboratory Press, Cold Spring Harbor, New York, Vols 1, 2 and 3

12. ASI ASI GenASIs Capture & Analysis. http://www.spectral-imaging.com/platforms/genasis-capture-a-analysis

INDEX

David J. Sharp (ed.), *Mitosis: Methods and Protocols*, Methods in Molecular Biology, vol. 1136,
DOI 10.1007/978-1-4939-0329-0, © Springer Science+Business Media New York 2014

Printed by Printforce, the Netherlands